Cyber-Secure Load Frequency Control for Power Systems

Cyber-Secure Load Frequency Control for Power Systems

Xinghua Liu

Xi'an University of Technology, Xi'an, China

Siwei Qiao

Hubei Normal University, Huangshi, China

Gaoxi Xiao

Nanyang Technological University, Singapore, Singapore

Peng Wang

Nanjing Normal University, Nanjing, China

IEEE Press

WILEY

Published by John Wiley & Sons, Inc., Hoboken, New Jersey.

Library of Congress Cataloging-in-Publication Data has been applied for:

Hardback ISBN: 9781394346363
ePDF ISBN: 9781394346387
ePub ISBN: 9781394346370
oBook ISBN: 9781394346394

Cover Design: Wiley
Cover Image: ImageFlow/Shutterstock

To those who are interested in the LFC issue.

Contents

Foreword

Power systems have long been inseparable from human society. In recent years, renewable energy, electric vehicles, and battery energy systems have been seen as the future of multi-area power systems. How to ensure the stable operation of modern power systems in a complex environment is an important issue. Besides, in the load frequency control power system, a communication network is used to transmit decision or control signals. The open nature of the network makes it vulnerable to cyberattacks in multi-area power systems. The significant importance of the normal operation of the power system makes it necessary to study the cybersecurity problems in complex and crowded environments, where the robustness and security of the power systems may suffer from cyberattacks. This book aims to deal with the problem from two perspectives: (1) to propose new defense methods with consideration of attack modelling; and (2) to describe the actual behavior of the attacks, some new attack models are proposed.

To ensure the safe and stable operation of the power system in complex environments. We have investigated a number of control approaches to balance load demand and generation. The proposed method solves the contamination of cyberattacks, the fluctuation of renewable energy, the connection with electric vehicles, and the integration of battery energy systems. We have also researched event-triggered devices with different trigger conditions, such that the high-pressure problem of the communication network can be solved. Overall, this book will be helpful for researchers working on load frequency control, cybersecurity issues, event-triggered approaches, stability analysis for modern power systems, and so on.

About the Authors

Xinghua Liu received the BSc from Jilin University, Changchun, China, in 2009; and the PhD degree in Automation from University of Science and Technology of China, Hefei, in 2014. From 2014 to 2015, he was invited as a Visiting Fellow at RMIT University in Melbourne, Australia. From 2015 to 2018, he was a Research Fellow at the School of Electrical and Electronic Engineering in Nanyang Technological University, Singapore. Dr. Liu has joined Xi'an University of Technology as a Professor since September 2018. His research interests include state estimation and control, intelligent systems, autonomous vehicles, cyber-physical systems, robotic systems, etc.

Siwei Qiao received the MS degree in the College of Mechatronics and Control Engineering from Hubei Normal University, Huangshi, China, in 2019. He received the PhD degree in the College of Electrical Engineering from Xi'an University of Technology, Xi'an, China, in 2025. Now, he is currently working as a Lecturer in the College of Electrical Engineering and Automation in Hubei Normal University, Huangshi, China. His research interests include networked control systems, robust control, and power systems.

Gaoxi Xiao received the BS and MS degrees in applied mathematics from Xidian University, Xi'an, China, in 1991 and 1994, respectively, and the PhD degree in computing from Hong Kong Polytechnic University, Hong Kong, in 1998. He was an Assistant Lecturer with Xidian University from 1994 to 1995. He was a Post-Doctoral Research Fellow with Polytechnic University, Brooklyn, NY, USA, in 1999, and a Visiting Scientist with The University of Texas at Dallas, Richardson, TX, USA, from 1999 to 2001. In 2001, he joined the School of Electrical and Electronic Engineering, Nanyang Technological University, Singapore, where he is currently an Associate Professor. His research interests include complex systems and complex networks, communication networks, smart grids, and system resilience and risk regulation.

Peng Wang is a National-level Leading Talent, an IEEE Fellow, a Professor at the School of Electrical and Automation Engineering, Nanjing Normal University, and a Tenured Professor at Nanyang Technological University (NTU), Singapore. He is a world-renowned expert in the fields of AC/DC hybrid microgrids and power system reliability. As a global pioneer in this domain, he was the first worldwide to propose the concept and architecture of the AC/DC hybrid microgrid. Serving as the Principal Investigator (PI), he has spearheaded multiple major international projects, securing cumulative research funding exceeding 15 million USD. He directed the development of a series of multi-port modular integrated products for hybrid microgrids and successfully established the world's first 8-node AC/DC hybrid microgrid, as well as Southeast Asia's first standalone AC/DC hybrid microgrid demonstration project. These accomplishments represent significant breakthroughs in key technologies and their industrial applications. Professor Wang has authored over a hundred papers in prestigious international journals, including IEEE Transactions, with numerous publications recognized as Highly Cited Papers. His accolades include the Outstanding Engineer Award from the IEEE Power & Energy Society (PES) Singapore Chapter and the Contribution Award from the CSEE Journal of Power and Energy Systems (JPES).

Preface

Load frequency control (LFC) is well known for balancing the load demand and frequency for a multi-area power system. Studies have proven that LFC can improve the global performance of multi-area power systems. In recent years, the increasing proportion of renewable energy, integration of EVs, and cyberattacks have become the main challenges in LFC power systems. Different strategies have been applied in the literature for LFC power systems and the possible impacts of renewable energy, EVs, and cyberattacks. This book is devoted to the research on directions in LFC multi-area power systems.

September, 2025 *Xi'an*

Acknowledgments

We would like to thank Prof. Shuzhi Sam Ge, Prof. Yu Kang, Prof. Yuanzhe Wang, Prof. Dianhui Wang, Prof. Meng Zhang, Prof. Shiping Wen, Prof. Badong Chen, Prof. Guoqing Yang, and Yuru Liang for their collaboration in this research area in recent years. We thank Prof. Gaoxi Xiao and Prof. Peng Wang for carefully reviewing the book and for the continuous support given to us in the entire publication process. We thank the authors in [Information Sciences, 676, No. 120793, 2024; Applied Mathematics and Computation, 472, No. 128636, 2024; Journal of the Franklin Institute, 362(7), No. 107644, 2025] for the technical and theoretical support. In addition, we would like to thank the National Natural Science Foundation of China (under Grant 62473309), and in part by the Shaanxi Outstanding Youth Science Fund Project (under Grant 2024JC-JCQN-68), and in part by the Xi'an Science and Technology Plan Project (under Grant 24GXFW0050), and in part by the Shaanxi Provincial Youth Innovation Team Scientific Research Program Project (under Grant 25JP116), and in part by the Xi'an Key Laboratory (under Grant 24ZDSY0015), and in part by the Future Resilient Systems Programme (FRS-II) at the Singapore-ETH Centre (SEC), funded by National Research Foundation (NRF), Singapore.

X.H.L.

Acronyms

LFC	Load Frequency Control
SMC	Sliding Mode Control
SMLFC	Sliding Mode Load Frequency Contrcl
AMD	Adaptive-memory-derivative
DFIG	Doubly-fed Induction Generator
DoS	Denial-of-Service
EVs	Electric Vehicles
BESSs	Battery Energy Storage Systems
LMI	Linear Matrix Inequality
V2G	Vehicle-to-Grid
AGC	Automatic Generation Control
ACE	Area Control Error
PI	Proportional-Integral
PID	Proportional-Integral-Derivative
MPC	Model Predictive Control
SOC	State-of-Charge

Introduction

In order to adapt to the development of human society, traditional power systems have gradually evolved into modern power systems. Modern power systems are also known as multi-area power systems, which connect several sub-areas via communication networks. In the next five years, the use of renewable energies, the integration of electric vehicles, the connection to battery energy systems, and cybersecurity will certainly be the basis of the emergency. Load frequency control has been widely investigated to solve the above problems and ensure the normal operation of the power system. Load frequency control can be used to maintain a balance between load and power fluctuations. Furthermore, new power electronics devices, communication network security protocols, and efficient control methods are also promoting the development of multi-area power systems.

An increasing number of different types of cyberattacks have occurred in the real world. Researchers must therefore consider potential cyberattack models in advance in order to develop effective preemptive strategies. Research into different attack models can help us describe the actual behavior of cyberattacks and provide timely responses when an attack occurs. In this book, we have described some new cyberattack models with a unique structure. The considered attack model will help in researching the damage of cyberattacks on power systems. Researchers can also develop defense strategies in advance to respond quickly to attacks.

To ensure the safe and stable operation of the power system in complex environments. We have investigated a number of control approaches to balance load demand and generation. The proposed method solves the contamination of cyberattacks, the fluctuation of renewable energy, the connection with electric vehicles, and the integration of battery energy systems. We have also researched event-triggered devices with different trigger conditions, such that the high-pressure problem of the communication network can be solved. Overall, this book will be helpful for researchers working on load frequency control, cybersecurity issues, event-triggered approaches, stability analysis for modern power systems, and so on.

1

Observer-based Sliding Mode Load Frequency Control of Power Systems Under Deception Attack

1.1 Background

Developments in power systems have greatly improved human life. Ensuring stability and sustainability of power systems, as an increasingly important research area, has attracted extensive attention [1–3].

One of the most popular goals of such studies is to ensure stable operation of the power system, to which the use of control methods helps provide a boost [4, 5]. Sliding mode control (SMC), with its advantages in overcoming system uncertainties and achieving strong robustness against disturbances, has been widely used in a variety of fields [6–8]. System robustness is enhanced by combining dynamic surface control with sliding mode [9–12]. It can be used in different systems like nonlinear systems [13], Markovian jump systems [14], power systems, and so on. Here, we focus on the power system. The robust stability of adaptive event-triggered load frequency control (LFC) with SMC for multi-area power systems is investigated in [4]. The authors Huang and Wang [15] proposed a fractional-order sliding mode controller for effectively stabilizing a nonlinear power system in a fixed time. In [16], a highly robust observer sliding mode-based load frequency and tie-power control to compensate for primary frequency control of multi-area interconnected power systems was designed. While the power system can be kept stable by SMC, the modern power system relies on the critical cyber infrastructure, which makes it vulnerable to hostile cyber threats.

Some robust control methods are applied to the LFC of the multi-area power system. It should be pointed out that in the analysis of multi-area power system problems, the attacks from outside the system may be ignored. There are many types of cyberattacks that can act on the power systems, among them include denial-of-service attack, deception attack, and data integrity attack, etc [17–21]. A deception attack is to compromise the trustworthiness of data by manipulating the packets transmitted over the communication networks. When under a deception attack, the flow of information in the system will be changed [22, 23], which may destabilize or even crash the system.

We pay special attention to the deception attack in this chapter. In [24], a new malicious data deception attack by considering a practical attacking situation was investigated. An active synchronous detection method was proposed in [25] to detect deception attack in microgrids without impeding system operations. In [26], the authors investigated the viability of machine learning methods in detecting cyberattacks. Considering the impact of damaged information, the observer-dependent event-triggered scheme is applied in this chapter. The event-trigger control works only when a certain set of event-trigger conditions are met. Using this method can not only make effective use of communication resources, but also prolong the service time of the controller. This makes it be rapidly developed in the system control, especially in dealing with complex networked dynamic systems. A properly designed event-triggering scheme may also help us find out whether the system is under malicious attack. In [27], a dynamic event-triggered mechanism was proposed to modulate the transmission of data packets. The cooperative control problem has been investigated based on a novel event-triggered scheme in [28]. The random deception attack is well handled according to a memory-based event-triggered scheme proposed in [29]. In [30], a mathematical model concerning the event-triggered LFC under deception attack was developed. The collaborative design of both the event-trigger strategy and the adaptive robust controller is investigated in [31]. The authors in [32] investigated the observer-based predictive event-triggered LFC for a smart grid incorporating electric vehicles. Benefiting from the use of an event-triggered mechanism, the negative impacts of deception attack on information transmission can be effectively detected. Based on which the control module may work in time to ensure the stable operation of the system.

We design an observer-based sliding mode LFC scheme for controlling multi-area power systems under deception attack. Firstly, by taking into account the characteristic of the deception attack, a model of attacked power system is built. Then, an observer is proposed to estimate the state of the damaged system. Further, the event trigger is introduced to judge whether the normal operation of the system is compromised using the designed event trigger conditions. The main contributions of this chapter can be generalized as follows:

1. A novel SMC is proposed for the multi-area power systems under deception attack. The proposed observer-based event-trigger SMC helps save the limited network channel resources, so that the desired control can be performed successfully.

2. The sliding mode load frequency problem in an interconnected power system with a deception attack is well-analyzed. Considering the designed SMC surface, the equivalent control under the basic SMC is obtained through analysis. The observer and the event triggering scheme are introduced to analyze the power system under deception attack. Using the linear matrix inequality (LMI) method, the conditions of the system's asymptotic stability are obtained. Finally, an appropriate SMC law is designed to ensure the reachability of the system trajectory.

1.2 Problem Formulation

1.2.1 Model Description

The multi-area power system is a type of complex control system. In this section, the control problem of a sliding mode LFC-based power system under deception attack is analyzed. The i-th area of a power system under deception attack model is shown in Figure 1.1, which consists of a generator, a governor and a turbine, and so on. The meanings of the notations in Figure 1.1 are shown in Table 1.1.

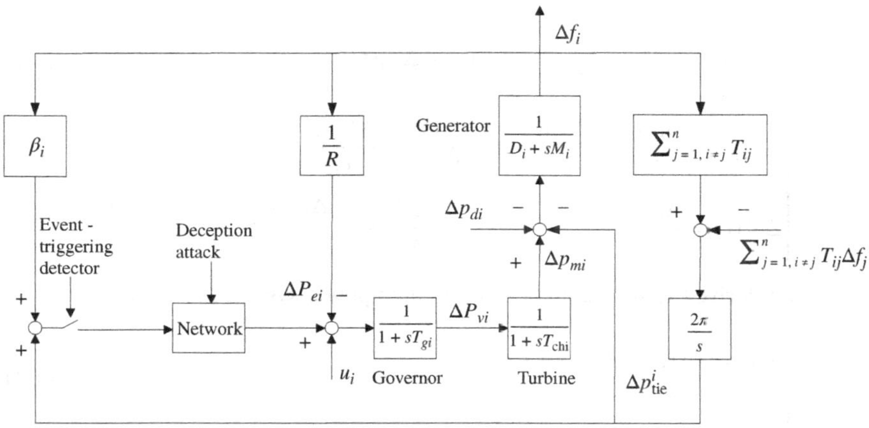

Figure 1.1 The model of the i-th area of the power system.

Table 1.1 Definition of notations.

Δf_i	Frequency deviation
ΔP^i_{tie}	Deviation of tie-line power exchange between areas
ΔP_{mi}	Mechanical output deviation
ΔP_{di}	Load deviation
D_i	Damping coefficient
M_i	Inertia of the generator
ΔP_{vi}	The deviation value
R_i	Droop coefficient
T_{gi},	Time constant of the governor
u_i	Control input
T_{chi}	Time constant of the turbine
T_{ij}	Synchronization coefficient between areas

Based on the system model and information transfer, the dynamic equation of the system can be obtained as follows

$$\begin{cases} \Delta\dot{P}_{mi}(t) = -\dfrac{\Delta P_{mi}(t)}{T_{\text{chi}}} + \dfrac{\Delta P_{vi}(t)}{T_{\text{chi}}} \\ \Delta\dot{P}_{ei}(t) = \Delta P^i_{\text{tie}}(t) + \beta_i \Delta f_i(t) \\ \Delta\dot{P}^i_{\text{tie}}(t) = -2\pi \sum\limits_{j=1,i\neq j}^{n} T_{ij}(\Delta f_i(t) - \Delta f_j(t)) \\ \Delta\dot{P}_{vi}(t) = \dfrac{u_i(t)}{T_{gi}} - \dfrac{\Delta f_i(t)}{R_i T_{gi}} - \dfrac{\Delta P_{vi}(t)}{T_{gi}} - \dfrac{\Delta P_{ei}(t)}{T_{gi}} \\ \Delta\dot{f}_i(t) = \dfrac{\Delta P_{mi}(t)}{M_i} - \dfrac{\Delta P_{di}(t)}{M_i} - \dfrac{D_i \Delta f_i(t)}{M_i} - \dfrac{\Delta P^i_{\text{tie}}(t)}{M_i} \end{cases} . \tag{1.1}$$

We can derive the state equation of i-th area power system as

$$\begin{cases} \dot{x}(t) = Ax(t) + Bu(t) + Fw(t) \\ y(t) = Cx(t) \end{cases} \tag{1.2}$$

where $w(t)$ is the disturbance of the system and defined as $w(t) = \Delta P_{di}$, $w(t)$ is bounded and satisfies $\Delta P_{\text{di}} < h_i$, h_i is the positive constants,

$$x_i(t) = [x_1^T(t), x_2^T(t), \cdots, x_n^T(t)]^T,$$

$$u(t) = [u_1^T(t), u_2^T(t), \cdots, u_n^T(t)]^T,$$

$$y_i(t) = [y_1^T(t), y_2^T(t), \cdots, y_n^T(t)]^T,$$

$$w_i(t) = [\Delta P_{d1}(t), \Delta P_{d2}(t), \cdots, \Delta P_{dn}(t)]^T,$$

$$x(t) = [\Delta f_i, \Delta P_{mi}, \Delta P_{vi}, \Delta P_{ei}, \Delta P_{tie}^i]^T.$$

where A, B, C, and F are the parameter matrixes, $A = [A_{ij}]_{n \times n}$, $B = \text{diag}\underbrace{\{B_1, \cdots, B_i\}}_{n}$, $C = \text{diag}\underbrace{\{C_1, \cdots, C_i\}}_{n}$, $F = \text{diag}\underbrace{\{F_1, \cdots, F_i\}}_{n}$.

$$A = \begin{bmatrix} -\frac{D_i}{M_i} & \frac{1}{M_i} & 0 & 0 & -\frac{1}{M_i} \\ 0 & -\frac{1}{T_{\text{chi}}} & \frac{1}{T_{\text{chi}}} & 0 & 0 \\ -\frac{1}{R_i T_{gi}} & 0 & -\frac{1}{T_{gi}} & 0 & 0 \\ \beta_i & 0 & 0 & 0 & 1 \\ -2\pi \sum\limits_{j=1, j\neq i}^{n} T_{ij} & 0 & 0 & 0 & 0 \end{bmatrix},$$

$$B = \begin{bmatrix} 0 & 0 & \frac{1}{T_{gi}} & 0 & 0 \end{bmatrix}^T, F = \begin{bmatrix} 0 & 0 & 0 & 0 & -\frac{1}{M_i} \end{bmatrix}^T.$$

1.2.2 Event-triggered Control

The data information of the power system is transmitted by the shared network to the corresponding sliding mode controller. In this section, a sliding mode observer-based load frequency controller for the ith-area power systems is given. An event-triggered scheme will be put forward taking into account the consideration of the deception attack

$$\tilde{y}(t) = (1 - \alpha(t))y(t) + \alpha(t)\varepsilon(t), \tag{1.3}$$

where $\varepsilon(t)$ is the deception attack from outside of the system and $\alpha(t)$ is an energy-bounded signal belonging to $L_2 \in [0, +\infty)$. $\alpha(t)$ is a stochastic variable subject to a Bernoulli distribution satisfying

$$\begin{cases} P\{\alpha(t) = 1\} = \mathbb{E}\{\alpha(t)\} := \alpha_0 \\ P\{\alpha(t) = 0\} = 1 - \mathbb{E}\{\alpha(t)\} := 1 - \alpha_0 \end{cases} \tag{1.4}$$

When the deception attack happens, the information transmitted will be changed, the system equation of state can be expressed as

$$\begin{cases} \dot{x}(t) = Ax(t) + Bu(t) + Fw(t) \\ \tilde{y}(t) = (1 - \alpha(t))Cx(t) + \alpha(t)\varepsilon(t) \end{cases} \tag{1.5}$$

Using the observed information as the transmitted information. To facilitate the analysis of the problem, the state of the observed information is defined as $\hat{x}(t)$, then

$$\begin{cases} \dot{\hat{x}}(t) = A\hat{x}(t) + Bu(t) + L(\tilde{y}(t) - \hat{y}(t)) \\ \hat{y}(t) = (1 - \alpha(t))C\hat{x}(t) \end{cases}. \tag{1.6}$$

To facilitate the detection of the deception attack on the system, the event-trigger is used in the system model 1.1. In the absence of deception attack, a common event-triggered condition [33] is given as

$$\begin{aligned} &t_{k+1}h = t_k h + \\ &\min_{d \in \mathbb{R}^+} \{dh \| e^T(j_k h)\phi e(j_k h) \geq \delta(t_k h)\hat{x}^T(t_k h)\phi\hat{x}(t_k h)\} \end{aligned}, \tag{1.7}$$

where $\phi = \eta C^T C, \eta$ is a positive definite matrix, and $\phi = \text{diag}\{\phi_1, \cdots, \phi_n\}$ is defined as a weighting matrix. We denote the information error as $e(j_k h) = \hat{x}(j_k h) - \hat{x}(t_k h)$, $j_k h = t_k h + dh$, $d \in \mathbb{R}_0^+$. δ is a predefined scalar confined in the interval $[0, 1)$.

The system error is $\Theta(t) = x(t) - \hat{x}(t)$. Then

$$\begin{aligned} \dot{\Theta}(t) &= A\Theta(t) + Fw(t) \\ &\quad - L(\alpha(t)\varepsilon(t) + (1 - \alpha(t))C\Theta(t)) \end{aligned}. \tag{1.8}$$

1.2.3 Sliding Mode Design

To make the system working stable, the SMC method is introduced. We construct the following sliding mode surface function

$$\sigma(t) = G\hat{x}(t) - \int_0^t G(A + BK)\hat{x}(\tau)d\tau, \tag{1.9}$$

where G is a constant matrix and it is selected to ensure that GB stays nonsingular, $0 \leq \tau(t) \leq h + \tau_{\max}$, $\tau_{\max} > 0$.

We all know that the switch surface function should satisfy

$$\sigma(t) = 0 \text{ and } \dot{\sigma}(t) = 0. \tag{1.10}$$

The derivative of the sliding surface is

$$\dot{\sigma}(t) = G\dot{\hat{x}}(t) - GA\hat{x}(t) - GBK\hat{x}(t). \tag{1.11}$$

According to (1.6) and (1.11), we have

$$\begin{aligned}\dot{\sigma}(t) &= G[A\hat{x}(t) + Bu(t) + L(\tilde{y}(t) - \hat{y}(t))] \\ &\quad - GA\hat{x}(t) - GBK\hat{x}(t) \\ &= -GBK\hat{x}(t) + GBu(t) + GL(\tilde{y}(t) - \hat{y}(t))\end{aligned}. \tag{1.12}$$

Noticing the function (1.10), we have

$$-GBK\hat{x}(t) + GBu(t) + GL(\tilde{y}(t) - \hat{y}(t)) = 0. \tag{1.13}$$

Thus, we get the equivalent controller

$$u_{\text{eq}} = K\hat{x}(t) - (GB)^{-1}GL(\tilde{y}(t) - \hat{y}(t)). \tag{1.14}$$

Then, the controlled system and the error system are reformed as

$$\begin{cases}\dot{\hat{x}}(t) = (A + BK)\hat{x}(t) \\ + (L - B(GB)^{-1}GL)(\alpha(t)\varepsilon(t) + (1 - \alpha(t))C\Theta(t)) \\ \dot{\Theta}(t) = A\Theta(t) + Fw(t) \\ - L(\alpha(t)\varepsilon(t) + (1 - \alpha(t))C\Theta(t))\end{cases}. \tag{1.15}$$

1.3 Main Results

In this section, sufficient conditions are established for the closed-loop system (1.15) to be globally asymptotically stable with an L-∞ norm bound. Before presenting the main results, the following useful lemma is introduced.

Definition 1.1 Considering the power system (14), if $w(t) = 0$ and $\varepsilon(t) = 0$, there exists a positive definite continuous first-order partial derivative $V(\hat{x}(t))$ and negative definite $\mathbb{E}\{\dot{V}(\hat{x}(t))\}$. One can conclude that the system is asymptotically stable. Besides, under the zero initial condition of $\hat{x}(t) = 0, t \in [-\tau, 0]$, any nonzero $w(t), \varepsilon(t) \in L_2[0, +\infty)$. The disturbance rejection level of H_∞ is given as γ, we have the following condition: $\mathbb{E}\{\|\hat{y}(t)\|_2\} \leq \gamma\mathbb{E}\{\|w(t)\|_2 + \|\varepsilon(t)\|_2\}$.

Lemma 1.1 *For any real vectors u, v and symmetric positive matrix Q with compatible dimensions, the following inequality holds:*

$$u^T v + v^T u \leq u^T Q u + v^T Q^{-1} v.$$

1.3.1 Stability Analysis

Theorem 1.1 *For the given conditions $1 > \alpha_0 \geq 0$. System (1.5) is asymptotically stable for the given positive constants γ and real symmetric matrices $Q > 0$, $W > 0$ satisfying the following matrix inequality*

$$\begin{bmatrix} \varpi_{11} & \varpi_{12} \\ \varpi_{12}^T & \varpi_{22} \end{bmatrix} < 0, \tag{1.16}$$

where

$$\varpi_{11} = \varpi_a - \varpi_b, \varpi_a = [a_{7*7}], \varpi_b = \Gamma^T\Gamma,$$

$$a_{11} = \text{sym}[X(A + BK)] + Q,$$

$$a_{13} = (1 - \alpha_0)XLC, a_{17} = \alpha_0 XL,$$

$$a_{22} = \delta(j_k h)\phi, a_{25} = -\delta(j_k h)\phi,$$

$$a_{33} = \text{sym}[XA] + W - (1 - \alpha_0)XLC,$$

$$a_{36} = XF, a_{37} = -\alpha_0 XL,$$

$$a_{55} = (\delta(j_k h) - 1)\phi, a_{66} = a_{77} = -\gamma^2 I,$$

$$\Gamma = [(1 - \alpha_0)C \ \ 0^{1*6}],$$

$$\varpi_{12} = \begin{bmatrix} XB & 0 & 0 \\ 0 & 0 & 0 \\ 0 & (1 - \alpha_0)C^T L^T X & 0 \\ 0^{3\times 1} & 0^{3\times 1} & 0^{3\times 1} \\ 0 & 0 & \alpha_0 L^T X \end{bmatrix},$$

$$\varpi_{22} = \begin{bmatrix} -GB & 0 & 0 \\ 0 & -(1 - \alpha_0)X & 0 \\ 0 & 0 & -\alpha_0 X \end{bmatrix}.$$

Proof. Proposing a Lyapunov functional

$$V(t) = V_{\hat{x}}(t) + V_{\Theta}(t), \tag{1.17}$$

where

$$\begin{cases} V_{\hat{x}}(t) = \hat{x}^T(t)X\hat{x}(t) + \int_{t-\tau(t)}^{t} \hat{x}^T(v)Q\hat{x}(v)dv \\ V_{\Theta}(t) = \Theta^T(t)X\Theta(t) + \int_{t-\tau(t)}^{t} \Theta^T(v)W\Theta(\iota)dv \end{cases}.$$

Calculating the derivative of (1.17), we have

$$\dot{V}(t) = \dot{V}_{\hat{x}}(t) + \dot{V}_{\Theta}(t), \tag{1.18}$$

$$\begin{cases} \dot{V}_{\hat{x}}(t) = 2\hat{x}^T(t)X\dot{\hat{x}}(t) + \hat{x}^T(t)Q\hat{x}(t) \\ \dot{V}_{\Theta}(t) = 2\Theta^T(t)X\dot{\Theta}(t) + \Theta^T(t)W\Theta(t) \end{cases}$$

The expectation of (1.18) is

$$E\{\dot{V}(t)\} = E\{\dot{V}_{\hat{x}}(t) + \dot{V}_{\Theta}(t)\}, \tag{1.19}$$

for

$$\begin{aligned} E\{\dot{V}_{\hat{x}}(t)\} = {} & 2\hat{x}^T(t)X(A+BK)\hat{x}(t) \\ & + 2\alpha_0\hat{x}^T(t)X(L - B(GB)^{-1}GL)\varepsilon(t) \\ & + 2(1-\alpha_0)\hat{x}^T(t)X(L - B(GB)^{-1}GL)C\Theta(t) \\ & + 2\Theta^T(t)XF\omega(t) + \hat{x}^T(t)Q\hat{x}(t) \end{aligned},$$

$$\begin{aligned} E\{\dot{V}_{\Theta}(t)\} = {} & 2\Theta^T(t)XA\Theta(t) - 2\alpha_0\Theta^T(t)XL\varepsilon(t) \\ & - 2(1-\alpha_0)\Theta^T(t)XLC\Theta(t) - \Theta^T(t)W\Theta(t). \\ & + e^T(j_kh)\phi e(j_kh) - e^T(j_kh)\phi e(j_kh) \end{aligned}$$

Noticing that there exist some complex items in (1.19). According to Lemma 1.1, these complex items can be handled like

$$\begin{aligned} & -2\alpha_0\hat{x}^T(t)XB(GB)^{-1}GL\varepsilon(t) \\ & \quad \leq \alpha_0\hat{x}^T(t)XB(GB)^{-1}G\hat{x}(t) + \alpha_0\varepsilon^T(t)L^TXL\varepsilon(t) \end{aligned}$$

Also, we have

$$-2(1-\alpha_0)\hat{x}^T(t)XB(GB)^{-1}GLC\Theta(t)$$
$$\leq (1-\alpha_0)\hat{x}^T(t)XB(GB)^{-1}G\hat{x}(t)+(1-\alpha_0)\cdot$$
$$\times \Theta^T(t)C^TL^TXLC\Theta(t).$$

Take the event-triggered condition into considering

$$e^T(j_kh)\phi e(j_kh)$$
$$< \delta(j_kh)(\hat{x}(t-\tau(t))-e(j_kh))^T\phi(\hat{x}(t-\tau(t))-e(j_kh))$$

For convenience, we define a matrix as follows

$$\xi(t)=\mathrm{col}[\hat{x}^T(t)\hat{x}^T(t-\tau(t))\Theta^T(t)\Theta^T(t-\tau(t))$$
$$e^T(j_kh)\omega(t)\varepsilon^T(t)]$$

Then

$$\begin{aligned}&\mathbb{E}\{V(t)\}+\mathbb{E}\{\hat{y}^T(t)\hat{y}(t)\}-\gamma^2\mathbb{E}\{\omega^T(t)\omega(t)+\varepsilon^T(t)\varepsilon(t)\}\\&\leq \xi^T(t)(\varpi_{11}-\varpi_{12}\varpi_{22}^{-1}\varpi_{12}^T)\xi(t)\end{aligned}\tag{1.20}$$

Using the Schur theorem given in [34], (1.20) is concluded as

$$\mathbb{E}\{V(t)\}\leq \gamma^2\mathbb{E}\{\omega^T(t)\omega(t)-\hat{y}^T(t)\hat{y}(t)+\varepsilon^T(t)\varepsilon(t)\}.$$

Integrate the deduction above, we have

$$\begin{aligned}\mathbb{E}\{V(+\infty)\}-\mathbb{E}\{V(0)\}\leq \mathbb{E}\{\int_0^{+\infty}[-\hat{y}^T(t)\hat{y}(t)\\+\gamma^2(\omega^T(t)\omega(t)+\varepsilon^T(t)\varepsilon(t))]dt\}\end{aligned}$$

Under zero initial condition, we can further obtain

$$\mathbb{E}\{\|\hat{y}(t)\|_2\}\leq \gamma\mathbb{E}\{\|\omega(t)\|_2\}+\mathbb{E}\{\|\varepsilon(t)\|_2\}.\tag{1.21}$$

When $\omega(t)=0$ and $\varepsilon(t)=0$, the inequality is concluded as

$$\mathbb{E}\{\dot{V}(t)\}\leq -\mathbb{E}\{\hat{y}^T(t)\hat{y}(t)\}<0.\tag{1.22}$$

Then, there exists a positive scalar $\epsilon>0$ that makes the following inequality holds

$$\mathbb{E}\{\dot{V}(t)\}\leq -\epsilon\mathbb{E}\{\|\hat{x}(t)\|^2\}.\tag{1.23}$$

Based on the proposed control scheme, the multi-area power system to be robustly stable with an optimal attenuation level γ. It completes the proof.

Corollary 1.1 *System (1.5) is asymptotically stable for the given conditions $\alpha_0 \geq 0$, if there exist matrixes $Q > 0$ and $W > 0$, appropriate known positive number $\gamma > 0$, real symmetric matrixes $\chi > 0$, and an arbitrary matrix Y satisfying the following LMI*

$$\begin{bmatrix} \tilde{\varpi}_{11} & \tilde{\varpi}_{12} & \tilde{\varpi}_{13} \\ * & \tilde{\varpi}_{22} & 0^{1*3} \\ * & * & -I \end{bmatrix} < 0, \tag{1.24}$$

where

$$\tilde{\varpi}_{11} = [\tilde{a}_{7*7}]\,,\, \tilde{a}_{11} = \mathrm{sym}[A\chi + BY] + \chi Q \chi,$$

$$\tilde{a}_{13} = (1 - \alpha_0)LC\chi,\, \tilde{a}_{17} = \alpha_0 L,$$

$$\tilde{a}_{22} = \delta(j_k h)\phi,\, \tilde{a}_{25} = -\delta(j_k h)\phi,$$

$$\tilde{a}_{33} = \mathrm{sym}[A\chi] + \chi W \chi - (1 - \alpha_0)LC\chi,$$

$$\tilde{a}_{36} = F,\, \tilde{a}_{37} = -\alpha_0 L,$$

$$\tilde{a}_{55} = (\delta(j_k h) - 1)\phi,\, \tilde{a}_{66} = \tilde{a}_{77} = -\gamma^2 I,$$

$$\tilde{\varpi}_{13} = [(1 - \alpha_0)C\chi \;\; 0^{1*6}]^T,$$

$$\tilde{\varpi}_{12} = \begin{bmatrix} B & 0 & 0 \\ 0 & 0 & 0 \\ 0 & (1 - \alpha_0)\chi^T C^T N^T & 0 \\ 0^{3\times 1} & 0^{3\times 1} & 0^{3\times 1} \\ 0 & 0 & \alpha_0 L^T \end{bmatrix},$$

$$\tilde{\varpi}_{22} = \begin{bmatrix} -GB & 0 & 0 \\ 0 & -(1-\alpha_0)\chi & 0 \\ 0 & 0 & -\alpha_0\chi \end{bmatrix}.$$

Proof. Define $\chi = X^{-1}$ and $Y = K\chi$, pre- and post-multiplying both sides of inequality (1.16) with diag$\{\chi, I, \chi, I, I, I, I, I, \chi, \chi\}$.

We can notice that the matrix in Theorem 1.2 cannot be solved directly. The reason is for the nonlinear terms $LC\chi$, $\chi W\chi$ and $\chi Q\chi$. To deal with the nonlinear term $LC\chi$ in Theorem 1.2, we have added the following theorem. So that the nonlinear problem is solvable. Besides, $\chi W\chi$ and $\chi Q\chi$ are defined as $\chi W\chi = \tilde{W}$ and $\chi Q\chi = \tilde{Q}$, respectively.

Theorem 1.2 *Based on Corollary 1.1, for the given full-rank matrix M and any proper matrix N with appropriate dimensions, such that (1.24) holds, and the following LMI*

$$\begin{bmatrix} \tilde{\varpi}_{11}' & \tilde{\varpi}_{12}' & \tilde{\varpi}_{13} \\ * & \tilde{\varpi}_{22} & 0^{1*3} \\ * & * & -I \end{bmatrix} < 0, \tag{1.25}$$

$$\begin{bmatrix} \xi I & C^T M^T - X^T C^T \\ * & -I \end{bmatrix} < 0, \tag{1.26}$$

where ξ is approaching to 0, $\tilde{\varpi}_{11}'$ and $\tilde{\varpi}_{12}'$ are obtained from $\tilde{\varpi}_{11}$ and $\tilde{\varpi}_{12}$ by replacing $LC\chi$ with NC and replacing $\chi W\chi$, $\chi Q\chi$ with $\tilde{W}$, $\tilde{Q}$ respectively, the system is asymptotically stable with observer gains $L = NM^{-1}$. It completes the proof.

Remark 1.1: Based on the proposed sliding mode LFC scheme, Theorem 1.2 could guarantee the attacked multi-area power system to be robustly stable with an optimal attenuation level γ. In Theorem 1.2, the nonlinear coupling terms have been replaced by processable terms.

1.3.2 Control Law Design

For SMC theory, a proper sliding control law is of great importance. The trajectory of the power system shall be reachable to the sliding surface in a limited time under the supplied controller. For the controlled multi-area power system (1.5), the corresponding reachability condition for the i-th area can be described by the following theorem.

Theorem 1.3 *For the considered sliding mode surface (1.9), a proper control law is given as*

$$u(t) = K\hat{x}(t) - (1-\alpha(t))(GB)^{-1}GLC\Theta(t) \\ - h_i(GB)^{-1}\sigma_i(t) - h_i(GB)^{-1}\|GF\|\mathrm{sgn}(\sigma_i(t)), \tag{1.27}$$

where h_i is a positive constant, and $\mathrm{sgn}(\sigma_i(t))$ *is the switching function that*

$$\mathrm{sgn}(\sigma_i(t)) = \begin{cases} -1, & \text{if } \sigma_i(t) < 0 \\ 0, & \text{if } \sigma_i(t) = 0. \\ 1, & \text{if } \sigma_i(t) > 0 \end{cases}$$

Proof. The Lyapunov function is constructed as

$$V_L = \frac{1}{2}\sigma_i^T(t)\sigma_i(t). \tag{1.28}$$

From (1.12), (1.27), and (1.28), the derivative of $V_L(t)$ is

$$\begin{aligned} \dot{V}_L &= \frac{1}{2}\dot{\sigma}_i^T(t)\sigma_i(t) + \frac{1}{2}\sigma_i^T(t)\dot{\sigma}_i(t) \\ &= \sigma_i^T(t)\dot{\sigma}_i(t) \\ &= \sigma_i(t)[-GBK\hat{x} + GBu(t) + (1-\alpha(t))GLC\Theta(t)] \\ &= \sigma_i(t)[-GBK\hat{x} + (1-\alpha(t))GLC\Theta(t) \\ &\quad + GB(K\hat{x}(t) - (1-\alpha(t))(GB)^{-1}GLC\Theta(t) \\ &\quad - h_i(GB)^{-1}\sigma_i(t) - h_i(GB)^{-1}\|GF\|\mathrm{sgn}(\sigma_i(t)))] \\ &= -h_i\|\sigma_i(t)\|^2 - h_i\sigma_i(t)\|GF\|\mathrm{sgn}(\sigma_i(t)) \\ &< -h_i\|\sigma_i(t)\|^2 - h_i\|\sigma_i(t)\|\|GF\| < 0 \end{aligned}$$

Therefore, one can conclude that the system can be kept stable by the designed control law (1.27).

Remark 1.2: The variable structure control has the advantages of fast response and being insensitive to parameter changes and disturbances. The way of SMC can be designed and independent of plant parameters and disturbances. This method can be simply implemented in physical. The main disadvantage, however, is that the imperfections of all kinds of components of the actual system may contribute to causing the control switching delay. Specifically, when the state trajectory reaches the sliding surface, it is difficult to slide strictly along the sliding surface toward the equilibrium point. Instead, it may pass back and forth on both sides of the sliding surface, resulting in chattering. The error of the operation in MATLAB. Besides, the use of SMC will certainly cause some buffeting phenomena.

1.4 Numerical Examples

In this section, numerical examples are provided to demonstrate the effectiveness of the proposed design scheme. Take a three-area interconnected power system as an example. The simulation results verify the effectiveness and correctness of the proposed SMC scheme. By analyzing the simulation diagram, one can conclude that the system can reach a stable state under the designed SMC.

Parameter values of the three-area interconnected power system are presented in Table 1.2. We consider a three-area interconnected power system. By using the MATLAB LMI control toolbox, the stability condition could be verified effectively.

The controller gain and observer gain can be calculated as

$$K = \begin{bmatrix} -223.783 & -5.012 & -3.280 & 91.864 & 1.795 \\ -157.956 & -2.073 & -2.597 & 67.533 & 1.631 \\ -160.963 & -2.360 & -2.448 & 69.519 & 1.510 \end{bmatrix},$$

$$L = \begin{bmatrix} 0.272 & 1.219 & -0.280 & 0.075 & -0.116 \\ 0.108 & -0.04 & -0.224 & 0.019 & -0.295 \\ 0.207 & -0.017 & -0.292 & 0.063 & -0.135 \end{bmatrix}^T.$$

Under the control scheme, the frequency derivative of the power system is shown in Figure 1.2. It can be seen that with the designed control scheme, the frequency derivative value of the deception-attacked power system gradually tends to a stable state. This demonstrates that the proposed controller can effectively

Table 1.2 Parameters of three-area interconnected power system.

D_1, D_2, D_3	1, 1.5, 1.8
$T_{ch1}, T_{ch2}, T_{ch3}$	0.3, 0.17, 0.2
T_{g1}, T_{g2}, T_{g3}	0.37, 0.4, 0.35
R_1, R_2, R_3	0.05, 0.05, 0.05
beta1, beta2, beta3	21, 21.5, 21.8
P_{d1}, P_{d2}, P_d3	0.1, 0.1, 0.1
$\gamma_1, \gamma_2, \gamma_3,$	2, 2, 2

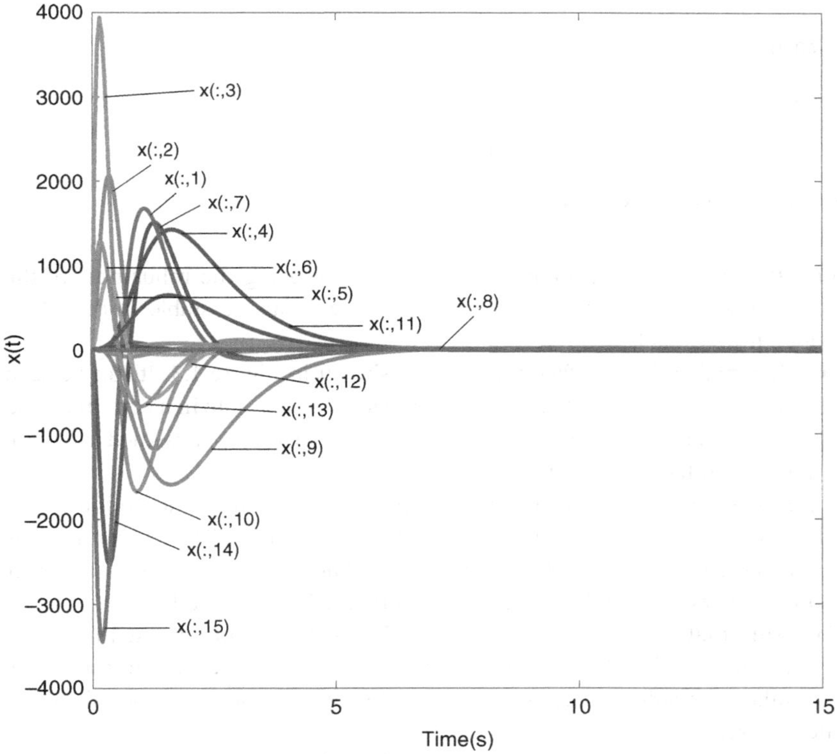

Figure 1.2 Frequency deviation of power system.

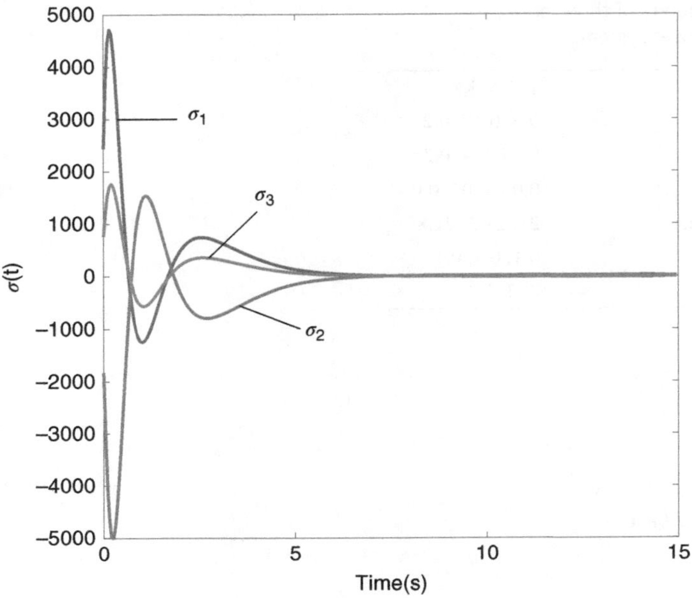

Figure 1.3 Designed sliding mode surface (SMC).

reduce the frequency deviation of the system, improving the robustness of the attacked power system. This is, the power system can be stable under SMC, although it was attacked.

The designed sliding surface trajectory is shown in Figure 1.3. It can be seen from the figure that the sliding surface can reach a stable state in a certain period and keep running. One can prove that the sliding mode that we designed can work normally and satisfies our purpose.

It can be seen from Figure 1.4 that the power system under deception attack with equivalent control can operate to a stable state over time. The test condition is listed in Table 1.2 and the controller gain is calculated as K. This means the power system can work in a stable state under the provided control method.

To ensure that the system can reach the sliding surface in a limited time, the state trajectory of the SMC law is shown in Figure 1.5. According to the curve, the simulation results can easily show the effectiveness of the proposed sliding mode LFC scheme.

We test on the case where something is random, the event-trigger frequency and probability are equally random. Figure 1.6 shows the event interval during the deception attack. The disorder event is shown in Figure 1.6, conforming to our requirements for random attacks.

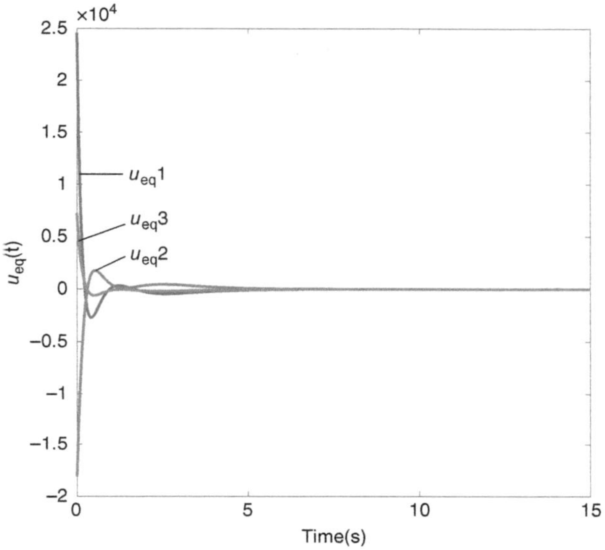

Figure 1.4 Equivalent control of sliding mode controlled power system.

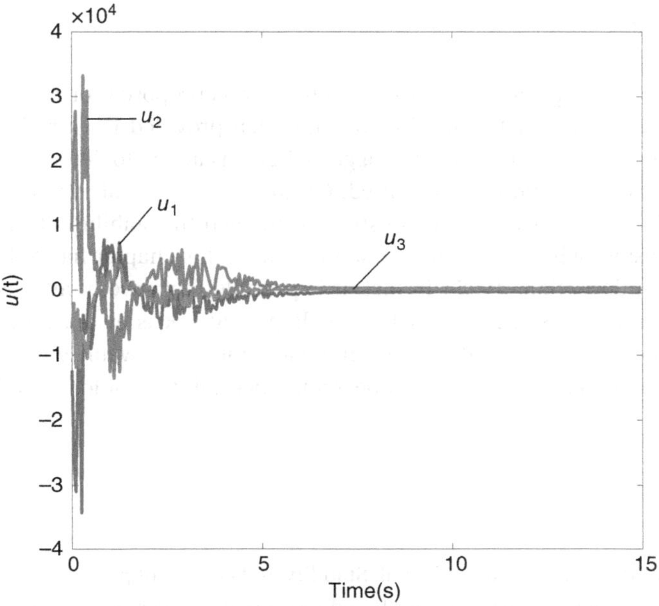

Figure 1.5 State of SMC law.

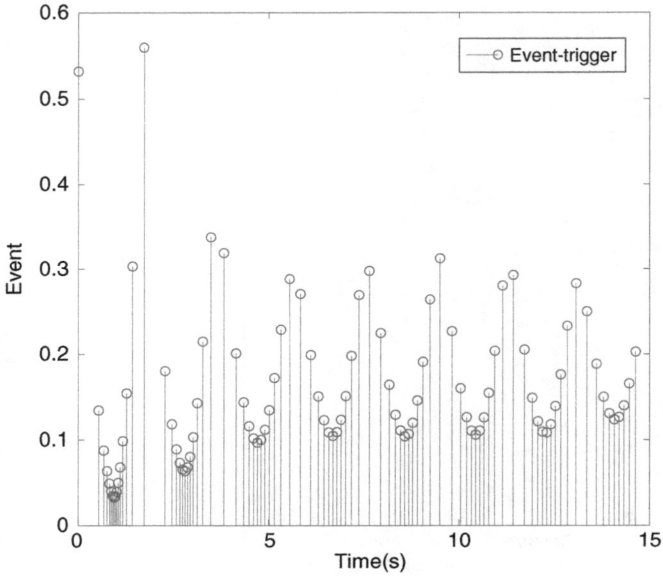

Figure 1.6 Event interval during deception attack interval.

1.5 Conclusion

An observer-based event-triggered transmission scheme was proposed for multi-area power systems. To be specific, an observer has been provided to gain disturbed transmission information. An event-trigger scheme is added to determine whether the information should be transmitted. On the basis of the above analysis, a sliding mode controller has been constructed to keep the stability of the multi-area power system which is injured by some attack. This chapter has only considered a kind of deception attack. In fact, it is possible that several attacks influence the system at the same time. The modeling of attack is also a problem worthy of consideration. In addition, with more and more renewable energy access to the power grid, its inherent randomness and uncertainty should be paid attention to.

References

1 R. W. Kenyon, M. Bossart, M. Markovi, et al. Stability and control of power systems with high penetrations of inverter-based resources: An accessible review of current knowledge and open questions. *Solar Energy*, 210:149–168, 2020.

2 H. Ahsan and M. D. Mufti. Comprehensive power system stability improvement with rocof controlled smes. *Electric Power Components and Systems*, 48:162–173, 2020.

3 A. C. Xue, J. H. Zhang, L. J. Zhang, et al. Transient frequency stability emergency control for the power system interconnected with offshore wind power through VSC-HVDC. *IEEE Access*, 8:53133–53140, 2020.

4 X. Lv, Y. Sun, Y. Wang, and V. Dinavahi. Adaptive event-triggered load frequency control of multi-area power systems under networked environment via sliding mode control. *IEEE Access*, 8:86585–86594, 2020.

5 Y. M. Alsmadi, A. Alqahtani, R. Giral, et al. Sliding mode control of photovoltaic based power generation systems for microgrid applications. *International Journal of Control*, 94:1704–1715, 2021.

6 M. Y. Silaa, M. Derbeli, O. Barambones, and A. Cheknane. Design and implementation of high order sliding mode control for PEMFC power system. *Energies*, 13:4317–4332, 2020.

7 A. T. Tran, B. Le Ngoc Minh, P. T. Tran, et al. Adaptive integral second-order sliding mode control design for load frequency control of large-scale power system with communication delays. *Complexity*, 2021:5564184, 2021.

8 R. Nie, S. He, F. Liu, and X. L. Luan. Sliding mode controller design for conic-type nonlinear semi-Markovian jumping systems of time-delayed Chua's circuit. *IEEE Transactions on Systems*, 51:2467–2475, 2021.

9 G. Q. Zhu, L. Nie, Z. Lv, et al. Adaptive fuzzy dynamic surface sliding mode control of large-scale power systems with prescribe output tracking performance. *ISA Transactions*, 99:305–321, 2020.

10 X. Lv, Y. Sun, W. Hu, and V. Dinavahi. Robust load frequency control for networked power system with renewable energy via fractional-order global sliding mode control. *IET Renewable Power Generation*, 15:1046–1057, 2021.

11 A. T. Tran, B. Le Ngoc Minh, V. V. Huynh, et al. Load frequency regulator in interconnected power system using second-order sliding mode control combined with state estimator. *Energy*, 14:863–880, 2021

12 M. Elsisi, N. Bazmohammadi, J. M. Guerrero, and M. A. Ebrahim. Energy management of controllable loads in multi-area power systems with wind power penetration based on new supervisor fuzzy nonlinear sliding mode control. *Energy*, 221:119867–119881, 2021.

13 S. P. He, W. Z. Lyu, and F. Liu. Robust sliding mode controller design of a class of time-delayed discrete conic-type nonlinear systems. *IEEE Transactions on Systems, Man, and Cybernetics: Systems*, 51:885–892, 2021.

14 W. H. Qi, G. D. Zong, and W. X. Zheng. Adaptive event-triggered smc for stochastic switching systems with semi-Markov process and application to boost converter circuit model. *IEEE Transactions on Circuits and Systems I: Regular Papers*, 68:786–796, 2020.

15 S. H. Huang and J. Wang. Fixed-time fractional-order sliding mode control for nonlinear power systems. *Journal of Vibration and Control*, 26:1425–1434, 2020.

16 V. V. Huynh, B. Le Ngoc Minh, E. N. Amaefule, et al. Highly robust observer sliding mode based frequency control for multi area power systems with renewable power plants. *Electronics*, 10:274–294, 2021.

17 N. K. Singh and V. Mahajan. Analysis and evaluation of cyber-attack impact on critical power system infrastructure. *Smart Science*, 9:1–13, 2021.

18 X. C. ShangGuan, Y. He, C. K. Zhang, et al. Switching system-based load frequency control for multi-area power system resilient to denial-of-service attacks. *Control Engineering Practice*, 107:104678–104689, 2021.

19 L. Zhao and W. Li. Co-design of dual security control and communication for nonlinear CPS under DoS attack. *IEEE Access*, 8:19271–19285, 2020.

20 B. Hu, H. Wang, Y. Zhao, et al. A brief overview of optimal robust control strategies for a benchmark power system with different cyber-physical attacks. *Complexity*, 2021(1):6646799, 2021

21 W. H. Qi, Y. K. Hou, G. D. Zong, and C. Ki Ahn. Finite-time event-triggered control for semi-Markovian switching cyber-physical systems with FDI attacks and applications. *IEEE Transactions on Circuits and Systems I: Regular Papers*, 68:2665–2674, 2021.

22 Z. H. Pang and G. P. Liu. Design and implementation of secure networked predictive control systems under deception attacks. *IEEE Transactions on Control Systems Technology*, 20:1334–1342, 2011.

23 M. R. Habibi, H. R. Baghaee, F. Blaabjerg, and T. Dragievi. False data injection cyber-attacks mitigation in parallel DC/DC converters based on artificial neural networks. *IEEE Transactions on Circuits and Systems II: Express Briefs*, 68:717–721, 2020.

24 D. J. Du, R. Chen, X. Li, et al. Malicious data deception attacks against power systems: A new case and its detection method. *Transactions of the Institute of Measurement and Control*, 41:1590–1599, 2019.

25 Y. Li, P. Zhang, B. Wang, and L. Zhang. Active synchronous detection of deception attacks in microgrid control systems. *IEEE Transactions on Smart Grid*, 8:373–375, 2016.

26 R. C. Borges Hink, J. M. Beaver, M. A. Buckner, et al. Machine learning for power system disturbance and cyber-attack discrimination. *7th International Symposium on Resilient Control Systems*, pages 1–8, 2014.

27 D. Zhao, Z. D. Wang, G. L. Wei, et al. A dynamic event-triggered approach to observer-based PID security control subject to deception attacks. *Automatica*, 120:109128–109141, 2020.

28 J. Shi. Cooperative control for nonlinear multi-agent systems based on event-triggered scheme. *IEEE Transactions on Circuits and Systems II: Express Briefs*, 68:1977–1981, 2020.

29 E. Tian and C. Peng. Memory-based event-triggering h_∞ load frequency control for power systems under deception attacks. *IEEE Transactions on Cybernetics*, 50:4610–4618, 2020.

30 D. J. Wang, F. Chen, B. Meng, et al. Event-based secure h_∞ load frequency control for delayed power systems subject to deception attacks. *Applied Mathematics and Computation*, 394:125788–125801, 2021.

31 W. Shen, S. Liu, and M. Liu. Adaptive sliding mode control of hydraulic systems with the event-trigger and finite-time disturbance observer. *Information Sciences*, 569:55–69, 2021.

32 Md. M. Hossain and C. Peng. Observer-based event-triggering h_∞ lfc for multi-area power systems under dos attacks. *Information Sciences*, 543:437–453, 2021.

33 D. Yue, E. Tian, and Q. L. Han. A delay system method for designing event-triggered controllers of networked control systems. *IEEE Transactions on Automatic Control*, 58:475–481, 2013.

34 S. Boyd, L. El Ghaoui, E. Feron, et al. *Linear Matrix Inequalities in System and Control Theory*. Philadelphia: SIAM, 1994.

2

Event-triggered Sliding Mode Load Frequency Control of Multi-area Power Systems Under Periodic Denial-of-service Attacks

After reading this chapter, you should be able to:

- Explain the main factors of the multi-area power system and the main relationship among equipments.
- Analysis the harm of cyberattacks and the characteristics of deception attacks.
- Clarify the design of the controller and the conditions for power system stability.

2.1 Background

Supported by the developments of science and technology, the scale of electric power systems is getting larger and larger. The multi-area interconnected power systems are gradually formed. The interconnected power systems are adding a shared network for information exchange. Due to the limitations of the network itself, safety and other issues are becoming increasingly prominent. The impacts of attacks on the multi-area power systems are often large-scale. Such attacks have caused significant damage to economic and social life [1, 2]. To handle such challenges, extensive studies have been conducted, generating some encouraging results [3–5]. The diversity of attacks on the multi-area power systems, however, makes the corresponding prevention a major challenge. Such attacks may include DoS attacks [6–9], dynamic load-altering attacks [10], false-data attacks [11–13], etc. Different attacks need to be discussed separately. This chapter mainly discusses on the control problem of multi-area power systems against DoS attacks.

DoS attacks impose interferences against communication networks. The purpose of DoS attacks is typically to interrupt the data transmission, causing time delay, packet dropout, and other problems. It is worth noting that even for the same type of attack, there are different attack modes. Unlike aperiodic DoS attacks [14–16], periodic DoS attacks are repeated periodically. The periodic or pulse width-modulated DoS attacks are relatively simple in structure and easy to implement, while still can effectively interrupt information transmission. Also, as such attacks adopt a periodic working mode, there exists a certain period of rest time for the attack signals. Hence, with a given amount of energy, periodic DoS attacks may affect the normal operation of the systems for a longer time. Protecting the system against periodic DoS attacks with a continuous response plan will inevitably lead to a large amount of wasted resources. When a DoS attack occurs, there would be missing inputs [17], imposing a major threat to normal work of the system. Therefore, exploring the periodic DoS attacks is meaningful. Various studies have been carried out to counter DoS attacks. The authors in [18] proposed an event-based plan to handle the DoS jamming attacks. The system stability while taking into account the influences of quantization and attacks was discussed in [19]. In [20] the authors focused on the security consistency of multi-agent systems, which may be contaminated by periodic DoS attacks. By designing the observer and feedback control method, the stability problem under multi-periodic DoS attacks was studied in [21]. In [22], nonhomogeneous agent systems under periodic DoS attacks were discussed, where a control method is proposed to handle the event-triggered output synchronization problem. In order to recover the damage caused by DoS attacks, a proper control method is a good and feasible idea.

Load frequency control (LFC) method is to meet the requirements of frequency balance between interconnection area power systems [23]. And, sliding mode control (SMC) has already been extensively used in the control field and engineering practice. This powerful control strategy has a good performance in helping build up stable closed-loop control systems, even under disturbances or uncertainties [24–27]. The sliding mode load frequency control (SMLFC) methods therefore naturally provide an efficient approach for solving the stability problem of interconnected multi-area power systems. A decentralized sliding mode load frequency controller was designed to solve the problem of matching and unmatched uncertainties in [28]. The authors in [29] proposed a new event-triggered SMLFC method in multi-area interconnected power systems. An SMLFC strategy was proposed for reducing frequency deviations of the overall power system in [30]. To use the SMLFC method in the power systems under attack, however, we need to know the system state information. To get the system state, the contamination caused by periodic DoS attacks has to be tackled. Along this direction, we shall propose an event-triggered scheme for system information transmission, of which the triggering conditions should be fully discussed.

The main advantage of event-triggered control methods is that they help reduce information exchanges and save resources in communication channels [31, 32]. In many interconnected engineering systems, operations of the samplers are time-triggered. An event-triggered mechanism may be applied to determine whether the sampled signals are useful, such that only useful signals are transmitted. By applying an event-triggered mechanism, we may also reduce the computation cost of the controller and the occupation of the sensors and actuators [33]. In [34], an adaptive torus-event-based control method was proposed to enhance the security of a communication network disturbed by external attacks. By setting multiple sampling values as a set of data, the authors in [35] proposed to use the average value as the event triggering condition. Event-triggered control is particularly suitable for a power system with external interferences since it timely responds to the changes of state by generating a triggering signal when the system state deviation value is greater than a certain preset threshold. Typically, we use the system instantaneous states in event-triggered scheme [36–40]. When we need applying event-triggered control in power systems under periodic DoS attacks, the actual triggering will be carried out periodically. Therefore, the triggering strategy needs to be redesigned. We design a suitable event-triggered scheme to apply the SMLFC methods under periodic DoS attacks.

In this chapter, we investigate the event-triggered SMLFC problem for multi-area power system under periodic DoS attacks. The main contributions are outlined as follows:

1. Unlike aperiodic DoS attacks [14–16] that appear randomly, periodic DoS attacks continually interrupt the information transmission of power systems for a long period of time. To deal with the unstable problem caused by periodic attacks, the SMLFC method is applied to establish exponential stability criterion.
2. Due to the influence of DoS attacks, the continuous event triggering strategy can no longer meet the periodic change. A periodic event triggering scheme is designed to conform the characteristics of periodic DoS attacks [31, 38, 39], which has a good performance in handling the periodic attacks and saving resources.
3. Different from the typical single Lyapunov function approach [23], a multiple Lyapunov function is constructed to make the result in a less conservative situation. Theoretical analysis results show that the multi-systems are stable under the proposed control scheme, and the corresponding controller gains are obtained.

2.2 Problem Formulation

2.2.1 Model Description

To analyze the SMLFC problem, we need to simplify the system model. The i-th subsystems under periodic attack is illustrated in Figure 2.1. The main composition power system, controller and components under attack are introduced.

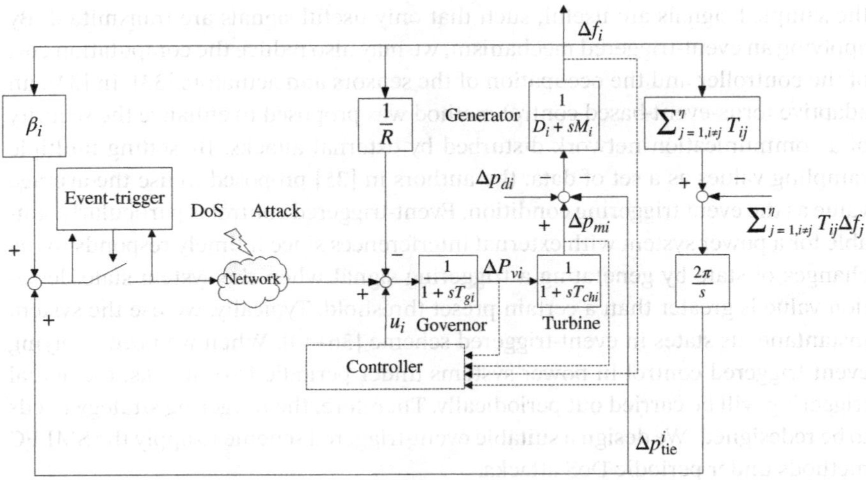

Figure 2.1 Model of the i-th subsystem under periodic DoS attack.

Consider the signal transmission inside the simplified model. The state equations of the attacked area are listed as

$$\begin{cases} \Delta \dot{P}_{mi}(t) = -\dfrac{\Delta P_{mi}(t)}{T_{\text{chi}}} + \dfrac{\Delta P_{vi}(t)}{T_{\text{chi}}}, \\ \Delta \dot{P}^i_{\text{tie}}(t) = 2\pi \displaystyle\sum_{j=1,i\neq j}^{n} T_{ij}(\Delta f_i(t) - \Delta f_j(t)), \\ \Delta \dot{P}_{vi}(t) = \dfrac{u_i(t)}{T_{gi}} - \dfrac{\Delta f_i(t)}{R_i T_{gi}} - \dfrac{\Delta P_{vi}(t)}{T_{gi}}, \\ \Delta \dot{f}_i(t) = \dfrac{\Delta P_{mi}(t)}{M_i} - \dfrac{D_i \Delta f_i(t)}{M_i} - \dfrac{\Delta P^i_{\text{tie}}(t)}{M_i} - \dfrac{\Delta P_{di}(t)}{M_i}. \end{cases} \tag{2.1}$$

Therefore, the following system state equations are obtained

$$\begin{cases} \dot{x}_i(t) = A_{ii}x_i(t) + B_i u_i(t) + \displaystyle\sum_{\substack{j=1, \\ i\neq j}}^{n} A_{ij}x_j(t) + F_i w_i(t), \\ y_i(t) = C_i x_i(t), \end{cases} \tag{2.2}$$

where system disturbance $w(t)$ is defined with generality as $w(t) = \Delta P_{\text{di}} < \epsilon$, where ϵ is a positive constants. And, we have the following definition

$$x(t) = [x_1(t), x_2(t), \cdots, x_\eta(t)]^\top,$$

$$x_i(t) = [\Delta P_{mi}, \Delta P_{vi}, \Delta P^i_{\text{tie}}, \Delta f_i]^\top,$$

$$u(t) = [u_1(t), u_2(t), \cdots, u_\eta(t)]^\top,$$

$$w(t) = [\Delta P_{d1}(t), \Delta P_{d2}(t), \cdots, \Delta P_{d\eta}(t)]^\top,$$

$$y(t) = [y_1(t), y_2(t), \cdots, y_\eta(t)]^\top,$$

$$A_{ii} = \begin{bmatrix} -\frac{1}{T_{\text{chi}}} & \frac{1}{T_{\text{chi}}} & 0 & 0 \\ 0 & -\frac{1}{T_{gi}} & 0 & -\frac{1}{R_i T_{gi}} \\ 0 & 0 & 0 & 2\pi \sum\limits_{j=1, j\neq i}^{\eta} T_{ij} \\ \frac{1}{M_i} & 0 & -\frac{1}{M_i} & -\frac{D_i}{M_i} \end{bmatrix},$$

$$A_{ij} = \begin{bmatrix} 0 & 0 & 0 & 0 \\ 0 & 0 & 0 & 0 \\ 0 & 0 & 0 & -2\pi T_{ij} \\ 0 & 0 & 0 & 0 \end{bmatrix}, \quad B = \begin{bmatrix} 0 & \frac{1}{T_{gi}} & 0 & 0 \end{bmatrix}^\top, \quad F = \begin{bmatrix} 0 & 0 & 0 & -\frac{1}{M_i} \end{bmatrix}^\top.$$

The entire interconnected power system is expressed

$$\dot{x}(t) = \tilde{A}x(t) + \tilde{B}u(t) + \tilde{F}w(t), \tag{2.3}$$

where

$$\tilde{A} = \begin{bmatrix} A_{11} & A_{12} & \dots & A_{1\eta} \\ A_{21} & A_{22} & \dots & A_{2\eta} \\ \vdots & \vdots & \vdots & \vdots \\ A_{\eta 1} & A_{\eta 2} & \dots & A_{\eta\eta} \end{bmatrix},$$

$$\tilde{F} = \text{diag}\{\underbrace{F_1, \cdots, F_i}_{\eta}\}, \tilde{B} = \text{diag}\{\underbrace{B_1, \cdots, B_i}_{\eta}\}.$$

2.2.2 Sliding Mode Surface Design

The SMLFC method is applied on multi-area power systems in this chapter. Based on the information transmission and connections in the multi-area power systems, considering the balance of load and frequency, we construct the state-space function (2.3). Then a proper sliding mode surface is designed with the system state $x(t)$. Specifically, we construct the following sliding surface function

$$\sigma(t) = Lx(t) - L\int_0^t (\tilde{A} + \tilde{B}K)x(\tau)d\tau, \tag{2.4}$$

where the matrix L is designed as $L = \tilde{B}^\top P$, $P \in \mathbb{R}^{\eta\times\eta}$ is positive and guarantees the nonsingularity of $L\tilde{B}$.

On account of the core principles of the SMC method, the designed surface should fulfill $\dot{\sigma}(t) = 0$ and $\sigma(t) = 0$. Therefore, we can derive the equivalent SMC law as

$$u_{\text{eq}}(t) = Kx(t) - (L\tilde{B})^{-1}L\tilde{F}w(t). \tag{2.5}$$

2.2.3 Event-triggered Strategy Under Periodic DoS Attacks

An event-triggered device is proposed to judge whether the sampled signal should be transmitted, based on which the amount of transmission data in the correspondence channels can be reduced.

Some essential definitions are given to allow study on contamination caused by periodic DoS attacks on the system. We take the periodic DoS attack signal in the following forms

$$S_{\text{DoS}}(t) = \begin{cases} 0,\ t \in [nT, nT + T_{\text{off}}), \\ 1,\ t \in [nT + T_{\text{off}}, nT + T), \end{cases} \tag{2.6}$$

where $T \in \mathbb{R}^+$ is the given attack period, $T_{\text{off}} \in \mathbb{R}^+$ and $T_{\text{off}} < T$, denote the period of time with normal transmission.

Thus, we have the following fact

$$S_{\text{DoS}}(t) = \begin{cases} \text{communication is successful},\ t \in [nT, nT + T_{\text{off}}), \\ \text{communication is impeded},\ t \in [nT + T_{\text{off}}, nT + T). \end{cases} \tag{2.7}$$

Figure 2.2 clearly conveys the pattern of periodic DoS attacks. T is the attacks period, T_{off} is defined as the sleeping interval.

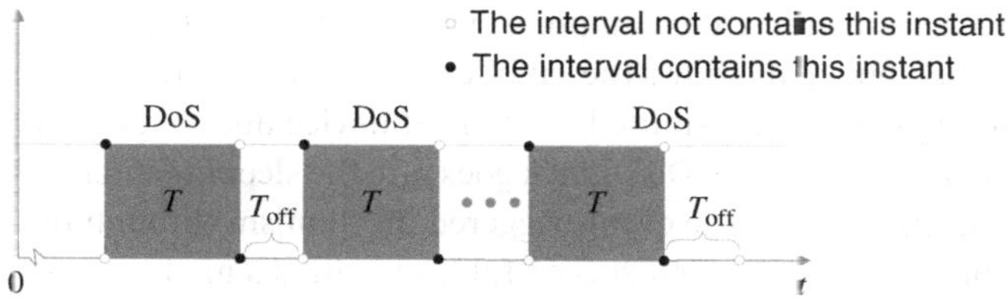

Figure 2.2 Pattern of DoS attacks.

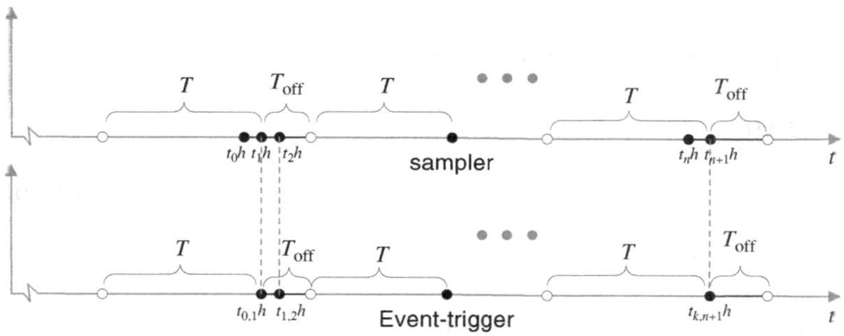

Figure 2.3 Sample instants and event-triggered instants.

The working principle of event-triggered scheme is to define the system error $e(\tilde{t}_k h) = x(\tilde{t}_k h) - x(t_k h)$, where $\tilde{t}_k h = t_k h + lh$. The basic form is

$$e^{\mathsf{T}}(\tilde{t}_k h) W e(\tilde{t}_k h) > \nu x^{\mathsf{T}}(t_k h) W x(t_k h), \tag{2.8}$$

where W is the given positive weighting matrix, $\nu \in (0, 1)$ is the triggering parameter. $l \in \mathbb{N}$ is the sampling times, $0 < h < T_{\text{off}}$ is the sampling interval. In a word, the event-triggered scheme (2.8) is used to reduce the amount of data transmission and make it convenient for derivation.

Figure 2.3 illustrates the relation between the sampler and the event-triggered scheme. The releasing instants of the sampler is $\{t_{k,n}h\}$, where n is the sampling times, k is the triggering times.

Considering the attack mode in Figure 2.2, the triggering time under attacks shown in Figure 2.3 is redefined as

$$t_{k,n+1}h = \{t_k h \text{ satisfying}(2.8) | t_k \in \Gamma_{1,n}\} \cup \{(n+1)T\}, \tag{2.9}$$

where k stands for the k-th triggering in $(n+1)$-th attacks, $n \in \mathbb{N}$, $k \in \{0, 1, \dots, k(n)\}$, $t_{0,n+1}h \triangleq nT$, and $k(n) = \sup\{k \in \mathbb{N} | nT + T_{\text{off}} \geq t_{k,n+1}h\}$. $x(t_{k,n+1}h)$ is defined as the status of current sampling time.

Remark 2.1: Figure 2.3 illustrates the sampling instants and event-triggered instants. The figure in the upper part shows the sample instants. It is worth noting that not all the sample signals are transmitted to the next device due to the existence of DoS attacks. When the periodic DoS attack goes into the sleeping interval, the sampling signal is transmitted to the event-triggered mechanism through the communication network. Besides, the first successful transmitted signal $t_{0,h}$ cannot cause event triggering, as the event triggering requires two adjacent successful transmitted values. An event-triggered mechanism is used to determine whether the sample signal should be further transmitted.

The equivalent control under attack is expressed as

$$u_{eq} = \begin{cases} Kx(t_{k,n+1}h) - (L\tilde{B})^{-1}L\tilde{F}w(t), \\ t \in [t_{k,n+1}h, t_{k+1,n+1}h) \cap [nT, nT + T_{\text{off}}), \\ 0, t \in [nT + T_{\text{off}}, nT + T). \end{cases} \tag{2.10}$$

For the convenience of analysis, we have the following definition

$$\begin{cases} H_{k,n} = [t_{k,n+1}h, t_{k+1,n+1}h), \\ \Gamma_{1,n} = [nT, nT + T_{\text{off}}), \\ \Gamma_{2,n} = [nT + T_{\text{off}}, nT + T). \end{cases} \tag{2.11}$$

Remark 2.2: The control signal cannot be transferred during the attack interval, which means the event-triggering control sequence generated by (2.8) is not able to achieve the stability of the power system. To handle the problem, we need to modify the triggering condition given in (2.8). From (2.9), $t_{k,n+1}h$ is the triggering instant generated by (2.8) in the $(n + 1)$-th attack mode. Triggering instant $t_{k,n+1}h$ belongs to the time interval $\Gamma_{1,n}$ or right endpoint $(n + 1)T$ of each interval $[nT, (n + 1)T)$. That is, if the triggering instant does not belong to $\Gamma_{1,n}$, then the triggering instant only happens at the right endpoint $(n + 1)T$. Besides, in the given triggered strategy, the Zeno phenomenon is excluded naturally.

Combining (2.3), (2.10), the system can be expressed as

$$\dot{x}(t) = \begin{cases} \tilde{A}x(t) + \tilde{B}Kx(t_{k,n+1}h) + \hat{F}w(t), t \in H_{k,n} \cap \Gamma_{1,n} \\ \tilde{A}x(t), t \in \Gamma_{2,n}. \end{cases} \tag{2.12}$$

where $\hat{F} = [I - B(B^\top PB)^{-1}B^\top P]\tilde{F}$.

Remark 2.3: Due to the characteristics of periodic DoS attacks, there exists a working interval for attacking the power system and a sleeping interval for allowing information transmission. When the periodic DoS attacks happen, the released sampled state cannot reach the controller successfully, while the control signal cannot arrive at the actuator either. In the attack sleeping interval, the signal is transmitted according to the event-triggered mechanism. We can know that the successful transmission signal is $x(t_{k,n+1})$. The control signal can also be transmitted during the sleeping interval. By introducing the equivalent control (2.10) into the system state equation (2.3), we get the equivalent control equation under SMLFC.

We then take the event-triggered device into consideration. The following subsets is used to represented the time of triggering interval

$$\begin{aligned} H_{k,n} &= \cup_{m=1}^{M}[t_{k,n+1}h+(m-1)h,\ t_{k,n+1}h+mh) \\ &\quad \cup[t_{k,n+1}h+Mh,\ t_{k+1,n+1}h), \end{aligned} \tag{2.13}$$

where $M \triangleq \sup\{m \in \mathbb{N} | t_{t,n+1} + mh < t_{k+1,n+1}h\}$ is the maximum positive integer number between events.

The handling is similar to that in [41]. We introduce the following definition

$$\begin{aligned} H_{k,n}^{m} &= [t_{k,n+1}h+(m-1)h,\ t_{k,n+1}h+mh), \\ H_{k,n}^{M} &= [t_{k,n+1}h+Mh,\ t_{k+1,n+1}h). \end{aligned} \tag{2.14}$$

Then, combining (2.13 and 2.14), we have the following

$$\Gamma_{1,n} = \cup_{k=0}^{k(n)}\cup_{m=1}^{M}\Omega_{k,n}^{m},\quad \Omega_{k,n}^{m} = H_{k,n}^{m} \cap \Gamma_{1,n}. \tag{2.15}$$

Next, we make the following definition for easy construction of system state equation

$$\xi_{k,n}(t) = \begin{cases} t - t_{k,n+1}h,\ t \in \Omega_{k,n}^{1}, \\ t - t_{k,n+1}h - h,\ t \in \Omega_{k,n}^{2}, \\ \vdots \\ t - t_{k,n+1}h - Mh,\ t \in \Omega_{k,n}^{M}, \end{cases} \tag{2.16}$$

$$\varpi_{k,n}(t) = \begin{cases} 0,\ \Omega_{k,n}^{1} \\ x(t_{k,n+1}h) - x(t_{k,n+1}h+h),\ t \in \Omega_{k,n}^{2}, \\ \vdots \\ x(t_{k,n+1}h) - x(t_{k,n+1}h+Mh),\ t \in \Omega_{k,n}^{M}, \end{cases} \tag{2.17}$$

where $\xi_{k,n}(t) = [0, h), t \in H_{k,n} \cap \Gamma_{1,n}$.

Further, the transmitted triggering sampled state is

$$x(t_{k,n+1}h) = x(t - \xi_{k,n}(t)) + \varpi_{k,n}(t). \tag{2.18}$$

After the above preparation, we have

$$\dot{x}(t) = \begin{cases} \tilde{A}x(t) + \tilde{B}Kx(t - \xi_{k,n}(t)) + \tilde{B}K\varpi_{k,n}(t) \\ \qquad +\hat{F}w(t), t \in H_{k,n} \cap \Gamma_{1,n}, \\ \tilde{A}x(t), t \in \Gamma_{2,n}. \end{cases} \tag{2.19}$$

From $\xi_{k,n}(t)$ and event-triggered scheme, we have

$$-e(t_{k,n+1}h) = x(t_{k,n+1}h) - x(t_{k,n+1}h + h) = \varpi_{k,n}(t).$$

Besides, $\varpi_{k,n}(t) = 0, t \in \Omega^1_{k,n}$, the triggering condition will be

$$\varpi^{\mathrm{T}}_{k,n}(t)W\varpi_{k,n}(t) \leq \nu x^{\mathrm{T}}(t_{k,n+1}h)Wx(t_{k,n+1}h),$$

Noticing $\xi_{k,n}(t)$, and $t_{k,n+1}h$ can be rewritten as $t - \xi_{k,n}(t)$, inequality can be expressed as

$$\begin{aligned} &\varpi^{T}_{k,n}(t)W\varpi_{k,n}(t) \leq \\ &\qquad \nu x^{T}(t - \xi_{k,n}(t))Wx(t - \xi_{k,n}(t)), t \in H_{k,n} \cap \Gamma_{1,n}. \end{aligned} \tag{2.20}$$

For convenience, the following definition is introduced

$$\zeta(t) = \begin{cases} 1, t \in [-h, 0) \cup \Gamma_{1,n}, \\ 2, t \in \Gamma_{2,n}. \end{cases} \tag{2.21}$$

Define $t_{\tilde{\imath},n} = \{^{nT,\tilde{\imath}=1;}_{nT+T_{\text{off}},\tilde{\imath}=2}$, for $\zeta(t) = \tilde{\imath} \in \{1, 2\}$. Further, let $\Gamma_{\tilde{\imath},n} = [t_{\tilde{\imath},n}, t_{3-\tilde{\imath},n+\tilde{\imath}-1})$, $\zeta_1(t) = \tilde{\imath}$, $\zeta_2(t) = 3 - \tilde{\imath}$. The system state equation is finally rewritten as

$$\begin{cases} \dot{x}(t) = \tilde{A}x(t) + \hat{B}_{\tilde{\imath}}\left[x(t - \xi_{k,n}(t)) + \varpi_{k,n}(t)\right] + \hat{F}w(t), \\ y(t) = \tilde{C}x(t). \end{cases} \tag{2.22}$$

for $\hat{B}_1 = \tilde{B}K, \hat{B}_2 = 0, t \in [t_{\tilde{\imath},n}, t_{3-\tilde{\imath},n+\tilde{\imath}-1})$.

Remark 2.4: The event-triggered SMLFC method is proposed in this chapter. An event-triggered device is applied to reduce the information transmission. The event-triggered condition is designed based on the state error. The state error is defined as the two status values of the most recent successful transmission. By comparing the state error and the current sampling value, the sampling signal is determined to be transmitted or not. The transmission signal determined by the event trigger device is used as the system state term in the design of the sliding mode controller.

The balance point's inner stability of the system is equivalent to the origin's stability with $w_i(t) = 0$. For a sufficient duration following a step load change, it does not influent the stability of power system [23]. Thus, the state-space model of subarea i studied in this chapter can be summarized as follows

$$\dot{x}(t) = \begin{cases} \tilde{A}x(t) + \tilde{B}Kx(t - \xi_{k,n}(t)) + \tilde{B}K\varpi_{k,n}(t), t \in H_{k,n} \cap \Gamma_{1,n}, \\ \tilde{A}x(t), t \in \Gamma_{2,n}. \end{cases}$$

2.3 Main Results

Some necessary corollaries and theorems are proved and discussed in this section. Specifically, the globally exponential stability is proved in Section 2.3.1. Based on the proposed condition, a design procedure is presented in Section 2.3.2 to determine the controller gains. The SMC law is given in Section 2.3.3 to ensure the reachability of the system.

2.3.1 Exponential Stability Analysis

Corollary 2.1 *For $\bar{\imath} \in \{1, 2\}$, appropriate dimensions matrices $M_{\bar{\imath}}$, $S_{\bar{\imath}}$, $N_{\bar{\imath}}$, prescribed positive scalars $\alpha_{\bar{\imath}}$, h and $\nu \in (0, 1)$, some positive definite matrices W, $P_{\bar{\imath}}$, $Q_{\bar{\imath}}$, $R_{\bar{\imath}}$, and $Z_{\bar{\imath}}$ can be found to satisfy the following matrix inequality*

$$\Theta_{\bar{\imath}} = \begin{bmatrix} \Theta_{11}^{\bar{\imath}} & * & * & * & * \\ \sqrt{h}S_{\bar{\imath}}^T & \Theta_{22}^{\bar{\imath}} & * & * & * \\ \sqrt{h}N_{\bar{\imath}}^T & 0 & \Theta_{33}^{\bar{\imath}} & * & * \\ \sqrt{h}M_{\bar{\imath}}^T & 0 & 0 & \Theta_{44}^{\bar{\imath}} & * \\ \sqrt{h}(R_{\bar{\imath}} + Z_{\bar{\imath}})G_{\bar{\imath}} & 0 & 0 & 0 & -R_{\bar{\imath}} - Z_{\bar{\imath}} \end{bmatrix} < 0. \tag{2.23}$$

1. for $\tilde{i}=1$
$\Theta^1_{11}=\Delta_{11}+\Delta_1, \Theta^1_{22}=\Theta^1_{33}=-e^{-2\alpha_1 h}R_1, \Theta^1_{44}=-e^{-2\alpha_1 h}Z_1, G_1=[\tilde{A}\ \hat{B}_1\ 0\ \hat{B}_1]$,
$M_1=\left[M^{\top}_{11}\ M^{\top}_{12}\ M^{\top}_{13}\ M^{\top}_{14}\right]^{\top}, S_1=\left[S^{\top}_{11}\ S^{\top}_{12}\ S^{\top}_{13}\ S^{\top}_{14}\right]^{\top}$,
$N_1=\left[N^{\top}_{11}\ N^{\top}_{12}\ N^{\top}_{13}\ N^{\top}_{14}\right]^{\top}$

$$\Delta_{11}=\begin{bmatrix} P_1\tilde{A}+\tilde{A}^{\top}P_1 & * & * & * \\ K^{\top}\tilde{B}^{\top}P_1 & \nu W & * & * \\ 0 & 0 & -e^{2\alpha_1 h}Q_1 & * \\ K^{\top}\tilde{B}^{\top}P_1 & 0 & 0 & -W \end{bmatrix},$$

$$\Delta_1=\begin{bmatrix} M^{\top}_{11}+N^{\top}_{11} & S^{\top}_{11}-N^{\top}_{11} & -M^{\top}_{11}-S^{\top}_{11} & 0 \\ M^{\top}_{12}+N^{\top}_{12} & S^{\top}_{12}-N^{\top}_{12} & -M^{\top}_{12}-S^{\top}_{12} & 0 \\ M^{\top}_{13}+N^{\top}_{13} & S^{\top}_{13}-N^{\top}_{13} & -M^{\top}_{13}-S^{\top}_{13} & 0 \\ M^{\top}_{14}+N^{\top}_{14} & S^{\top}_{14}-N^{\top}_{14} & -M^{\top}_{14}-S^{\top}_{14} & 0 \end{bmatrix}.$$

2. for $\tilde{i}=2$
$\Theta^2_{11}=\Delta_{21}+\Delta_2, \quad \Theta^2_{22}=\Theta^2_{33}=-R_2, \quad \Theta^2_{44}=-Z_2, \quad G_2=[\tilde{A}\ 0\ 0]$,
$M_2=\left[M^{\top}_{21}\ M^{\top}_{22}\ M^{\top}_{23}\right]^{\top}, N_2=\left[N^{\top}_{21}\ N^{\top}_{22}\ N^{\top}_{23}\right]^{\top}, S_2=\left[S^{\top}_{21}\ S^{\top}_{22}\ S^{\top}_{23}\right]^{\top}$,

$$\Delta_{21}=\begin{bmatrix} P_2\tilde{A}+\tilde{A}^{\top}P_2-2\alpha_2P_2 & * & * \\ 0 & 0 & * \\ 0 & 0 & -Q_2 \end{bmatrix},$$

$$\Delta_2=\begin{bmatrix} M^{\top}_{21}+N^{\top}_{21} & -N^{\top}_{21}+S^{\top}_{21} & -M^{\top}_{21}-S^{\top}_{21} \\ M^{\top}_{22}+N^{\top}_{22} & -N^{\top}_{22}+S^{\top}_{22} & -M^{\top}_{22}-S^{\top}_{22} \\ M^{\top}_{23}+N^{\top}_{23} & -N^{\top}_{23}+S^{\top}_{23} & -M^{\top}_{23}-S^{\top}_{23} \\ M^{\top}_{24}+N^{\top}_{24} & -N^{\top}_{24}+S^{\top}_{24} & -M^{\top}_{24}-S^{\top}_{24} \end{bmatrix},$$

Then, we have

$$e^{2(-1)^{\tilde{i}}\alpha_{\tilde{i}}(t-t_{\tilde{i},n})}V_{\tilde{i}}(t_{\tilde{i},n})\geq V_{\tilde{i}}(t). \tag{2.24}$$

Proof. Considering the multiple Lyapunov function below

$$V_{\zeta(t)}(t)=V_{\tilde{i}1}+V_{\tilde{i}2}+V_{\tilde{i}3}+V_{\tilde{i}4}, \tag{2.25}$$

where

$$\begin{cases} V_{\tilde{i}1} = x^{\top}(t)P_{\zeta(t)}x(t), \\ V_{\tilde{i}2} = \int_{t-h}^{t} x^{\top}(\theta)f(t)Q_{\zeta(t)}x(\theta)d\theta, \\ V_{\tilde{i}3} = \int_{-h}^{0}\int_{t+\theta}^{t} \dot{x}^{\top}(\varrho)f(t)R_{\zeta(t)}\dot{x}(\varrho)d\varrho d\theta, \\ V_{\tilde{i}4} = \int_{-h}^{0}\int_{t+\theta}^{t} \dot{x}^{\top}(\varrho)f(t)Z_{\zeta(t)}\dot{x}(\varrho)d\varrho d\theta, \end{cases}$$

here $P_{\zeta(t)}$, $Q_{\zeta(t)}$, $R_{\zeta(t)}$, and $Z_{\zeta(t)}$are all positive, $f(t) = e^{2(-1)^{\zeta(t)}\alpha_{\zeta(t)}(t-\theta)}$.

For any given $k \in \mathbb{K}$, $\tilde{i} = 1$, the derivation of $V_1(t)$ for $t \in H_{k,n} \bigcap \Gamma_{1,n}$. Using the free weighting matrix method, we have the following

$$\begin{aligned} \dot{V}_1(t) \le & -2\alpha_1 V_1(t) + x^{\top}(t)Q_1x(t) + 2x^{\top}(t)P_1\dot{x}(t) \\ & + 2\alpha_1 x^{\top}(t)P_1x(t) + 2\psi^{\top}(t)[M_1x_{M_1} + N_1x_{N_1} + S_1x_{S_1}] \\ & + h\dot{x}^{\top}(t)R_1\dot{x}(t) - \int_{t-\xi_{k,n}(t)}^{t} \dot{x}^{\top}(\varrho)e^{-2\alpha_1 h}R_1\dot{x}(\varrho)d\varrho \\ & - \int_{t-h}^{t-\xi_{k,n}(t)} \dot{x}^{\top}(\varrho)e^{-2\alpha_1 h}R_1\dot{x}(\varrho)d\varrho - h\dot{x}^{\top}(t)Z_1\dot{x}(t) \\ & - \int_{t-h}^{t} \dot{x}^{\top}(\varrho)e^{-2\alpha_1 h}Z_1\dot{x}(\varrho)d\varrho \\ & - 2\psi^{\top}(t)N_1 \int_{t-\xi_{k,n}(t)}^{t} \dot{x}(\varrho)d\varrho - 2\psi^{\top}(t)M_1 \int_{t-h}^{t} \dot{x}(\varrho)d\varrho \\ & - 2\psi^{\top}(t)S_1 \int_{t-\xi_{k,n(t)}}^{t-h} \dot{x}(\varrho)d\varrho - x^{\top}(t-h)e^{-2\alpha_1 h}Q_1x(t-h), \end{aligned}$$

and $x_{S_1} = x(t-\xi_{k,n}(t)) - x(t-h)$, $x_{N_1} = x(t) - x(t-\xi_{k,n}(t))$, $x_{M_1} = x(t) - x(t-h)$, $\psi(t) = [x^{\top}(t)\ x_{k,n}^{\top}\ x^{\top}(t-h)\ \varpi_{k,n}^{\top}(t)]^{\top}$.

According to [18], it can be derived that

$$\begin{aligned} & -2\psi^{\top}(t)M_1 \int_{t-h}^{t} \dot{x}(\varrho)d\varrho \le h\psi^{\top}(t)M_1e^{2\alpha_1 h}Z_1^{-1}M_1^{\top}\varrho(t) \\ & + \int_{t-h}^{t} \dot{x}^{\top}(\varrho)e^{-2\alpha_1 h}Z_1\dot{x}(\varrho)d\varrho, \end{aligned} \tag{2.26}$$

$$
\begin{aligned}
&-2\psi^{\top}(t)N_1\int_{t-\xi_{k,n}(t)}^{t}\dot{x}(\varrho)d\varrho \le h\psi^{\top}(t)N_1e^{2\alpha_1 h}R_1^{-1}N_1^{\top}\varrho(t)\\
&+\int_{t-h}^{t}\dot{x}^{\top}(\varrho)e^{-2\alpha_1 h}R_1\dot{x}(\varrho)d\varrho,
\end{aligned}
\tag{2.27}
$$

$$
\begin{aligned}
&-2\psi^{\top}(t)S_1\int_{t-\xi_{k,n}(t)}^{t-h}\dot{x}(\varrho)d\varrho \le h\psi^{\top}(t)S_1e^{2\alpha_1 h}R_1^{-1}S_1^{\top}\varrho(t)\\
&+\int_{t-\xi_{k,n}(t)}^{t-h}\dot{x}^{\top}(\varrho)e^{-2\alpha_1 h}R_1\dot{x}(\varrho)d\varrho.
\end{aligned}
\tag{2.28}
$$

After some deduction and arrangement, we have

$$
\begin{aligned}
\dot{V}_1(t) \le& -2\alpha_1V_1(t)+\psi^{\top}(t)[\Theta_{11}^1+hM_1e^{2\alpha_1 h}Z_1^{-1}M_1^{\top}\\
&+hN_1e^{2\alpha_1 h}R_1^{-1}N_1^{\top}+hS_1e^{2\alpha_1 h}R_1^{-1}S_1^{\top}\\
&+hG_1^{\top}(R_1+Z_1)G_1]\psi(t),
\end{aligned}
\tag{2.29}
$$

where $G_1=[\tilde{A},\tilde{B}_1,0,\tilde{B}_1]$, $\Theta_{11}^1=\Delta_{11}+\Delta_1$,

$$
\Delta_{11}=\begin{bmatrix}
P_1\tilde{A}+\tilde{A}^{\top}P_1 & * & * & *\\
K^{\top}\tilde{B}^{\top}P_1 & \nu W & * & *\\
0 & 0 & -e^{2\alpha_1 h} & *\\
K^{\top}\tilde{B}^{\top}P_1 & 0 & 0 & -W
\end{bmatrix},
$$

$$
\Delta_1=\begin{bmatrix}
M_{11}^{\top}+N_{11}^{\top} & -N_{11}^{\top}+S_{11}^{\top} & -M_{11}^{\top}-S_{11}^{\top} & 0\\
M_{12}^{\top}+N_{12}^{\top} & -N_{12}^{\top}+S_{12}^{\top} & -M_{12}^{\top}-S_{12}^{\top} & 0\\
M_{13}^{\top}+N_{13}^{\top} & -N_{13}^{\top}+S_{13}^{\top} & -M_{13}^{\top}-S_{13}^{\top} & 0\\
M_{14}^{\top}+N_{14}^{\top} & -N_{14}^{\top}+S_{14}^{\top} & -M_{14}^{\top}-S_{14}^{\top} & 0
\end{bmatrix}.
$$

Based on the Schur complement, we have

$$
\begin{aligned}
&\Theta_{11}^1+hM_1e^{2\alpha_1 h}Z_1^{-1}M_1^{\top}+hN_1e^{2\alpha_1 h}R_1^{-1}N_1^{\top}\\
&+hS_1e^{2\alpha_1 h}R_1^{-1}S_1^{\top}+hG_1^{\top}(R_1+Z_1)G_1<0.
\end{aligned}
\tag{2.30}
$$

Therefore, inequality $\dot{V}_1 \leq -2\alpha_1 V_1(t)$ for $t \in [t_{1,n}, t_{2,n})$ is equivalent to $V_1 \leq e^{-2\alpha_1(t-t_{1,n})} V_1(t_{1,n})$.

By similar calculations, the derivation of $V_2(t)$ for $t \in [t_{2,n}, t_{1,n+1})$ is as follows

$$\begin{aligned}\dot{V}_2(t) \leq{} & -2\alpha_2 V_2(t) + \psi^{\top}(t)[\Theta_{11}^2 + hM_1 e^{2\alpha_2 h} Z_2^{-1} M_2^{\top} \\ & + hN_2 e^{2\alpha_2 h} R_2^{-1} N_2^{\top} + hS_2 e^{2\alpha_2 h} R_2^{-1} S_2^{\top} \\ & + hG_2^{\top}(R_2 + Z_2)G_2]\psi(t),\end{aligned} \tag{2.31}$$

and $\dot{V}_2 \leq -2\alpha_2 V_2(t)$ for $t \in [t_{2,n}, t_{1,n+1})$, is equivalent to $V_2 \leq e^{-2\alpha_2(t-t_{2,n})} V_2(t_{2,n}, t_{1,n+1})$. This completes the proof.

Remark 2.5: According to [19]. T and T_{off} in $S_{\text{DoS}}(t)$ are known in advance. The power systems (2.22) have controller gains K and $\nu \in (0, 1)$, some known scalars $\alpha_{\tilde{i}} \in (0, +\infty)$, $h \in (0, T_{\text{off}})$ and $\mu_{\tilde{i}} \in (1, +\infty)$, $\tilde{i} \in 1, 2$, positive ς satisfying

$$\varsigma \triangleq 2\alpha_1 T_{\text{off}} - 2\alpha_2(T - T_{\text{off}}) - 2(\alpha_1 + \alpha_2)h - \ln(\mu_1 \mu_2). \tag{2.32}$$

Theorem 2.1 *The* SMLFC *system* (2.22) *under periodic* DoS *attacks is globally exponentially stable for decay rate* $\rho = \frac{\varsigma}{2T}$, *if there exist positive definite matrices* P_1, P_2, $Z_{\tilde{i}}$, $R_{\tilde{i}}$ *and* $Q_{\tilde{i}}$, *positive scalars* $h \in (0, T_{\text{off}})$, $\alpha_{\tilde{i}}$, $\mu_{\tilde{i}}$, *such that (2.32) and the following equations are established*

$$\begin{cases} P_1 \leq \mu_2 P_2,\ P_2 \leq e^{2(\alpha_1+\alpha_2)h} \mu_1 P_1, \\ R_{\tilde{i}} \leq \mu_{3-\tilde{i}} R_{3-\tilde{i}},\ Q_{\tilde{i}} \leq \mu_{3-\tilde{i}} Q_{3-\tilde{i}},\ Z_{\tilde{i}} \leq \mu_{3-\tilde{i}} Z_{3-\tilde{i}} \end{cases}. \tag{2.33}$$

Proof. According to (2.24), we have

$$V(t) \leq \begin{cases} e^{-2\alpha_1(t-t_{1,n})} V_1(t_{1,n}),\ t \in [t_{1,n}, t_{2,n}), \\ e^{2\alpha_1(t-t_{2,n})} V_2(t_{2,n}),\ t \in [t_{2,n}, t_{1,n+1}). \end{cases} \tag{2.34}$$

According to (2.33)

$$\begin{cases} \mu_2 V_2(t_{1,n}^-) \geq V_1(t_{1,n}), \\ e^{2(\alpha_1+\alpha_2)h} \mu_1 V_1(t_{2,n}^-) \geq V_2(t_{2,n}). \end{cases} \tag{2.35}$$

There exists an integer $n \in [0, +\infty)$ making $t \in [t_{1,n}, t_{2,n})$ or $t \in [t_{2,n}, t_{1,n+1})$. Combining corollary and (2.34), (2.35) for $t \in [t_{1,n}, t_{2,n})$ we have

$$\begin{aligned} V(t) &\le \mu_2 e^{-2\alpha_1(t-t_{1,n})} V_2(t_{1,n}^-), \\ &\le \mu_2 e^{-2\alpha_1(t-t_{1,n-1})} V_2(t_{1,n}^-), \\ &\vdots \\ &\le e^{-\varsigma n} V_1(0), t \in [t_{1,n}, t_{2,n}), \end{aligned} \tag{2.36}$$

where $t < t_{2,n} = nT + T_{\text{off}}$, which leads to $n > \frac{t - T_{\text{off}}}{T}$. $V_1(0)$ is the initial value of $V(t)$. Then (2.36) is rewritten

$$V(t) \le V_1(0) e^{\varsigma \frac{T_{\text{off}}}{T}} e^{-\frac{\varsigma}{T} t}. \tag{2.37}$$

For $t \in [t_{2,n}, t_{1,n+1})$, by similar calculations, we have

$$V(t) \le V_1(0) \frac{1}{\mu_2} e^{-\frac{\varsigma}{T} t}, t \in [t_{2,n}, t_{1,n+1}). \tag{2.38}$$

We define $d_0 = \max\{e^{\frac{\varsigma T_{\text{off}}}{T}}, \frac{1}{\mu_2}\}$, $d_1 = \min\{\varsigma_{\min}(P_1)\}$, $d_2 = \varsigma_{\max}(P_1) + h\varsigma_{\max}(Q_1) + \frac{h^2}{2}\varsigma_{\max}(R_1 + Z_1)$, and $\rho = \frac{\varsigma}{2T}$. Then we have

$$V(t) \le V_1(0) d_0 e^{-\frac{\varsigma}{T} t}. \tag{2.39}$$

From (2.25), the following inequalities are established

$$V(t) \ge d_1 \|x(t)\|^2, \ V_1(0) \le d_2 \|\rho\|_h^2. \tag{2.40}$$

Then

$$\|x(t)\| \le \sqrt{\frac{d_0 d_2}{d_1}} e^{-\rho t} \|\rho\|, t \ge 0, \tag{2.41}$$

where ρ is the decay rate. Hence, the system 2.22 can be globally exponential stable.

Remark 2.6: The multiple Lyapunov functions are constructed to solve the stability problem of multi-area power systems against periodic DoS attacks, which are helpful to obtain a less conservative result of single Lyapunov functions. As shown in **Theorem 2.1**, based on the multiple Lyapunov functions, the globally exponential stability problem can be solved under the condition of 2.32 and 2.33.

2.3.2 Strategy of Controller Gain

Lemma 2.1 [15] *For the symmetric positive matrices W and X, there exists a suitable scalar a_0 satisfying the following inequality*

$$-XW^{-1}X \leq a_0^2 W - 2a_0 X. \tag{2.42}$$

Theorem 2.2 *The following* LMIs *holds if there exist positive definite matrices $X_{\bar{i}}$, $\bar{Q}_{\bar{i}}$, $\bar{R}_{\bar{i}}$, and $\bar{Z}_{\bar{i}}$, appropriate size matrices $Y_{\bar{i}}$, $\bar{S}_{\bar{i}}$, $\bar{M}_{\bar{i}}$, $\bar{N}_{\bar{i}}$, and some known scalars $\bar{b}_{\bar{i}}$, $c_{\bar{i}}$, $d_{\bar{i}}$, $f_{\bar{i}} \in (0, +\infty)$, $\mu_{\bar{i}} \in (1, +\infty)$, $h \in (0, T_{\text{off}}]$, and $\nu \in (0, 1)$.*

$$\bar{\Theta}_{\bar{i}} = \begin{bmatrix} \Upsilon_{11}^{\bar{i}} & * & * \\ \Upsilon_{21}^{\bar{i}} & \Upsilon_{22}^{\bar{i}} & * \\ \Upsilon_{31}^{\bar{i}} & 0 & f_{\bar{i}}^2(\bar{R}_{\bar{i}} + \bar{Z}_{\bar{i}}) - 2f_{\bar{i}}X_{\bar{i}} \end{bmatrix} < 0,$$

$$\begin{bmatrix} -\mu_2 X_2 & X_2 \\ * & -X_1 \end{bmatrix} \leq 0, \begin{bmatrix} -\mu_1 e^{2(\alpha_1+\alpha_2)h} X_1 & X_1 \\ * & -X_2 \end{bmatrix} \leq 0,$$

$$\begin{bmatrix} -\mu_{3-\bar{i}}\bar{R}_{3-\bar{i}} & X_{3-\bar{i}} \\ * & c_{\bar{i}}^2\bar{R}_{\bar{i}} - 2c_{\bar{i}}X_{\bar{i}} \end{bmatrix} \leq 0, \tag{2.43}$$

$$\begin{bmatrix} -\mu_{3-\bar{i}}\bar{Q}_{3-\bar{i}} & X_{3-\bar{i}} \\ * & \bar{b}_{\bar{i}}^2\bar{Q}_{\bar{i}} - 2\bar{b}_{\bar{i}}X_{\bar{i}} \end{bmatrix} \leq 0,$$

$$\begin{bmatrix} -\mu_{3-\bar{i}}\bar{Z}_{3-\bar{i}} & X_{3-\bar{i}} \\ * & d_{\bar{i}}^2\bar{Z}_{\bar{i}} - 2d_{\bar{i}}X_{\bar{i}} \end{bmatrix} \leq 0.$$

Then, the controller gain is calculated as $K = YX_1^{-1}$, event-triggered matrix $\bar{W} = X_1 W X_1$.

Proof. Define matrices $X_{\bar{i}} = P_{\bar{i}}^{-1}$.

1. For $\bar{i} = 1$, define $J_1^1 = \text{diag}\{J_1^2, X_1, X_1, X_1, I\}$, $J_1^2 = \text{diag}\{X_1, X_1, X_1, X_1\}$. By multiplying J_1^1 on left and right of Θ_1, we have $\bar{Q}_1 = X_1 Q_1 X_1$, $\bar{R}_1 = X_1 R_1 X_1$, $\bar{Z}_1 = X_1 Z_1 X_1$, $\bar{S}_1 = X_1 S_1 X_1$, $\bar{M}_1 = X_1 M_1 X_1$, $\bar{N}_1 = X_1 N_1 X_1$.

$$\Upsilon_{11}^{1} = \begin{bmatrix} \gamma_{11} & * & * & * \\ \gamma_{21} & \gamma_{22} & * & * \\ \gamma_{31} & \gamma_{32} & \gamma_{33} & * \\ \gamma_{41} & \gamma_{42} & \gamma_{43} & \gamma_{44} \end{bmatrix},$$

where

$$\gamma_{11} = 2\alpha_1 X_1 + \text{sym}(\tilde{A}X_1) + \bar{Q}_1 + \text{sym}(\bar{M}_{11} + \bar{S}_{11}),$$

$$\gamma_{21} = Y^{\mathsf{T}}\tilde{B}^{\mathsf{T}} + \bar{M}_{12} + \bar{S}_{12} - \bar{M}_{11}^{\mathsf{T}} + \bar{N}_{11}^{\mathsf{T}},$$

$$\gamma_{22} = \nu\bar{W} + \text{sym}(\bar{N}_{12} - \bar{M}_{12}), \gamma_{31} = \bar{M}_{13} + \bar{S}_{13} - \bar{S}_{11}^{\mathsf{T}} - \bar{N}_{11}^{\mathsf{T}},$$

$$\gamma_{32} = \bar{N}_{13} - \bar{M}_{13} - \bar{N}_{12}^{\mathsf{T}} - \bar{S}_{12}^{\mathsf{T}}, \gamma_{41} = Y^{T}\tilde{B}^{\mathsf{T}} + \bar{S}_{14} + \bar{M}_{14},$$

$$\gamma_{33} = -e^{-2\alpha_1 h}\bar{Q}_1 - \text{sym}(\bar{S}_{13} - \bar{N}_{13}), \gamma_{42} = \bar{N}_{14} - \bar{M}_{14} + \nu\bar{W},$$

$$\gamma_{43} = -\bar{S}_{14} - \bar{N}_{14}, \bar{N}_1 = \begin{bmatrix} \bar{N}_{11}^{\mathsf{T}} & \bar{N}_{12}^{\mathsf{T}} & \bar{N}_{13}^{\mathsf{T}} & \bar{N}_{14}^{\mathsf{T}} \end{bmatrix}^{\mathsf{T}},$$

$$a_{44} = (\nu - 1)\bar{W}, \Upsilon_{22}^{1} = e^{-2\alpha_1 h}\text{diag}\{-\bar{R}_1, -\bar{R}_1, -\bar{Z}_1\},$$

$$\bar{S}_1 = \begin{bmatrix} \bar{S}_{11}^{\mathsf{T}} & \bar{S}_{12}^{\mathsf{T}} & \bar{S}_{13}^{\mathsf{T}} & \bar{S}_{14}^{\mathsf{T}} \end{bmatrix}^{\mathsf{T}}, \Upsilon_{21}^{1} = \sqrt{h}\begin{bmatrix} \bar{M}_1 & \bar{N}_1 & \bar{S}_1 \end{bmatrix}^{\mathsf{T}},$$

$$\Upsilon_{31}^{1} = \begin{bmatrix} \tilde{A}X_1 & \tilde{B}Y & 0 & \tilde{B}Y \end{bmatrix}, \bar{M}_1 = \begin{bmatrix} \bar{M}_{11}^{\mathsf{T}} & \bar{M}_{12}^{\mathsf{T}} & \bar{M}_{13}^{\mathsf{T}} & \bar{M}_{14}^{\mathsf{T}} \end{bmatrix}^{\mathsf{T}};$$

2. For $\tilde{\imath} = 2$, define $J_2^1 = \text{diag}\{J_2^2, J_2^2, I\}$, $J_2^2 = \text{diag}\{X_2, X_2, X_2\}$. By multiplying J_2^1 on both sides of Θ_2, we have $\bar{S}_2 = X_2 S_2 X_2$, $\bar{M}_2 = X_2 M_2 X_2$, $\bar{N}_2 = X_2 N_2 X_2$.

$$\Upsilon_{11}^{1} = \begin{bmatrix} \bar{\gamma}_{11} & * & * \\ \bar{\gamma}_{21} & \bar{\gamma}_{22} & * \\ \bar{\gamma}_{31} & \bar{\gamma}_{32} & \bar{\gamma}_{33} \end{bmatrix},$$

where

$$\bar{\gamma}_{11} = 2\alpha_2 X_2 + \text{sym}(\tilde{A}X_2) + \bar{Q}_2 + \text{sym}(\bar{M}_{21} + \bar{S}_{21}),$$

$$\bar{\gamma}_{21} = \bar{M}_{22} + \bar{S}_{22} - \bar{M}_{21}^{\mathrm{T}} + \bar{N}_{21}^{\mathrm{T}}, \bar{\gamma}_{22} = \text{sym}(\bar{N}_{22} - \bar{M}_{22}),$$

$$\bar{\gamma}_{31} = \bar{M}_{23} + \bar{S}_{23} - \bar{S}_{21}^{\mathrm{T}} - \bar{N}_{21}^{\mathrm{T}}, \bar{\gamma}_{32} = \bar{N}_{23} - \bar{M}_{23} - \bar{N}_{22}^{\mathrm{T}} - \bar{S}_{22}^{\mathrm{T}},$$

$$\bar{\gamma}_{33} = -\bar{Q}_2 - \text{sym}(\bar{S}_{23} - \bar{N}_{23}), \bar{M}_2 = \begin{bmatrix} \bar{M}_{21}^{\mathrm{T}} & \bar{M}_{22}^{\mathrm{T}} & \bar{M}_{23}^{\mathrm{T}} & \bar{M}_{24}^{\mathrm{T}} \end{bmatrix}^T,$$

$$\bar{S}_2 = \begin{bmatrix} \bar{S}_{21}^{\mathrm{T}} & \bar{S}_{22}^{\mathrm{T}} & \bar{S}_{23}^{\mathrm{T}} & \bar{S}_{24}^{\mathrm{T}} \end{bmatrix}^{\mathrm{T}}, \bar{N}_2 = \begin{bmatrix} \bar{N}_{21}^{\mathrm{T}} & \bar{N}_{22}^{\mathrm{T}} & \bar{N}_{23}^{\mathrm{T}} & \bar{N}_{24}^{\mathrm{T}} \end{bmatrix}^T,$$

$$\Upsilon_{21}^2 = \sqrt{h}[\bar{M}_2 \quad \bar{N}_2 \quad \bar{S}_2]^{\mathrm{T}}, \Upsilon_{31}^2 = [\tilde{A}X_2 \quad 0 \quad 0 \quad 0]$$

$$\Upsilon_{22}^2 = \text{diag}\{-\bar{R}_2, -\bar{R}_2, -\bar{Z}_2\}.$$

Combining with Lemma 2.1, inequality $X_{\bar{l}}(\bar{R}_{\bar{l}} + \bar{W}_{\bar{l}})^{-1}X_{\bar{l}} \geq 2f_{\bar{l}}X_{\bar{l}} - f_{\bar{l}}^2(\bar{R}_{\bar{l}} + \bar{W}_{\bar{l}})$ can be solved. When $\bar{\Theta}_{\bar{l}} < 0$ holds, $\Theta_{\bar{l}} < 0$ can be determined. The proof is completed.

Remark 2.7: The inequality (2.23) is transformed into (2.43), since the nonlinear terms $K^T B^T P$, which contains the controller gain, cannot be solved directly. By using MATLAB, we can solve the parameters $\bar{Y}$, $\bar{W}$, and X_1. The controller gain $K = YX_1^{-1}$ can be calculated easily. The existence of the controller gain means that the system is globally exponentially stable under SMLFC.

2.3.3 Control Law Design

Theorem 2.3 *For* SMLFC *power system (2.22), the sliding mode surface is given as (2.4). By using the* SMLFC *methods, the system can guarantee to meet the requirements in finite time with load disturbance $w(t) = \Delta P_d < \epsilon$. The stable performance of the controlled system can be guaranteed*

$$u(t) = Kx(t) - \epsilon L\tilde{F}\text{sgn}(\sigma), \tag{2.44}$$

Proof. Build the Lyapunov function as

$$V = \frac{1}{2}\sigma^{\mathrm{T}}(t)(L\tilde{F})^{-1}\sigma(t).$$

Take the state function and control law into derivative $\dot{V}(t)$,

$$\begin{aligned}\dot{V} &= \sigma^T(t)(L\tilde{F})^{-1}\dot{\sigma}(t)\\ &= \sigma^{\mathrm{T}}(t)(L\tilde{F})^{-1}(-L\tilde{B}Kx(t) + L\tilde{B}u(t) + L\tilde{F}w(t))\\ &= \sigma^{\mathrm{T}}(t)(L\tilde{F})^{-1}[L\tilde{B}(Kx(t) - \epsilon L\tilde{F}\mathrm{sgn}(\sigma))\\ &\quad + L\tilde{F}w(t) - L\tilde{B}Kx(t)]\\ &= \sigma^{\mathrm{T}}(t)(L\tilde{F})^{-1}(-\epsilon L\tilde{F}\mathrm{sgn}(\sigma) + L\tilde{F}w(t))\\ &= \sigma^{\mathrm{T}}(t)(-\epsilon\mathrm{sgn}(\sigma) + w(t)) \leq \|\sigma^{\mathrm{T}}(t)\|(-\epsilon + \|w(t)\|) < 0.\end{aligned}$$

Thus, $\dot{V}(t) < 0$ is true. The sliding mode surface $\sigma(t)$ is located in the desired location. Then, we can say that the SMLFC power system, by applying the proposed control law, can stay in the designed surface. This completes the proof.

2.4 Numerical Examples

In this section, three numerical examples are provided to demonstrate the superiority to the SMLFC method.

Example 2.1 *The* SMLFC *technique is simulated for the three-area interconnected power system under periodic* DoS *attacks. To verify the correctness of the* SMLFC *method, some simulation examples are given.*

Parameter values of each area are basically the same as those in [15, 23, 29] and listed in Table 2.1.

The other parameters are given as $\mu_1 = 1.01$, $b_1 = c_1 = d_1 = \hat{f}_1 = 0.1$, $b_2 = c_2 = d_2 = \hat{f}_2 = 0.3$, $b_3 = c_3 = d_3 = \hat{f}_3 = 0.5$, event-triggered probability ν_1, ν_2, ν_3 equal 0.1, 0.2, 0.4, respectively, periods of three areas are $T_1 = 2$, $T_2 = 3$, $T_3 = 5$. The weighting matrices of event trigger can be calculated from Theorem 2.2.

Table 2.1 Parameters for each area.

Area-1	Parameters	D_1	T_{ch1}	T_{g1}	R_1	T_{12}	h_1	M_1
	Value	0.1	0.3	0.1	0.3	0.34π	0.01	1.0
Area-2	Parameters	D_2	T_{ch2}	T_{g2}	R_2	T_{13}	h_2	M_2
	Value	0.2	0.3	0.3	0.5	0.3π	0.01	1.5
Area-3	Parameters	D_3	T_{ch3}	T_{g3}	R_3	T_{23}	h_3	M_3
	Value	0.3	0.3	0.5	0.8	0.35π	0.01	1.8

Controller gain of the SMLFC methods is calculated from the given LMIs in Theorem 2.2. We have

$$K_1 = \begin{bmatrix} -0.0023 & 0.0237 & 0.0023 & 0.0877 \end{bmatrix},$$

$$K_2 = \begin{bmatrix} -0.0008 & 0.0241 & 0.0013 & 0.0605 \end{bmatrix},$$

$$K_3 = \begin{bmatrix} 0.0026 & 0.0070 & 0.0019 & 0.0561 \end{bmatrix}.$$

The system state of SMLFC power system under the periodic DoS attack is shown in Figure 2.4. From Figure 2.4, we know that the final state remains stable, which means SMLFC method reduces the contamination of periodic DoS attacks. The relationship between DoS attacks and the power system state are revealed in Figure 2.5. We can observe that all indexes of the i-th regional fluctuate and change, which means that the power system becomes unstable. With the continuous intervention of event-triggered SMLFC, the system state gradually becomes stable. The change in simulation results are consisted with the previous analysis and derivation results. The states of the system before and after the control scheme are shown in Figure 2.6. The results show that the SMLFC method is effective in countering the DoS attack, the controller responses and attacks period have uniformity. Figure 2.7 shows the event-triggering instant.

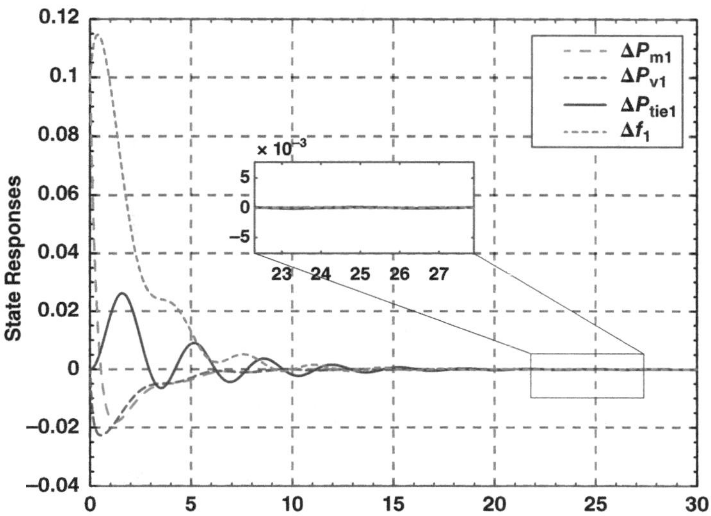

Figure 2.4 State responses of SMLFC scheme under periodic DoS attacks.

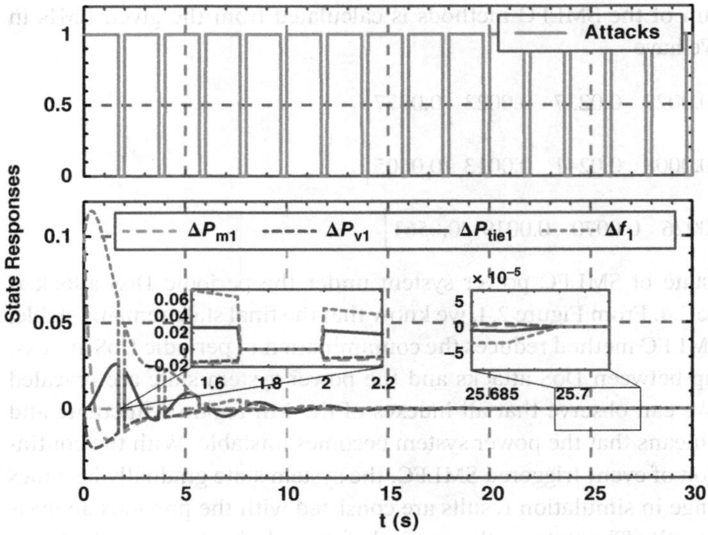

Figure 2.5 System state under periodic DoS attacks.

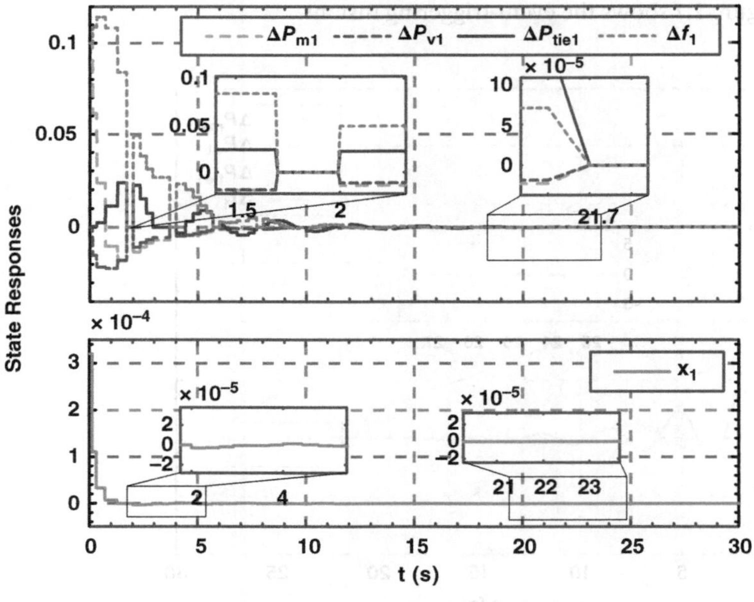

Figure 2.6 The performance of control scheme to the system states.

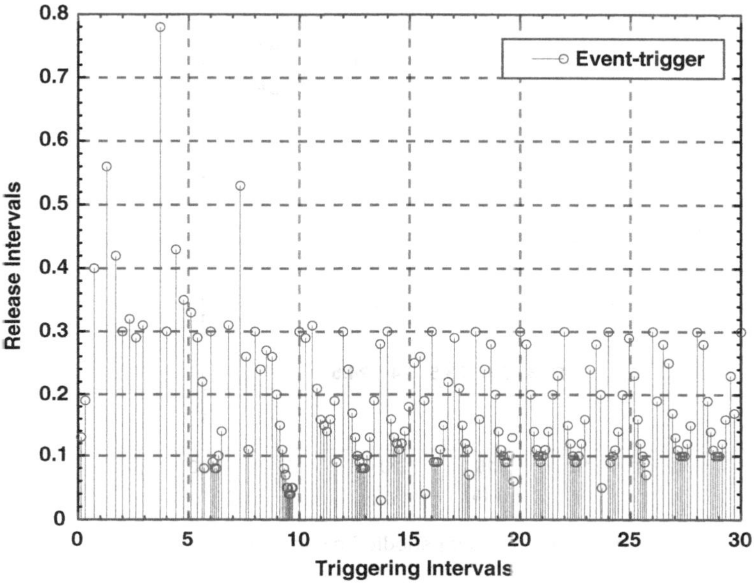

Figure 2.7 Event-triggered instants.

Remark 2.8: For an ideal SMC system, the process of control structure switching has no time and space lag. The measurement of the system has no error, and the control quantity has no amplitude limit. Then the system will be a motion process that gradually tends to and stabilizes at the origin. Thus, there will be no chattering problem. On the other hand, in the simulation part, all the work is done by Matlab and Simulink soft. The data is processing and analyzed by the computer, the lag in time and space is very small. So the chattering phenomenon is not obvious in the simulation.

The states of $\Delta P_{\text{tie}i}$ and Δf_i for the interconnected power system are given in Figures 2.8 and 2.9, respectively. Simulation results show that the SMLFC method can make the system stable, the $\Delta P_{\text{tie}i}$ and Δf_i can arrive at the unbiased state even under periodic DoS attacks.

Example 2.2 *Under the periodic* DoS *attacks, the following example shows the performance among proposed* SMC, SMC [42], proportional–integral (PI) *control* [23], *and* proportional–integral–derivative (PID) *control* [43]. *Simulations are carried out with the same parameters.*

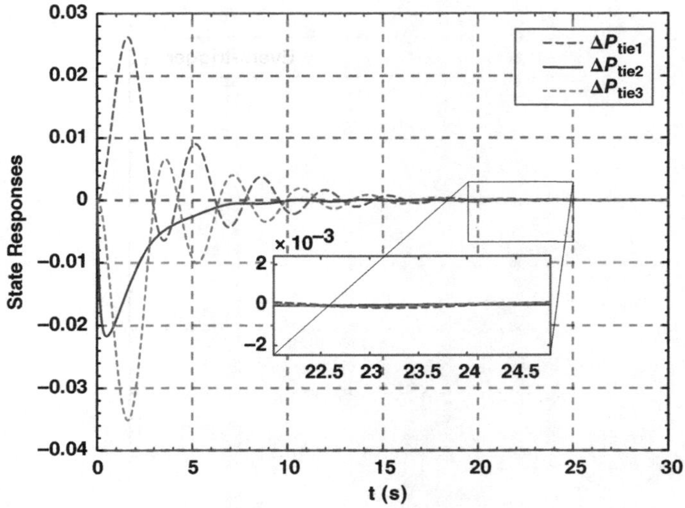

Figure 2.8 Tie-line power deviation under periodic DoS attacks.

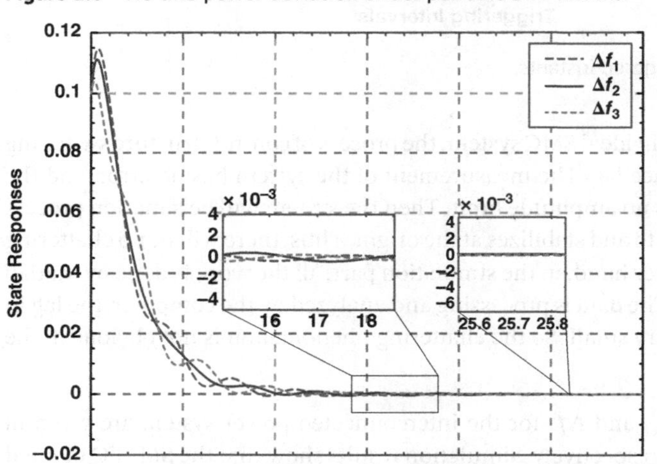

Figure 2.9 Frequency deviation under periodic DoS attacks.

Most of the parameters are the same as **Example 2.1**, the other parameters of **Example 2.2** are given as $D_i = 0.001$, $T_{\text{chi}} = 0.3$, $T_{gi} = 0.3$, $R_i = 1$, $h_i = 0.01$, $M_i = 0.7$. To evaluate the efficiencies of different methods in the LFC for power systems under periodic DoS attack, we test the cases adopting the proposed SMLFC, SMLFC [42], PI [23], and PID [43], respectively. The controller gains K_{SMC}, $K_{\text{SMC}-[42]}$, K_{PI}, and K_{PID} are calculated as follows:

$$K_{\mathrm{SMC}} = \begin{bmatrix} -0.0782 & -0.0799 & -0.0022 & 0.1397 \end{bmatrix},$$

$$K_{\mathrm{SMC-[42]}} = \begin{bmatrix} -0.3578 & -0.5496 & 1.6065 & 0.6086 \end{bmatrix},$$

$$K_{\mathrm{PI}} = \begin{bmatrix} 0.0733 & 0.0128 \end{bmatrix}, K_{\mathrm{PID}} = \begin{bmatrix} 0.0695 & 0.0014 & 0.0103 \end{bmatrix}.$$

Figures 2.10–2.12 illustrate the trajectories of system state for attacked i-th area power system with different control methods. The frequency deviation Δf of the system is shown in Figure 2.10. The deviation value is eliminated to be zero by the SMC method at $t = 8.5$, while the SMC [42], PI control and PID control fail to converge to zero. The mechanical output deviation P_m is shown in Figure 2.11. By utilizing the SMC method, the deviation value becomes zero at $t = 9.5$, but while there are fluctuations under SMC [42], PI control and PID control. The governor valve position deviation P_v is shown in Figure 2.12, the deviation value becomes zero with the SMC method at $t = 9$, while SMC [42], PI control and PID control cannot keep the value to be zero. Schematic diagram of different control inputs are given in Figure 2.13. It can be concluded from Figures 2.10–2.13 that the performance of the proposed SMLFC is better than the SMC [42], and the SMC [42] outperforms the PID-based LFC or PI-based LFC.

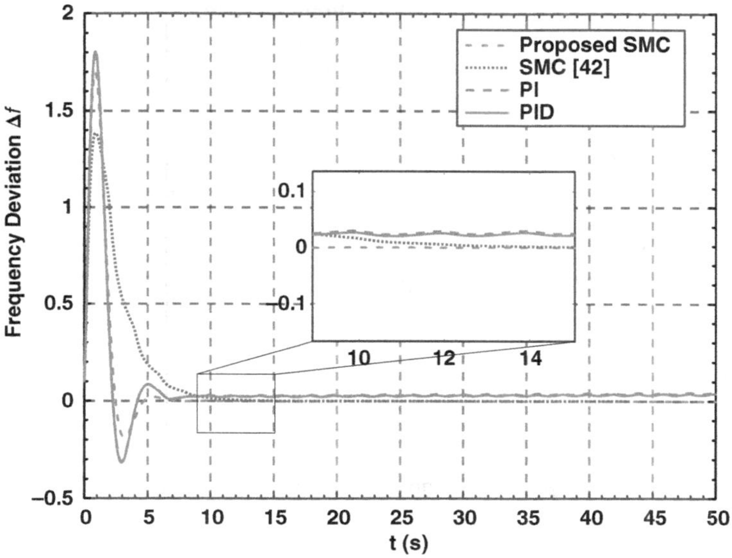

Figure 2.10 Δf of the i-th area power system with different control methods.

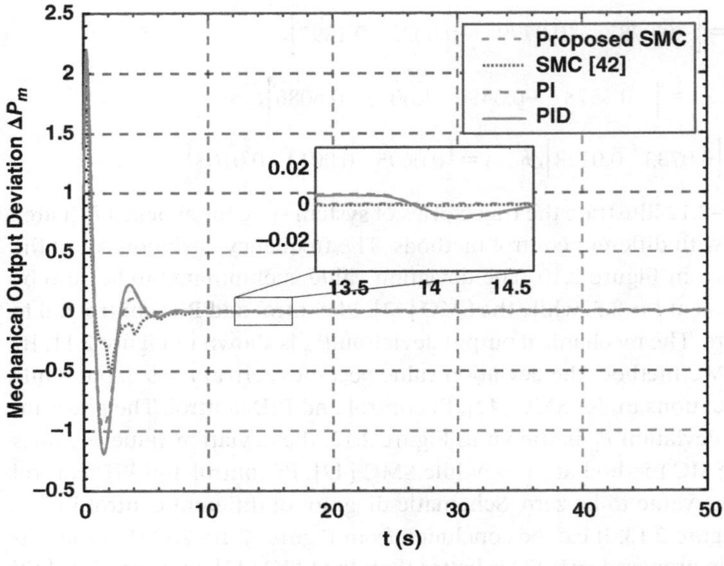

Figure 2.11 ΔP_m of the i-th area power system with different control methods.

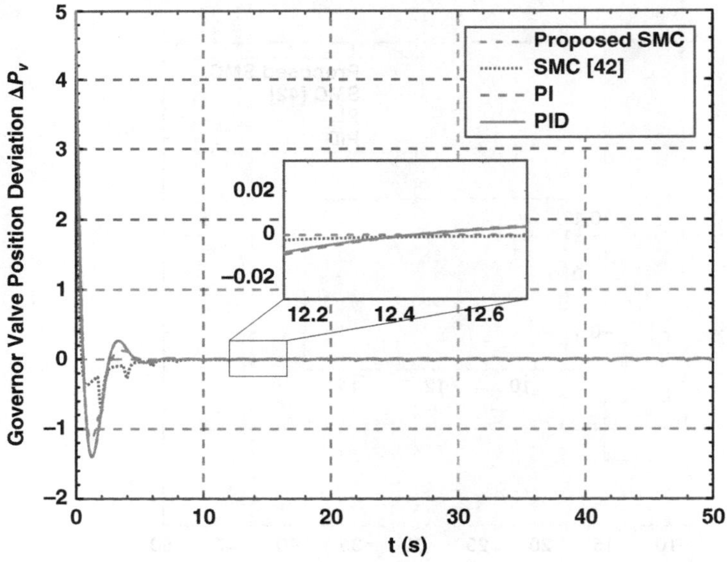

Figure 2.12 ΔP_v of the i-th area power system with different control methods.

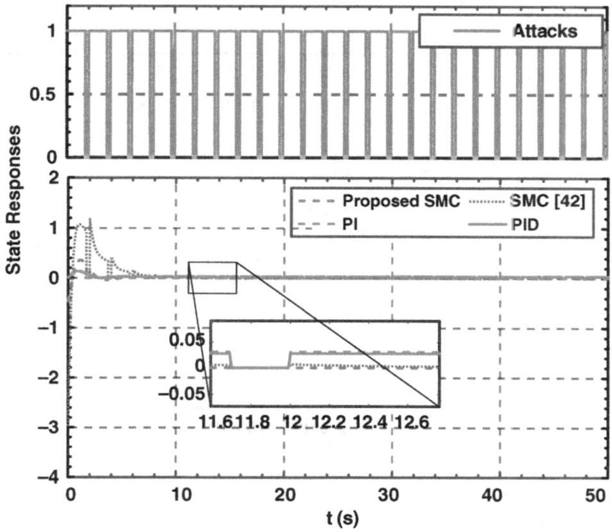

Figure 2.13 Different control input under periodic DoS attacks.

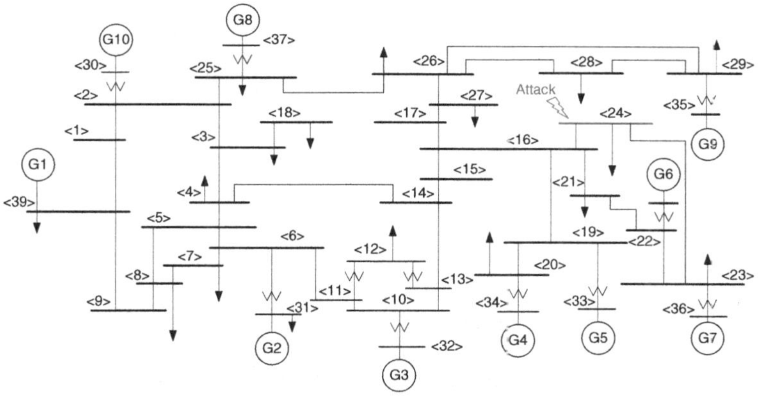

Figure 2.14 IEEE-39-bus system under periodic attacks.

Example 2.3 *In the following example, we carry out simulations on an* IEEE *10-generator, 39-bus system. The rated frequency of the system is 60* Hz. *The model of generator, transformer, and other devices is the same as* [44]. *Due to space limitations, relevant parameters can be seen in* [44].

The 10-generator, 39-bus system is shown in Figure 2.14. The proposed method is applied in the 39-bus system. To be convenient, we suppose that the periodic

Table 2.2 Resistance of lines from bus i to j ($\times 10^{-2}$).

1-2	1-39	2-3	2-25	3-4	3-18	4-5	4-14	5-6
2.55	1.20	2.58	24.42	1.83	2.48	1.88	1.86	2.31
5-8	6-7	6-11	7-8	8-9	9-39	10-11	10-13	13-14
2.14	1.96	2.56	2.61	1.90	1.20	2.79	2.79	2.67
14-15	15-16	16-17	16-19	16-21	16-24	17-18	17-27	21-22
2.49	2.87	2.36	2.46	1.78	1.53	2.56	2.25	1.71
22-23	23-24	25-26	26-27	26-28	26-29	28-29		
1.88	1.89	2.97	2.86	2.72	2.74	2.78		

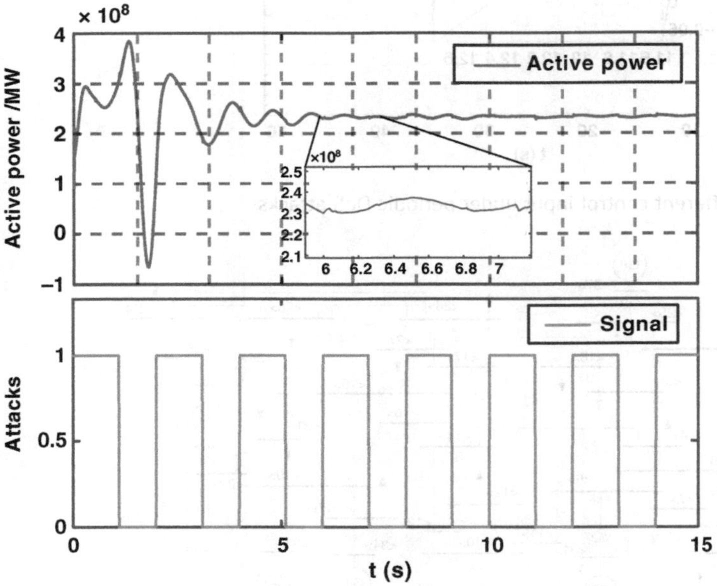

Figure 2.15 Active power of IEEE-39-bus system under periodic DoS attacks.

DoS attacks appear at the bus-24. Table 2.2 is given to show the resistance of lines from bus i to j. The reactance is 0.000796.

Simulation results are shown in Figures 2.15 and 2.16. Figure 2.15 presents the active power of the event-triggered SMC 39-bus system under periodic DoS attacks. From Figure 2.15, we can see that the fluctuation of active power to the power system is related to periodic DoS attacks. Meanwhile, with the continuous operation of the event-triggered SMC, the 39-bus power system gradually becomes stable. The simulation results show the effectiveness of the proposed event-triggered SMC method.

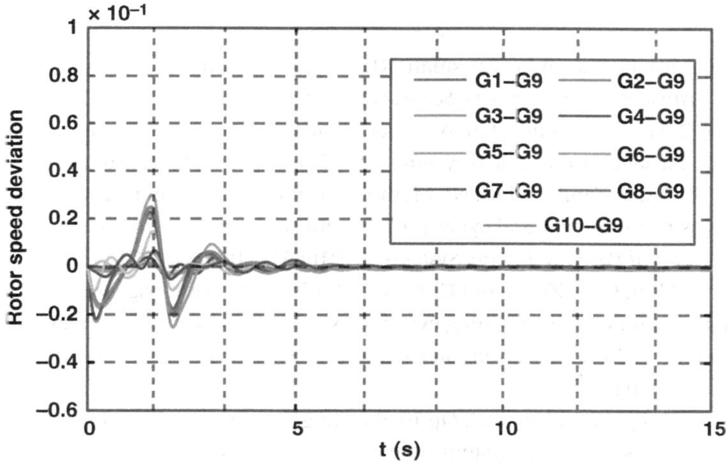

Figure 2.16 Rotor speed deviation of 10-generator 39-bus system under periodic DoS attacks.

Besides, we made the simulation of the rotor speed deviation. As the simulation is processed under a 10-generator, 39-bus system, we arbitrarily select the rotor speed of G9 as the reference object. Figure 2.16 is to show the rotor speed deviation of the generator. From the results, it can be found that the rotor speed has big oscillations caused by the periodic DoS attacks. With the intervention of event-triggered SMC, the rotor speed becomes stable. Based on the above simulation results, the proposed event-triggered SMC is proved to be effective.

2.5 Conclusion

To maintain effective control of the multi-area power systems under periodic DoS attacks, this chapter proposed an event-based SMLFC method. Specifically, we proposed a suitable sliding surface to deduce the state-space equation for the controlled power system. To determine whether the sampled state information should be transmitted, an event-triggered device has been designed. We derived a criterion for globally exponential stability, which relates the controller gains, the triggering matrix, and the periodic DoS attacks. By constructing the multiple Lyapunov stability function, we got the sufficient conditions of exponential decay rate of the power system. A suitable control law has been established to ensure the reachability of the power system. Numerical simulation results have been presented to demonstrate the superiority of event-based SMLFC.

References

1 S. Soltan, M. Yannakakis, and G Zussman. REACT to cyber attacks on power grids. *IEEE Transactions on Network Science and Engineering*, 6:459–473, 2018.

2 D. Zhang, Q. G. Wang, G. Feng, et al. A survey on attack detection, estimation and control of industrial cyber–physical systems. *ISA Transactions*, 116:1–16, 2021.

3 Q. Wang, X. P. Cai, Y. Tang, and M. Ni. Methods of cyber-attack identification for power systems based on bilateral cyber-physical information. *International Journal of Electrical Power & Energy Systems*, 125:106515–106125, 2021.

4 W. H. Qi, Y. K. Hou, G. D. Zong, and C. Ki Ahn. Finite-time event-triggered control for semi-Markovian switching cyber-physical systems with FDI attacks and applications. *IEEE Transactions on Circuits and Systems I: Regular Papers*, 68(6):2665–2674, 2021.

5 C. S. Liu, R. L. Deng, W. L. He, et al. Optimal coding schemes for detecting false data injection attacks in power system state estimation. *IEEE Transactions on Smart Grid*, 13(1):738–749, 2021. https://doi.org/10.1109/TSG.2021.3107972.

6 C. Peng, J. C. Wu, and E. G. Tian. Stochastic event-triggered H-∞ control for networked systems under denial-of-service attacks. *IEEE Transactions on Systems, Man, and Cybernetics: Systems*, 52(7):4200–4210, 2021. https://doi.org/10.1109/TSMC.2021.3090024.

7 A. Chawla, A. Singh, P. Agrawal, et al. Denial-of-service attacks pre-emptive and detection framework for synchrophasor based wide area protection applications. *IEEE Systems Journal*, 16(1):1570–1581, 2021. https://doi.org/10.1109/JSYST.2021.3093494.

8 J. W. Tian, B. H. Wang, T. Y. Li, et al. Coordinated cyber-physical attacks considering DoS attacks in power systems. *International Journal of Robust and Nonlinear Control*, 30(11):4345–4358, 2020.

9 A. Torabi, J. Zarei, R. Razavi-Far, and M. Saif. Decentralized resilient output-feedback control design for networked control systems under denial-of-service. *IEEE Systems Journal*, 16(4):5620–5629, 2022. https://doi.org/10.1109/JSYST.2021.3118333.

10 S. Lakshminarayana, S. Adhikari, and C. Maple. Analysis of IoT-based load altering attacks against power grids using the theory of second-order dynamical systems. *IEEE Transactions on Smart Grid*, 12(5):4415–4425, 2021.

11 X. L. Chen, S. L. Hu, Y. Li, et al. Co-estimation of state and FDI attacks and attack compensation control for multi-area load frequency control systems under FDI and DoS attacks. *IEEE Transactions on Smart Grid*, 13(3):2357–2368, 2022.

12 W. J. Bi, G. Chen, and K. F. Zhang. Profit-oriented false data injection attack against wind farms and countermeasures. *IEEE Systems Journal*, 16(3):3700–3710, 2022. https://doi.org/10.1109/JSYST.2021.3107910.

13 O. Boyaci, A. Umunnakwe, A. Sahu, et al. Graph neural networks based detection of stealth false data injection attacks in smart grids. *IEEE Systems Journal*, 16(2):2946–2957, 2022. https://doi.org/10.1109/JSYST.2021.3109082.

14 A. Amini, A. Asif, A. Mohammadi, and A. Azarbahram. Sampled-data dynamic event-triggering control for networked systems subject to DoS attacks. *IEEE Transactions on Network Science and Engineering*, 8(3):1978–1990, 2021.

15 C. Peng, J. C. Li, and M. R. Fei. Resilient event-triggering H-∞ load frequency control for multi-area power systems with energy-limited DoS attacks. *IEEE Transactions on Power Systems*, 32(5):4110–4118, 2016.

16 Md. M. Hossain, C. Peng, Y. L. Wang, and X. Du. Handshake logic-based event-triggered load frequency control for smart grids under dos attacks. *IEEE Transactions on Industrial Informatics*, 18(6):3863–3872, 2022. https://doi.org/10.1109/TII.2021.3116165.

17 W. A. Zhang and L. Yu. Stabilization of sampled-data control systems with control inputs missing. *IEEE Transactions on Automatic Control*, 55(2):447–452, 2010.

18 S. L. Hu, D. Yue, X. P. Xie, et al. Resilient event-triggered controller synthesis of networked control systems under periodic DoS jamming attacks. *IEEE Transactions on Cybernetics*, 49(12):4271–4281, 2018.

19 X. L. Chen, Y. G. Wang, and S. L. Hu. Event-based robust stabilization of uncertain networked control systems under quantization and denial-of-service attacks. *Information Sciences*, 459:369–386, 2018.

20 Y. Xu, M. Fang, P. Shi, and Z.-G. Wu. Event-based secure consensus of mutiagent systems against DoS attacks. *IEEE Transactions on Cybernetics*, 50(8):3468–3476, 2019.

21 Y. Z. Zhu and W. X. Zheng. Observer-based control for cyber-physical systems with periodic DoS attacks via a cyclic switching strategy. *IEEE Transactions on Automatic Control*, 65(8):3714–3721, 2019.

22 Y. Xu, M. Fang, Y. J. Pan, et al. Event-triggered output synchronization for nonhomogeneous agent systems with periodic denial-of-service attacks. *International Journal of Robust and Nonlinear Control*, 31(6):1851–1865, 2021.

23 X. C. ShangGuan, Y. He, C. K. Zhang, et al. Switching system-based load frequency control for multi-area power system resilient to denial-of-service attacks. *Control Engineering Practice*, 107:104678–104689, 2021.

24 M. Saeedi, J. Zarei, R. Razavi-Far, and M. Saif. Event-triggered adaptive optimal fast terminal sliding mode control under denial-of-service attacks. *IEEE Systems Journal*, 16(2):2684–2692, 2022. https://doi.org/10.1109/JSYST.2021.3073816.

25 W. H. Qi, X. Yang, J. H. Park, et al. Fuzzy SMC for quantized nonlinear stochastic switching systems with semi-Markovian process and application. *IEEE Transactions on Cybernetics*, 52(9):9316–9325, 2022. https://doi.org/10.1109/TCYB.2021.3069423.

26 C. W. Wu, L. G. Wu, J. X. Liu, and Z. P. Jiang. Active defense-based resilient sliding mode control under denial-of-service attacks. *IEEE Transactions on Information Forensics and Security*, 15:237–249, 2019.

27 R. J. Ma, P. Shi, and L. G. Wu. Dissipativity-based sliding-mode control of cyber-physical systems under denial-of-service attacks. *IEEE Transactions on Cybernetics*, 51(5):2306–2318, 2020.

28 Y. L. Cui, L. L. Xu, M. R. Fei, and Y. Shen. Observer based robust integral sliding mode load frequency control for wind power systems. *Control Engineering Practice*, 65:1–10, 2017.

29 X. J. Su, X. X. Liu, and Y. D. Song. Event-triggered sliding-mode control for multi-area power systems. *IEEE Transactions on Industrial Electronics*, 64(8):6732–6741, 2017.

30 P. C. Chen, L. Yu, and D. Zhang. Event-triggered sliding mode control of power systems with communication delay and sensor faults. *IEEE Transactions on Circuits and Systems I: Regular Papers*, 68(2):797–807, 2020.

31 W. P. M. H. Heemels, M. C. F. Donkers, and A. R. Teel. Periodic event-triggered control for linear systems. *IEEE Transactions on Automatic Control*, 58(4):847–861, 2012.

32 X. N. Song, R. Z. Zhang, C. K. Ahn, and S. Song. Dissipative synchronization of semi-Markov jump complex dynamical networks via adaptive event-triggered sampling control scheme. *IEEE Systems Journal*, 16(3):4653–4663, 2022. https://doi.org/10.1109/JSYST.2021.3124082.

33 X. P. Xie, Q. Zhou, D. Yue, and H. Y. Li. Relaxed control design of discrete-time Takagi–Sugeno fuzzy systems: An event-triggered real-time scheduling approach. *IEEE Transactions on Systems, Man, and Cybernetics: Systems*, 48(12):2251–2262, 2017.

34 S. X. Zhu, E. G. Tian, D. H. Xu, and J. L. Liu. An adaptive torus-event-based controller design for networked T-S fuzzy systems under deception attacks. *International Journal of Robust and Nonlinear Control*, 32(6):3425–3441, 2022.

35 Z. Gu, S. Yan, J. H. Park, and X. P. Xie. Event-triggered synchronization of chaotic lur'e systems via memory-based triggering approach. *IEEE Transactions on Circuits and Systems II: Express Briefs*, 69(3):1427–1431, 2021.

36 N. Zhao, P. Shi, W. Xing, and R. K. Agarwal. Resilient event-triggered control for networked cascade control systems under denial-of-service attacks and actuator saturation. *IEEE Systems Journal*, 16(1):1114–1122, 2022. https://doi.org/10.1109/JSYST.2021.3066540.

37 E. Tian and C. Peng. Memory-based event-triggering h_{∞} load frequency control for power systems under deception attacks. *IEEE Transactions on Cybernetics*, 50:4610–4618, 2020.

38 A. K. Behera, B. Bandyopadhyay, and X. H. Yu. Periodic event-triggered sliding mode control. *Automatica*, 96:61–72, 2018.

39 Z. S. Wang, X. F. Fan, and Z. Shi. Periodic event-triggered integral sliding-mode control for t–s fuzzy systems. *IEEE Transactions on Cybernetics*, 52(8):7669–7681, 2022. https://doi.org/10.1109/TCYB.2020.3036888.

40 G. D. Qi, W. H. Zong, and W. X. Zheng. Adaptive event-triggered SMC for stochastic switching systems with semi-Markov process and application to boost converter circuit model. *IEEE Transactions on Circuits and Systems I: Regular Papers*, 68:786–796, 2020.

41 D. Yue, E. Tian, and Q. L. Han. A delay system method for designing event-triggered controllers of networked control systems. *IEEE Transactions on Automatic Control*, 58:475–481, 2013.

42 Y. Mi, Y. Fu, C. S. Wang, and P. Wang. Decentralized sliding mode load frequency control for multi-area power systems. *IEEE Transactions on Power Systems*, 28(4):4301–4309, 2013.

43 H. Shabani, B. Vahidi, and M. Ebrahimpour. A robust PID controller based on imperialist competitive algorithm for load-frequency control of power systems. *ISA Transactions*, 52(1):88–95, 2013.

44 I. Hiskens. IEEE PES task force on benchmark systems for stability controls. *Technical Report*, pages 1–23, 2013.

3

H_∞ Load Frequency Control of Power System Integrated with EVs Under DoS Attacks: Non-fragile Output Sliding Mode Control Approach

After reading this chapter, you should be able to:

- Figure out the difference among different participations of EVs in the power system.
- Clarify the effect of non-fragile on the controller performance.
- Understand the characteristics of the considered real-time frequency-triggered DoS attacks.

3.1 Background

In recent years, growing environmental concerns have spurred the extensive adoption of electric vehicles (EVs) owing to their emission-free nature [1–3]. In the multi-area interconnected power system integrated with EVs, keeping load and frequency in balance is an important and basic issue. To realize a balance between load command and power generation, load frequency control (LFC) strategies have been widely considered [4, 5]. It is worth noting that EVs can enhance the dynamic performance of LFC power systems because EVs have faster response characteristics in power output than conventional generators. Therefore, many studies are proposed for LFC power systems integrated with EVs. The topic of frequency regulation of a single-area power system connected to an EV aggregator over a nonideal communication network is investigated in [6]. The participation ability of the EVs aggregator in the LFC power system is investigated in [7].

In addition to the above results, wide range of use of EVs brings about fresh challenges on the stability to power systems, such as decreasing charging device lifetime along with the increasing number of charging or discharging, and modeling the uncertain nature of EV users' behavior [8, 9]. Besides, cyberattacks cause serious threat to the stability of multi-area power systems integrated with EVs.

As multi-area power systems continue to advance, the processing and transmission of information are becoming increasingly pivotal. To expedite seamless and efficient signal transmission, communication networks are employed to establish connections within the framework of a multi-area interconnected power system. LFC signals or decisions are transmitted from the control center to the devices through the communication network. While enjoying the convenience and efficiency of the network, we have to face a series of threats, including the risk of cyberattacks brought by the openness of the network [10–12]. Cyberattacks pose a significant risk to system availability, integrity, and confidentiality. These traits represent prime targets for cyber threats, leading to grave repercussions such as information leakages and the potential destruction of essential infrastructure [13]. Thus, the stability issue of multi-area power systems integrated with EVs under cyberattacks is the key point for multi-area power system development.

As detailed in [13], the LFC system faces primary threats in the form of denial-of-service (DoS) attacks that compromise data availability and data integrity attacks, which manipulate transmission data to jeopardize system integrity. Particularly, DoS attacks lead to communication channel disruptions [14], halting the successful transmission of system signals and consequently undermining data validity. This situation is exacerbated by the erroneous reception of signals by generation units, resulting in a cascading degradation of overall system performance. Significant efforts have been made to study cyberattacks on the LFC power systems, and many encouraging results have been generated. Considering the event-based secure control scheme, the authors in [15] designed an H_∞ load-frequency controller for power systems suffering from deception attacks. Adopting a memory-event-triggered mechanism [16], the authors solved the H_∞ LFC of multi-area power systems with cyberattacks and communication delays. Most of the existing studies have been focusing on finding appropriate control methods to reduce the impacts of cyberattacks. In such studies, cyberattacks are typically modeled as stochastic processes occurring at a given stochastic probability [17–20]. In other words, it is assumed that cyber attacks can happen at any moment with a given probability, constructed against the networked control system. Although the multi-area power system can indeed be seen as a complex networked control system, the cyberattack is not directly carried out on the power system itself. An attacker may be able to exploit certain characteristics of the power system to devise a particular attack against the stability of the power system. Considering this, a novel cyberattack model is proposed for the first time, which is related to the frequency of the multi-area power system.

From the attacker's perspective, cyberattacks are willing to destroy the stable state of the power system. Cyberattacks may lead to unstable system issues by affecting the normal transmission signals of the power system. When we are trying to solve the stability problem for a multi-area LFC power system under cyberattacks, a widely used method is to build the system state function. The state function is constructed with the frequency deviation, the deviation of generator mechanical output, the valve position deviation, the deviation of load, and so on. In the above feature parameters, the most easily accessible one is the frequency of the interconnected power system. Frequency information is available to the attack, say, by connecting a frequency sensor to the power system. Therefore, a frequency sensor is assumed to be used by the attacker to complete the acquisition of frequency information. Based on the real-time frequency, a new kind of DoS attack that relates to the power system frequency can be easily developed.

It is not difficult to understand that stochastic cyberattacks may occur in both stable and unstable states of the system [20–23]. Since the energy of cyberattack is limited, cyberattack that occur in an unstable state cause little impact on system stability. The purpose of the cyberattack is to disrupt the normal operation of the system, then it is not very meaningful to carry out the attack when the system is in an unstable state. This caused us to think about whether it is possible to make the cyberattack not occur at the time of the unstable state. Then, a real-time frequency-triggered DoS attack model is proposed and first considered in LFC power system. In this chapter, the attacker is assumed to obtain the frequency of the system by a sensor device. Therefore, based on the real-time frequency information, the system state can be separated into stable region and unstable region. Also, based on the real-time frequency information and event-triggered strategy [15, 16], a frequency-triggered strategy is designed to decide whether to carry out a cyberattack. Once the frequency-triggered condition is satisfied, a cyberattack is carried out. Therefore, the designed cyberattacks are named real-time-frequency-triggered DoS attacks. Furthermore, an accurate controller is developed for the stabilization of the power system integrated with EVs under real-time frequency-triggered DoS attacks in this work.

Sliding mode control (SMC) emerges as a robust control strategy, particularly well-suited to addressing disturbances and unaccounted dynamics. Its potency has found a significant application in resolving LFC challenges. In [24], disturbances and parameter uncertainties are incorporated into the LFC model, leading to the design of an adaptive SMC scheme for power system frequency regulation. In [25], the authors design a fuzzy-based integral sliding mode control scheme, which resolves the LFC problem of multi-area interconnected power systems integrated with a wind farm. In existing studies [26–28], the sliding mode surface is designed by using the system state or system output which can only ensure the state or output stable. An output sliding mode surface is designed to ensure the stability of the system state and system output stable concurrently.

However, in reality, immature technology, aging equipment, and other issues lead to disturbances, and parameter uncertainties can also appear in the design of the controller. To handle the influence of uncertainties, studies have been carried out on the non-fragile control problem. One of the main features of a non-fragile controller is that it is insensitive to uncertainties such that well control performance can be obtained. In [29], a non-fragile proportional integral control strategy is applied in dynamic event-triggered H_∞ LFC approach for multi-area nonlinear power systems. In [30], the authors designed a non-fragile sliding mode controller for discrete singular systems with external disturbances. In [31], a robust non-fragile H_∞ fault-tolerant control method is proposed to accommodate the simultaneous presence of performance limitations and disturbances. In the above studies, the sliding mode surface is designed only to use system state or output. In this study, the non-fragile output sliding mode controller is designed with the system state and output at the same time.

In this study, we comprehensively explore the stability problem of a multi-area LFC power system that integrates EVs, operating amidst real-time frequency-triggered DoS attacks. Firstly, we propose an innovative model for real-time frequency-triggered DoS attacks, which is crafted by combining real-time frequency sensing, an event-triggering strategy, and the concept of DoS attacks. Secondly, a non-fragile output SMC approach is proposed for power systems with EV integration, targeted at tackling the challenges posed by real-time frequency-triggered DoS attacks. Our approach involves the design of an output sliding mode controller, which hinges upon the system's state and output to ensure the stability of both aspects once the system is stabilized. Thirdly, we incorporate considerations of non-fragility in the controller design, making it more aligned with real-world scenarios. Furthermore, we analyze the intricate impact of real-time frequency-triggered DoS attacks and EV participation to the power system.

Key contributions are summarized as:

1. We propose a novel real-time frequency-triggered DoS attack model, which uniquely combines real-time frequency dynamics, event-triggering mechanisms, and the attributes of DoS attacks. This innovative approach establishes a more profound relationship between the proposed attack model and the power system compared to existing literature [17–20].
2. To address the challenges of power system stability under frequency-triggered DoS attacks, we put forth a non-fragile OSMLFC approach. Distinguished from previous studies [26–28], our approach develops a sliding mode surface based on both system state and output. Additionally, we account for fluctuations in the controller gain, which enhances the robustness of the system.
3. Based on the Lyapunov stability theory, we derive a sufficient condition for non-fragile H_∞ stability in power systems. By adeptly employing inequality treatment methods [32–34], we achieve results that are less conservative compared to previous works [35–37].

3.2 Problem Formulation

The multi-area power system integrated with EVs is a kind of complex control system. We focus on a simplified subsystem i comprising a set of governor, turbine, power generator and EVs and so on. The power system model integrated with EVs is given in Figure 3.1.

3.2.1 Model Description

Dynamic performance of EVs can be expressed by

$$\Delta\dot{P}_{\mathrm{evi}}(t) = \frac{\alpha_e k_e u_i(t)}{T_{ei}} - \frac{\Delta P_{\mathrm{evi}}}{T_{ei}} \tag{3.1}$$

where $P_{\mathrm{evi}}(t)$ is the output power of EVs, T_{ei} and k_e are time constant and EV gain, respectively. α_e is the participation of EVs.

The dynamic equation of the system base on the system model and information transfer can be described as

$$\begin{cases} \Delta\dot{P}_{mi}(t) = -\dfrac{\Delta P_{mi}(t)}{T_{\mathrm{chi}}} + \dfrac{\Delta P_{vi}(t)}{T_{\mathrm{chi}}} \\ \Delta\dot{P}^{i}_{\mathrm{tie}}(t) = 2\pi \displaystyle\sum_{j=1, i\neq j}^{n} T_{ij}(\Delta f_i(t) - \Delta f_j(t)) \\ \Delta\dot{P}_{vi}(t) = -\dfrac{\Delta f_i(t)}{R_i T_{gi}} - \dfrac{\Delta P_{vi}(t)}{T_{gi}} - \dfrac{\Delta P_{ei}(t)}{T_{gi}} \\ \Delta\dot{f}_i(t) = \dfrac{\Delta P_{mi}(t)}{M_i} - \dfrac{D_i \Delta f_i(t)}{M_i} - \dfrac{\Delta P^{i}_{\mathrm{tie}}(t)}{M_i} - \dfrac{\Delta P_{di}(t)}{M_i} \\ \Delta\dot{P}_{ei}(t) = \beta_i \Delta f_i(t) + \Delta P^{i}_{\mathrm{tie}}(t) \end{cases} \tag{3.2}$$

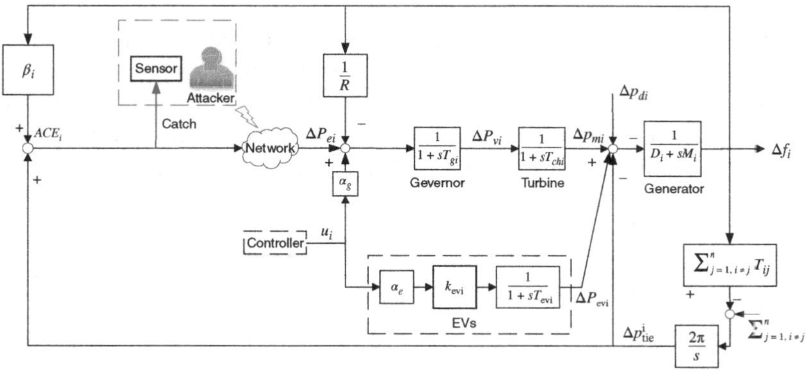

Figure 3.1 Model of i-th area power system integrated with EVs under real-time frequency-triggered DoS attacks.

Hence, the state vector space model of the multi-area power system can be represented as

$$\dot{x}_i(t) = A_i x_i(t) + B_i u_i(t) + F_i w_i(t) \tag{3.3}$$

where

$$x_i(t) = [\Delta P_{mi} \ \ \Delta P_{vi} \ \ \Delta P^i_{\text{tie}} \ \ \Delta f_i \ \ \Delta P_{ei} \ \ \Delta P_{\text{evi}}]^\text{T},$$

$$A_i = \begin{bmatrix} -\frac{1}{T_{\text{chi}}} & \frac{1}{T_{\text{chi}}} & 0 & 0 & 0 & 0 \\ 0 & -\frac{1}{T_{gi}} & 0 & -\frac{1}{R_i T_{gi}} & -\frac{1}{T_{gi}} & 0 \\ 0 & 0 & 0 & 2\pi \sum_{j=1, j\neq i}^{n} T_{ij} & 0 & 0 \\ \frac{1}{M_i} & 0 & -\frac{1}{M_i} & -\frac{D_i}{M_i} & 0 & 0 \\ 0 & 0 & 1 & \beta_i & 0 & 0 \\ 0 & 0 & 0 & 0 & 0 & -\frac{1}{T_{\text{evi}}} \end{bmatrix},$$

$$B_i = \begin{bmatrix} 0 & \frac{\alpha_g}{T_{gi}} & 0_{1\times 3} & \frac{\alpha_e k_e}{T_{\text{evi}}} \end{bmatrix}, F_i = \begin{bmatrix} 0_{1\times 3} & -\frac{1}{M_i} & 0 & 0 \end{bmatrix}.$$

$w_i(t)$ is the bounded disturbance of the system, and generality defined as $w_i(t) = \Delta P_{di} < d_i$, d_i is the positive constants and α_g is the participation of the governor, $\alpha_g + \alpha_e = 1$.

3.2.2 Real-time Frequency-triggered Mechanism

In this chapter, we consider a kind of triggered DoS attacks. The attackers are assumed to get the real-time frequency of the power system by a frequency sensor. Then, based on the obtained frequency and event-triggering mechanism, attacks are reconfigured. For the purpose of clarifying the description of the problem. Let's review the description of event-triggering scheme, which is given as

$$t_{k+1} = t_k + \sup\{t \in R | g_i(t) < 0\}, \tag{3.4}$$

where $g_i(t)$ is the triggering condition. Obviously, a new event is generated once $g_i(t) < 0$.

Then, we extend the results to the power system. In an ideal power system, the frequency deviation of a normal operating is seen as zero. The frequency deviation is given as

$$\Delta\bar{f}_i = f_s - f_N, \tag{3.5}$$

where f_s is defined as the real-time frequency measured by frequency sensor, f_N is the frequency under normal operating of power system.

Combined with the above analysis, the real-time frequency-triggered mechanism is described as

$$\begin{cases} |\Delta\bar{f}_i| < \Delta f_A, & \text{triggered} \\ |\Delta\bar{f}_i| \geq \Delta f_A, & \text{not triggered} \end{cases} \tag{3.6}$$

where Δf_A is defined as the weighting parameters, and used to judge whether the cyberattacks happening.

Remark 3.1: The real-time frequency-triggered mechanism is given as (3.6). The event triggered once $|\Delta\bar{f}_i| < \Delta f_A$, which means the DoS attacks happening. Different from the Bernoulli distribution DoS attacks [20, 21, 35]. The real-time frequency-triggered DoS attacks are connected to the power system characteristics. By monitoring the frequency of the power system, the attacker gives the DoS attacks that affect the normal signal transmission of the power system.

The design of the triggering mechanism as presented in (3.6) may be justified as follows. For an ideal operating power system, the stable state is confirmed when $\Delta\bar{f}_i = 0$. However, from the attacker's perspective, the DoS attack signal is to affect or even destroy the normal operation of the power system. Combining the criterion of power system stability, the purpose of the DoS attacks, and (3.4), we have that the system state can be divided into two different regions as follows

$$\begin{cases} |\Delta\bar{f}_i| < \Delta f_A, & \text{stable region} \\ |\Delta\bar{f}_i| \geq \Delta f_A, & \text{unstable region} \end{cases} \tag{3.7}$$

An example is proposed in Figure 3.2 to explain (3.7) and real-time frequency-triggered DoS attacks contamination region (stable region).

Obviously, only the sampling signal which belongs to the stable region will be influenced by real-time frequency-triggered DoS attacks.

Remark 3.2: Based on the design of (3.7), the proposed real-time frequency-triggered DoS attacks happened when the system frequency satisfies $|\Delta\bar{f}_i| < \Delta f_A$. This means the stable region of the power system equals the attack region to the

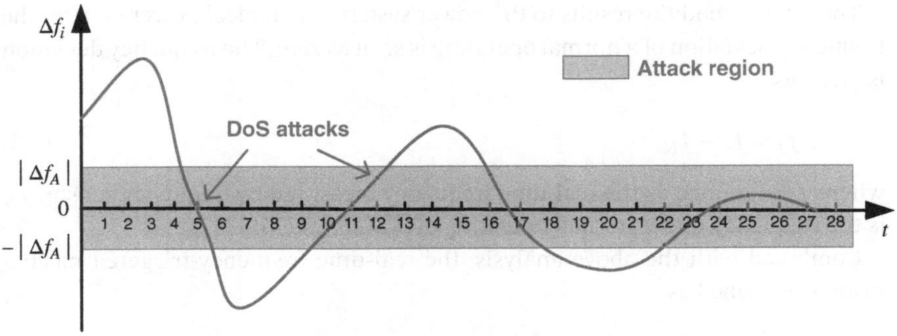

Figure 3.2 Example of real-time frequency-triggered DoS attacks.

attacker. Simply, all the real-time frequency-triggered DoS attacks happened in the stable region. The purpose of real-time frequency-triggered DoS attacks is to avoid the system state close to stability or becoming stable.

3.2.3 Real-time Frequency-triggered DoS Attacks

Assumption 1: (Attack frequency) [35] For a fixed sampling period, define l_1 as the number of DoS attacks on the power system, and l_2 is the number of normal transmissions without DoS attacks. Therefore, the frequency of DoS attacks is defined as $l_1/(l_1 + l_2)$.

Assumption 2: (Attack duration) [35] For a fixed sampling period T, the system is subjected to a DoS attack under $[kT, (k+m)T)$ with $m \in \{1, 2, \cdots, m_{\max}\}$, the sampling time $\{kT, (k+1)T, \cdots, (k+m-1)T\}$ is prevented from being transmitted. Therefore, the durations of DoS attacks are defined as mT, and the largest duration is $m_{\max}T$.

Firstly, we recall the DoS attacks which are considered in [21, 35]. DoS attack refers to preventing the control signals from being executed by obstructing the communication channel. Once the target network suffers a DoS attack, it will be unable to provide normal service or resource access. This means the state signal and control signal cannot be normally transmitted in the whole time t.

In order to analyze the contamination of DoS attacks [21, 35] to the power system. The state space function is built by considering the influence of DoS attacks. Power system model (3.3) under DoS attacks can be rewritten as

$$\begin{cases} \dot{x}(t) = Ax(t) + (1 - \beta(k))Bu_i(t_k) \\ \qquad +\beta(t)Bu_i(t_{k-1}) + F_i w_i(t), \\ y(t) = Cx(t). \end{cases} \tag{3.8}$$

where $\beta(k) \in \{0, 1\}$ is Bernoulli distribution related to occurrences of DoS attack, that is

$$E(\beta(k)) = \beta, E((\beta(k) - \beta)^2) = \beta(1 - \beta) = \varepsilon^2 \tag{3.9}$$

where β is the probability of DoS attacks happening. $\beta = 0$ denotes that the power system is not contaminated by DoS attacks, and $\beta = 1$ means DoS attacks happening. The details analysis about Bernoulli distributed DoS attacks can be seen in [21, 35].

Remark 3.3: In the previous study [21, 35], the DoS attacks should explicit some characteristics. They are 1) Attack duration; 2) Attack frequency; 3) Initial attack time. Different attack duration and frequencies will affect the stability of the power system to varying degrees. A larger DoS attack duration will lead to more changes [18]. The initial attack time is simply seen as the injection time of the attack.

Remark 3.4: For an operated power system, the actuator uses the control signal $u_i(t_k)$. When a DoS attacks happens, the control signal cannot be transmitted normally. The actuator has to use the latest received control signal $u_i(t_{k-1})$.

Obviously, every sampled signal may be contaminated by Bernoulli-distributed DoS attacks. But, not all sampled data are important for power system stability. In the event-triggered scheme [37], neighbored signals with little difference will be given up. In the memory-event-triggered scheme [16, 36], the latest released signal is seen to be more important. In this chapter, the sampled signal which is close to the stable state is seen to be more important and will be the goal of the attacker.

Therefore, the frequency-triggered scheme (3.5) is applied to determine whether the DoS attacks should happen. The real-time frequency-triggered DoS attacks can be given as

$$\begin{cases} |\Delta \bar{f}_i| < \Delta f_A, & \text{DoS attacks} \\ |\Delta \bar{f}_i| \geq \Delta f_A, & \text{without attacks} \end{cases} \tag{3.10}$$

Combining (3.6), (3.7), and (3.10), all the real-time frequency-triggered DoS attacks guarantee to contaminate the sampled signal which is located in the defined stable region. Then, we have the following

$$\hat{u}_i(t) = \begin{cases} u_i(t_{k-1}), & \text{DoS attacks} \\ u_i(t_k), & \text{without attacks} \end{cases} \tag{3.11}$$

Remark 3.5: Power system state function under Bernoulli distributed DoS attacks and real-time frequency-triggered DoS attacks are given as (3.8) and (3.12), respectively. The happening of DoS attacks in (3.8) is the given probability (3.9) while the real-time frequency-triggered DoS attacks in (3.12) is dependent on the predetermined threshold (3.10).

Thus, state space function of power system integrated with EVs under real-time frequency-triggered DoS attack by non-fragile output sliding mode LFC can be rewritten as

$$\begin{cases} \dot{x}_i(t) = A_i x_i(t) + B_i \hat{u}_i(t) + \hat{F}_i w_i(t) \\ y_i(t) = C_i x_i(t) \end{cases} \tag{3.12}$$

Remark 3.6: As mentioned in **Remark 3.3**, stochastic DoS attacks exist some characteristics. Similarly, in the considered real-time frequency-triggered DoS attacks, attack duration can be seen as a time-varying delay. Differently, due to the use of a triggering mechanism, attack frequency is dependent on the triggering rules. Also, the first real-time frequency-triggered DoS attack can be ensured by the triggering mechanism. The initial attack time can be seen as the first happened attack.

The frequency of the transmitted signal will be caught by the attacker to determine whether a DoS attack shall be interjected. By comparing the current sampling frequency with the system frequency under normal operation, the DoS attacker shall decide whether to contaminate the power system or not. Under the

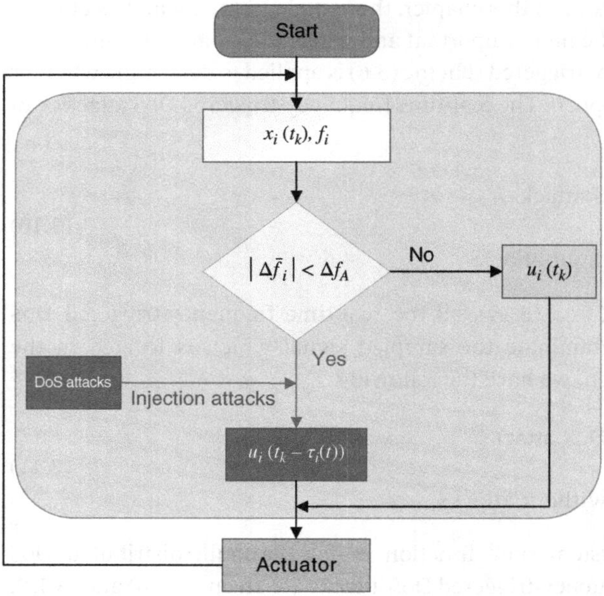

Figure 3.3 Flowchart of real-time frequency-triggered DoS attacks.

condition (3.10) is fulfilled, data transmission will be interrupted and the actuator use the latest received signal. The overall real-time frequency-triggered DoS attacks flowchart is shown in Figure 3.3.

3.3 Non-fragile Output SMC Scheme

PI-type sliding mode surface [26–28] is widely discussed, in which the sliding mode surface is designed based on the system state. In this chapter, a non-fragile output SMC is developed for the power system integrated with EVs.

If the system output can be considered into sliding mode surface design, good performance can be expected. Once the trajectory of sliding mode surface is reached, the state and output can be both ensured to be stable. Since $\Delta P_{ei}(t) = \int_0^t ACE_i(t)dt$, and $y_i(t) = ACE_i(t)$, we have

$$\Delta \dot{P}_{ei}(t) = ACE_i(t) = y_i(t). \tag{3.13}$$

Thus, the output sliding mode surface can be developed as

$$\sigma = G_i x_i(t) - \int_0^t G_i(A_i - B_i K_i) x_i(s) ds - K_i P_{ei}(t), \tag{3.14}$$

where G_i is selected to ensure $G_i B_i$ is nonsingular.

Taking the deviation of (3.14), we have

$$\dot{\sigma} = G_i \dot{x}_i(t) - G_i(A_i - B_i K_i) x_i(t) - K_i y_i(t). \tag{3.15}$$

Based on the SMC theorem, a proper sliding mode surface should satisfies $\sigma = \dot{\sigma} = 0$. Therefore, we have the following equation

$$G_i \dot{x}_i(t) - G_i(A_i - B_i K_i) x_i(t) - K_i y_i(t) = 0. \tag{3.16}$$

Mostly, the uncertainty or disturbance is considered in the power system state. Differently, we suppose that there exist gain fluctuations in the controller gain matrix. Combining with system state function (3.3), the equivalent non-fragile output sliding mode controller can be derived as

$$u_{eq}(t) = (\hat{K}_i + \Delta \hat{K}_i) x_i(t) - (G_i B_i)^{-1} G_i F_i w_i(t), \tag{3.17}$$

where $\hat{K}_i = (G_i B_i)^{-1} K_i C_i - K_i$, $\Delta \hat{K}_i$ is defined as some perturbed matrices appearing in the controller.

Remark 3.7: Complex external environments lead to adverse effects on the control performance. The non-fragile issue is seen as the uncertain in process of the controller design [21, 32]. The controller gain fluctuation $\Delta \hat{K}_i = M_i \xi(t) N_i$, in which M_i, and N_i are constant matrices, and $\xi(t)$ is defined as a time-varying matrix with $\xi^{\mathrm{T}}(t)\xi(t) < I$ [38].

Remark 3.8: The researchers have noted the applicability of the control methods in real systems. To illustrate the capability of the proposed PI-based controller in real-time LFC applications, an experimental study has been performed on a large-scale analog power system simulator at the research laboratory of the Kyushu Electric Power company in [39]. Real-world effects have been considered in the proposed frequency-triggered DoS attack model and the non-fragile OSMLFC approach.

Defining h as the fixed sampling period, $t_k h$ is the sampling state, and t_k is the successfully transmitted sampling signal. Then, we have $t_k - t_{k-1} = mh$, $m \in \{1, 2, \cdots, m_{\max}\}$. In the ordinary course of working, all sampling data are transmitted without external interference. Obviously, it can be seen as no DoS attacks with $m = 1$. Otherwise, the durations of real-time frequency-triggered DoS attacks is $m_A h$, $m_A \in \{2, 3, \cdots, m_{\max}\}$.

Therefore, we take the influence of real-time frequency-triggered DoS attacks as the time-varying delay. Defining the following

$$\tau_i(t) = m_A h, m_A \in \{2, 3, \cdots, m_{\max}\}. \tag{3.18}$$

Further define

$$\check{x}_i(t_k) = \begin{cases} x_i(t_k - \tau_i(t)), & \text{DoS attacks} \\ x_i(t), & \text{without attacks} \end{cases}. \tag{3.19}$$

Then, we can derive the non-fragile OSMLFC power system state space function as

$$\begin{cases} \dot{x}_i(t) = A_i x_i(t) + B_i(\hat{K}_i + \Delta\hat{K}_i)\check{x}_i(t_k) + \hat{F}_i w_i(t) \\ y_i(t) = C_i x_i(t) \end{cases} \tag{3.20}$$

where $\hat{F}_i = F_i - B_i(G_i B_i)^{-1} G_i F_i$.

3.4 Main Results

In this section, the non-fragile stability problem of the power system integrated with EVs under frequency-based DoS attacks is analyzed. Also, the reachability of the power system integrated with EVs under the proposed non-fragile OSMLFC is proved.

Definition 3.1: (Non-fragile H_∞ stability) [31] For a positive definite matrix P_i, define the Lyapunov function with $V = x_i^\top P_i x_i > 0$ for system model (3.20). There exists

$$\dot{V} = 2\dot{x}_i^\top P_i x_i < 0.$$

Such that, the system is non-fragile H_∞ stability via a non-fragile controller.

Lemma 3.1: [33] *For $0 < a < b < c \in R$ and $x_i(t) \in R^n$, there exist positive definite matrix P_i, such that the following inequality holds*

$$\begin{aligned}
&-(c-a)\int_{t-c}^{t-a} x_i^\top(s)P_i x_i(s)ds \leq \\
&-\int_{t-c}^{t-b} x_i^\top(s)dsP_i \int_{t-c}^{t-b} x_i(s)ds \\
&-\int_{t-b}^{t-a} x_i^\top(s)dsP_i \int_{t-b}^{t-a} x_i(s)ds \\
&-\frac{b-a}{c-a}\int_{t-c}^{t-b} x_i^\top(s)dsP_i \int_{t-c}^{t-b} x_i(s)ds \\
&-\frac{c-b}{c-a}\int_{t-b}^{t-a} x_i^\top(s)dsP_i \int_{t-b}^{t-a} x_i(s)ds.
\end{aligned}$$

Lemma 3.2: [33] *For $\tau_i \in R$ and $x_i(t) \in R^n$, there exist positive definite matrix P_i, such that the following inequality holds*

$$-\tau_i \int_{t-\tau_i}^{t} x_i^\top(s)P_i x_i(s)ds \leq -\int_{t-\tau_i}^{t} x_i^\top(s)dsP_i \int_{t-\tau_i}^{t} x_i(s)ds.$$

Lemma 3.3: [34] *For $0 < a < b < c \in R$ and $x_i(t) \in R^n$, there exist positive definite matrix P_i, such that the following inequality holds*

$$\begin{aligned}
&-\frac{c^2-a^2}{2}\int_{-c}^{-a}\int_{t+\beta}^{t} x_i^\top(s)P_i x_i(s)dsd\beta \leq \\
&-\int_{-b}^{-a}\int_{t+\beta}^{t} x_i^\top(s)dsP_i \int_{-b}^{-a}\int_{t+\beta}^{t} x_i(s)ds \\
&-\int_{-c}^{-b}\int_{t+\beta}^{t} x_i^\top(s)dsP_i \int_{-c}^{-b}\int_{t+\beta}^{t} x_i(s)ds \\
&-\frac{c^2-b^2}{c^2-a^2}\int_{-b}^{-a}\int_{t+\beta}^{t} x_i^\top(s)dsP_i \int_{-b}^{-a}\int_{t+\beta}^{t} x_i(s)ds \\
&-\frac{b^2-a^2}{c^2-a^2}\int_{-c}^{-b}\int_{t+\beta}^{t} x_i^\top(s)dsP_i \int_{-c}^{-b}\int_{t+\beta}^{t} x_i(s)ds.
\end{aligned}$$

Lemma 3.4: [32] *For a symmetrical matrix P, matrices $N_1>0, N_2>0$ with appropriate dimensions. For $\xi^\top\xi<I$, and $P+N_1\xi N_2+N_2^\top\xi^\top N_1^\top<0$, if and only if there exist constant $a>0$, such that*

$$P+aN_1N_1^\top+a^{-1}N_2^\top N_2<0.$$

3.4.1 Non-fragile H_∞ Stability Analysis

Theorem 3.1: *For given positive τ_1, τ_2, the multi-area power system is non-fragile H_∞ stabilizable if there exist symmetric matrices $P_1>0$, matrix $P_2>0, P_3>0$, $Q_b>0$, $R_c>0$, $S_c>0$, with appropriate dimensions, for $b=1,2$ and $c=1,2,3,4$, such that the following inequality hold*

$$J=\begin{bmatrix}\Psi_{11} & \Psi_{21}^\top \\ \Psi_{21} & \Psi_{22}\end{bmatrix}\prec 0 \tag{3.21}$$

where $\Psi_{11}=[\phi_{\hat{i}\hat{j}}]$, $\hat{i},\hat{j}=1,2,\cdots,11$, $\phi_{11}=A^\top P_1+P_1^\top A+Q_1+\tau^2R_1-R_2+(\tau_2-\tau_1)^2R_3+C_i^\top C_i+\frac{\tau_1^4}{4}S_1-m^2S_2+\frac{(\tau_2^2-\tau_1^2)^2}{4}S_3-(2+m+n)(\tau_2-\tau_1)^2S_4$,
$\phi_{21}=(\hat{K}_i+\Delta\hat{K}_i)^\top B_i^\top P_1$, $\phi_{31}=R_2$, $\phi_{51}=P_2+mS_2$, $\phi_{61}=(1+m)(\tau_2-\tau_1)S_4$, $\phi_{71}=(1+n)(\tau_2-\tau_1)S_4$, $\phi_{1,11}=F^\top P_1$, $\phi_{22}=-(2+n+m)R_4$, $\phi_{32}=(1+n)R_4$, $\phi_{42}=(1+m)R_4$, $\phi_{33}=Q_1+Q_2-(1+n)R_4+R_3$, $\phi_{53}=-P_2$, $\phi_{63}=P_3$, $\phi_{73}=P_3$, $\phi_{44}=-Q_2-(1+m)R_4$, $\phi_{64}=-P_3$, $\phi_{74}=-P_3$, $\phi_{55}=-R_1-S_2$, $\phi_{66}=-(1+m)R_3-(1+\tilde{m})S_4$, $\phi_{77}=-(1+n)R_4-(1+\tilde{n})S_4$, $\phi_{88}=-S_1$, $\phi_{99}=-(1+\tilde{m})S_3$, $\phi_{10,10}=-(1+\tilde{n})S_3$, $\phi_{11,11}=-\gamma^2I$, $m=\frac{\tau_i(t)-\tau_1}{\tau_2-\tau_1}$, $n=\frac{\tau_2-\tau_i(t)}{\tau_2-\tau_1}$, $\tilde{m}=\frac{\tau_i^2(t)-\tau_1^2}{\tau_2^2-\tau_1^2}$, $\tilde{n}=\frac{\tau_2^2-\tau_i^2(t)}{\tau_2^2-\tau_1^2}$, $\Psi_{21}=\mathrm{col}[\tau_1\Lambda,(\tau_2-\tau_1)\Lambda,\frac{\tau_1^2}{2}\Lambda,\frac{\tau_2^2-\tau_1^2}{2}\Lambda]$, $\Psi_{22}=-\mathrm{diag}\{R_2,R_4,S_2,S_4\}$, $\Lambda=[A,B_i(\hat{K}_i+\Delta\hat{K}_i),0,0,0,0,0,0,0,0,\hat{F}_i]$.

Proof. Considering the Lyapunov–Krasovskii candidate function as

$$V(t)=V_1(t)+V_2(t)+V_3(t), \tag{3.22}$$

where

$$\begin{aligned}V_1(t)=&\,x_i^\top(t)P_1x_i(t)+\int_{t-\tau_1}^{t}x_i^\top(s)dsP_2\int_{t-\tau_1}^{t}x_i(s)ds\\&+\int_{t-\tau_2}^{t-\tau_1}x_i^\top(s)dsP_3\int_{t-\tau_2}^{t-\tau_1}x_i(s)ds,\end{aligned}$$

$$V_2(t) = \int_{t-\tau_1}^{t} x_i^{\top}(s)Q_1x_i(s)ds + \int_{t-\tau_2}^{t-\tau_1} x_i^{\top}(s)Q_2x_i(s)ds,$$

$$\begin{aligned}V_3(t) = {} & \tau_1 \int_{-\tau_1}^{0}\int_{t+\alpha}^{t} x_i^{\top}(s)R_1x_i(s)dsd\alpha \\ & + \tau_1 \int_{-\tau_1}^{t}\int_{t+\alpha}^{t} \dot{x}_i^{\top}(t)R_2\dot{x}_i(t)dsd\alpha \\ & + (\tau_2-\tau_1)\int_{-\tau_2}^{-\tau_1}\int_{t+\alpha}^{t} x_i^{\top}(t)R_3x_i(t)dsd\alpha \\ & + (\tau_2-\tau_1)\int_{-\tau_2}^{-\tau_1}\int_{t+\alpha}^{t} \dot{x}_i^{\top}(t)R_4\dot{x}_i(t)dsd\alpha,\end{aligned}$$

$$\begin{aligned}V_4(t) = {} & \frac{\tau_1^2}{2}\int_{-\tau_1}^{0}\int_{\alpha}^{0}\int_{t+\nu}^{t} x_i^{\top}(s)S_1x_i(s)dsd\alpha d\nu \\ & + \frac{\tau_1^2}{2}\int_{-\tau_1}^{0}\int_{\alpha}^{0}\int_{t+\nu}^{t} \dot{x}_i^{\top}(s)S_2\dot{x}_i(s)dsd\alpha d\nu \\ & + \frac{\tau_2^2-\tau_1^2}{2}\int_{-\tau_2}^{-\tau_1}\int_{\alpha}^{0}\int_{t+\nu}^{t} x_i^{\top}(s)S_3x_i(s)dsd\alpha d\nu \\ & + \frac{\tau_2^2-\tau_1^2}{2}\int_{-\tau_2}^{-\tau_1}\int_{\alpha}^{0}\int_{t+\nu}^{t} \dot{x}_i^{\top}(s)S_4\dot{x}_i(s)dsd\alpha d\nu,\end{aligned}$$

Take deviation on $V(t)$, we have

$$\begin{aligned}\dot{V}1(t) = {} & sym\Bigg\{\dot{x}i^{\top}(t)P1x_i(t) \\ & + \left(x_i(t)-x_i(t-\tau_1)\right)^{\top}P_2\int_{t-\tau 1}^{t} x(s)ds \\ & + \left(x_i(t-\tau 1)-x_i(t-\tau_2)\right)^{\top}P_3\int_{t-\tau_2}^{t-\tau_1} x_i(s)ds\Bigg\},\end{aligned}$$

$$\begin{aligned}\dot{V}2(t) = {} & x_i^{\top}(t)Q_1x_i(t) - x^{\top}(t-\tau_1)Q_1x_i(t-\tau_1) \\ & + x_i^{\top}(t-\tau_1)Q_2x_i(t-\tau_1) - x_i^{\top}(t-\tau_2)Q_2x_i(t-\tau_2),\end{aligned}$$

$$\begin{aligned}\dot{V}_3(t) &= \tau_1^2 x_i^{\mathrm{T}}(t)R_1x_i(t) - \tau_1\int_{t-\tau_1}^{t} x_i^{\mathrm{T}}(s)R_1x(s)ds\\ &+\tau_1^2\dot{x}_i^{\mathrm{T}}(t)R_2\dot{x}_i(t) - \tau_1\int_{t-\tau_1}^{t}\dot{x}^{\mathrm{T}}(s)R_2\dot{x}_i(s)ds\\ &+(\tau_2-\tau_1)^2x_i^{\mathrm{T}}(t)R_3x_i(t)+(\tau_2-\tau_1)^2\dot{x}_i^{\mathrm{T}}(t)R_4\dot{x}_i(t)\\ &-(\tau_2-\tau_1)\int_{t-\tau_2}^{t-\tau_1}x_i^{\mathrm{T}}(s)R_3x_i(s)ds\\ &-(\tau_2-\tau_1)\int_{t-\tau_2}^{t-\tau_1}\dot{x}_i^{\mathrm{T}}(s)R_4\dot{x}_i(s)ds,\end{aligned}$$

$$\begin{aligned}\dot{V}_4(t) &= \frac{\tau_1^4}{4}x_i^{\mathrm{T}}(t)S_1x_i(t)+\frac{\tau_1^4}{4}\dot{x}_i^{\mathrm{T}}(t)S_2\dot{x}_i(t)\\ &-\frac{\tau_1^2}{2}\int_{-\tau_1}^{0}\int_{t+\alpha}^{t}x_i^{\mathrm{T}}(s)S_1x_i(s)dsd\alpha-\frac{\tau_1^2}{2}\int_{-\tau_1}^{0}\int_{t+\alpha}^{t}\dot{x}_i^{\mathrm{T}}(s)S_2\dot{x}_i(s)dsd\alpha\\ &+\frac{(\tau_2^2-\tau_1^2)^2}{4}x_i^{\mathrm{T}}(t)S_3x_i(t)-\frac{\tau_2^2-\tau_1^2}{2}\int_{-\tau_2}^{-\tau_1}\int_{t+\alpha}^{t}x_i^{\mathrm{T}}(s)S_3x_i(s)dsd\alpha\\ &+\frac{(\tau_2^2-\tau_1^2)^2}{4}\dot{x}_i^{\mathrm{T}}(t)S_4\dot{x}_i(t)-\frac{\tau_2^2-\tau_1^2}{2}\int_{-\tau_2}^{-\tau_1}\int_{t+\alpha}^{t}\dot{x}_i^{\mathrm{T}}(s)S_4\dot{x}_i(s)dsd\alpha.\end{aligned}$$

Applying Lemmas 3.1 and 3.2, we have

$$\begin{aligned}&-\tau_1\int_{t-\tau_1}^{t}\dot{x}^{\mathrm{T}}(s)R_2\dot{x}_i(s)ds\le\\ &-x_i^{\mathrm{T}}(t)R_2x_i(t)+x_i^{\mathrm{T}}(t-\tau_1)R_2x_i(t)\\ &+x_i^{\mathrm{T}}(t)R_2x_i(t-\tau_1)-x_i^{\mathrm{T}}(t-\tau_1)R_2x_i(t-\tau_2),\\ &-(\tau_2-\tau_1)\int_{t-\tau_2}^{t-\tau_1}x_i^{\mathrm{T}}(s)R_3x_i(s)ds\le\\ &-(1+m)\int_{t-\tau_2}^{t-\tau_i(t)}x_i^{\mathrm{T}}(s)dsR_3\int_{t-\tau_2}^{t-\tau_i(t)}x_i(s)ds\\ &-(1+n)\int_{t-\tau_i(t)}^{t-\tau_1}x_i^{\mathrm{T}}(s)dsR_3\int_{t-\tau_i(t)}^{t-\tau_1}x_i(s)ds\\ &-(\tau_2-\tau_1)\int_{t-\tau_2}^{t-\tau_1}\dot{x}_i^{\mathrm{T}}(s)R_4\dot{x}_i(s)ds\le\end{aligned}$$

$$-\varrho_i^{\mathrm{T}}(t)R_4\varrho_i(t)-\vartheta_i^{\mathrm{T}}(t)R_4\vartheta_i(t)$$

$$-m\varrho_i^{\mathrm{T}}(t)R_4\varrho_i(t)-n\vartheta_i^{\mathrm{T}}(t)R_4\vartheta_i(t),$$

$$-\frac{\tau_1^2}{2}\int_{-\tau_1}^{0}\int_{t+\alpha}^{t}x_i^{\mathrm{T}}(s)S_1x_i(s)dsd\alpha\leq$$
$$-\int_{-\tau_1}^{0}\int_{t+\alpha}^{t}x_i^{\mathrm{T}}(s)dsd\alpha S_1\int_{-\tau_1}^{0}\int_{t+\alpha}^{t}x_i^{\mathrm{T}}(s)dsd\alpha,$$

wherein $m=\frac{\tau_2-\tau_i(t)}{\tau_2-\tau_1}, n=\frac{\tau_i(t)-\tau_1}{\tau_2-\tau_1}$, $\varrho_i(t)=x(t-\tau_i(t))-x_i(t-\tau_2), \vartheta_i(t)=x(t-\tau_1)-x_i(t-\tau_i(t))$.

Similarly, we have

$$-\frac{\tau_1^2}{2}\int_{-\tau_1}^{0}\int_{t+\alpha}^{t}\dot{x}_i^{\mathrm{T}}(s)S_2\dot{x}_i(s)dsd\alpha\leq$$
$$-m^2x_i^{\mathrm{T}}(t)S_2x_i(t)+mx_i^{\mathrm{T}}(t)S_2\int_{t-\tau_1}^{t}x(s)ds$$
$$+m\int_{t-\tau_1}^{t}x^{\mathrm{T}}(s)dsS_2x_i(t)-\int_{t-\tau_1}^{t}x^{\mathrm{T}}(s)dsS_2\int_{t-\tau_1}^{t}x(s)ds,$$

Applying Lemma 3.3, we have

$$-\frac{\tau_2^2-\tau_1^2}{2}\int_{-\tau_2}^{-\tau_1}\int_{t+\alpha}^{t}x_i^{\mathrm{T}}(s)S_3x_i(s)dsd\alpha\leq$$
$$-(1+\tilde{n})\int_{-\tau_i(t)}^{-\tau_1}\int_{t+\alpha}^{t}x_i^{\mathrm{T}}(s)dsS_3\int_{-\tau_i(t)}^{-\tau_1}\int_{t+\alpha}^{t}x_i(s)ds$$
$$-(1+\tilde{m}\int_{-\tau_2}^{-\tau_i(t)}\int_{t+\alpha}^{t}x_i^{\mathrm{T}}(s)dsS_3\int_{-\tau_2}^{-\tau_i(t)}\int_{t+\alpha}^{t}x_i(s)ds$$

wherein $\tilde{m}=\frac{\tau_i^2(t)-\tau_1^2}{\tau_2^2-\tau_1^2}, \tilde{n}=\frac{\tau_2^2-\tau_i^2(t)}{\tau_2^2-\tau_1^2}$.

For further, we have

$$-\frac{\tau_2^2-\tau_1^2}{2}\int_{-\tau_2}^{-\tau_1}\int_{t+\beta}^{t}\dot{x}_i^{\mathrm{T}}(s)S_4\dot{x}_i(s)dsd\beta\leq$$
$$-\tilde{\varrho}_i^{\mathrm{T}}(t)S_4\tilde{\varrho}_i(t)-\tilde{\vartheta}_i^{\mathrm{T}}(t)S_4\tilde{\vartheta}_i(t)$$

$$-n\tilde{\varrho}_i^{\mathrm{T}}(t)S_4\tilde{\varrho}_i(t)-m\tilde{\vartheta}_i^{\mathrm{T}}(t)S_4\tilde{\vartheta}_i(t),$$

where $\tilde{\varrho}_i(t) = (\tau_2 - \tau_1)x_i(t) - \int_{t-\tau_i(t)}^{t-\tau_1} x_i(s)ds$, $\tilde{\vartheta}_i(t) = (\tau_2 - \tau_1)x_i(t) - \int_{t-\tau_2}^{t-\tau_i(t)} x_i(s)ds$.

By defining

$$\Phi^{\top}(t) = \Bigg[x_i(t), x_i(t-\tau_i(t)), x_i(t-\tau_1), x_i(t-\tau_2), \int_{t-\tau_1}^{t} x_i(s)ds, \int_{t-\tau_i(t)}^{t-\tau_1} x_i(s)ds, \int_{t-\tau_2}^{t-\tau_i(t)} x_i(s)ds, \int_{-\tau_1}^{0}\int_{t+\alpha}^{t} x_i(s)dsd\alpha, \int_{-\tau_i(t)}^{-\tau_1}\int_{t+\alpha}^{t} x_i(s)dsd\alpha, \int_{-\tau_2}^{-\tau_i(t)}\int_{t+\alpha}^{t} x_i(s)dsd\alpha\Bigg]^{\top}.$$

Then, we have

$$\dot{V}(t) \le \Phi^{\top}(t)(\tilde{\Psi}_{11} - \tilde{\Psi}_{21}^{\top}\tilde{\Psi}_{22}\tilde{\Psi}_{21})\Phi(t). \tag{3.23}$$

Based on [20], we have the following

$$\dot{V} \le -y_i^{\top}(t)y_i(t) + \gamma^2 w_i^{\top}(t)w_i(t). \tag{3.24}$$

Under condition of $V(0) = 0$, one can obtain

$$\int_0^{+\infty} y_i^{\top}(t)y_i(t) \le \int_0^{+\infty} \gamma^2 w_i^{\top}(t)w_i(t), \tag{3.25}$$

which means $\|y_i(t)\| \le \gamma\|w_i(t)\|$ for nonzero $w_i(t) \in L_2[0, +\infty)$. Therefore, the non-fragile output sliding mode LFC power system is non-fragile asymptotically stable with an H_∞ performance. The proof is completed.

Remark 3.9: A non-fragile output sliding mode LFC method is developed in Theorem 3.1 to ensure the stability of multi-area power systems under real-time frequency-triggered DoS attacks. A less conservative stability condition is obtained by using Lemmas 3.1–3.3. Under the designed Lyapunov function, a less conservative result can be obtained. The non-fragile controller gain will be solved in the following theorem.

Theorem 3.2: *For given positive τ_1, τ_2, θ_i, there exist symmetric matrices $X_1 > 0$, matrix $\tilde{P}_2 > 0$, $\tilde{P}_3 > 0$, $\tilde{Q}_b > 0$, $\tilde{R}_c > 0$, $\tilde{S}_c > 0$, $b = 1, 2$ and $c = 1, 2, 3, 4$, matrices M_i, N_i, Y_1 with appropriate dimensions, such that the following linear matrix inequality (LMI) is satisfied*

$$\begin{bmatrix} \hat{\Psi}_{11} & * & * & * & * \\ \hat{\Psi}_{21} & \hat{\Psi}_{22} & * & * & * \\ \hat{\Psi}_{31} & 0 & -I & * & * \\ \hat{\Psi}_{41} & 0 & 0 & -\theta_i I & * \\ \hat{\Psi}_{51} & 0 & 0 & 0 & -\theta_i I \end{bmatrix} \prec 0 \tag{3.26}$$

where $\hat{\Psi}_{11} = [\tilde{\phi}_{\hat{i}\hat{j}}]$, $\hat{i},\hat{j} = 1, 2, \cdots, 11$, $\tilde{\phi}_{11} = X_1 A_i^\top + A_i X_1 + \tilde{Q}_1 + \tau^2 \tilde{R}_1 - \tilde{R}_2 + (\tau_2 - \tau_1)^2 \tilde{R}_3 + \frac{\tau_1^4}{4}\tilde{S}_1 - m^2 \tilde{S}_2 + \frac{(\tau_2^2 - \tau_1^2)^2}{4}\tilde{S}_3 - (2 + m + n)(\tau_2 - \tau_1)^2 \tilde{S}_4, \tilde{\phi}_{21} = Y_1^\top B_i^\top$, $\tilde{\phi}_{31} = \tilde{R}_2$, $\tilde{\phi}_{51} = \tilde{P}_2 + m\tilde{S}_2$, $\tilde{\phi}_{61} = (1 + n)(\tau_2 - \tau_1)\tilde{S}_4$, $\tilde{\phi}_{71} = (1 + m)(\tau_2 - \tau_1)\tilde{S}_4$, $\tilde{\phi}_{11,1} = F^\top, \tilde{\phi}_{22} = -(2 + n + m)\tilde{R}_4, \tilde{\phi}_{32} = (1 + n)\tilde{R}_4, \tilde{\phi}_{42} = (1 + m)\tilde{R}_4, \tilde{\phi}_{33} = \tilde{Q}_1 + \tilde{Q}_2 -(1 + n)\tilde{R}_4 + \tilde{R}_3$, $\tilde{\phi}_{53} = -P_2$, $\tilde{\phi}_{63} = P_3$, $\tilde{\phi}_{73} = P_3$, $\tilde{\phi}_{44} = -\tilde{Q}_2 - (1 + m)\tilde{R}_4$, $\tilde{\phi}_{64} = -P_3$, $\tilde{\phi}_{74} = -P_3$, $\tilde{\phi}_{55} = -\tilde{R}_1 - \tilde{S}_2$, $\tilde{\phi}_{66} = -(1 + n)\tilde{R}_3 - (1 + \tilde{m})\tilde{S}_4$, $\tilde{\phi}_{77} = -(1 + m)\tilde{R}_4 - (1 + \tilde{n})\tilde{S}_4$, $\tilde{\phi}_{88} = -\tilde{S}_1$, $\tilde{\phi}_{99} = -(1 + \tilde{m})\tilde{S}_3$, $\tilde{\phi}_{10,10} = -(1 + \tilde{n})\tilde{S}_3$, $\tilde{\phi}_{11,11} = -\gamma^2 I$,

$$m = \frac{\tau_i(t) - \tau_1}{\tau_2 - \tau_1}, n = \frac{\tau_2 - \tau_i(t)}{\tau_2 - \tau_1}, \tilde{m} = \frac{\tau_i^2(t) - \tau_1^2}{\tau_2^2 - \tau_1^2}, \tilde{n} = \frac{\tau_2^2 - \tau_i^2(t)}{\tau_2^2 - \tau_1^2},$$

$$\hat{\Psi}_{21} = \mathrm{col}[\tau_1\Lambda, (\tau_2 - \tau_1)\Lambda, \frac{\tau_1^2}{2}\Lambda, \frac{\tau_2^2 - \tau_1^2}{2}\Lambda],$$

$$\tilde{\Lambda} = [A_i X_1, B_i Y_1, \underbrace{0, \cdots, 0}_{8}, \hat{F}_i],$$

$$\hat{\Psi}_{22} = -\mathrm{diag}\{\tilde{R}_2, \tilde{R}_4, \tilde{S}_2, \tilde{S}_4\}, \hat{\Psi}_{31} = [C_i X_1, \underbrace{0, \cdots, 0}_{10}],$$

$$\hat{\Psi}_{41} = B_i M_i X_1 [I, \underbrace{0, \cdots, 0}_{10}, \tau_1, (\tau_2 - \tau_1), \frac{\tau_1^2}{2}, \frac{\tau_2^2 - \tau_1^2}{2}],$$

$$\hat{\Psi}_{51} = [0, N_i, 0, 0, 0, 0, 0, 0, 0, 0, 0, 0, 0, 0, 0].$$

The non-fragile output sliding mode LFC gain can be calculated by $K_i = Y_1 X_1^{-1}$.

Proof. Define $X_1 = P_1^{-1}$, $Y_1 = K_i X_1$, pre- and post-multiply both sides of J with $\mathrm{diag}\{X_1, X_1, X_1, X_1, X_1, X_1, X_1, X_1, X_1, X_1, I, R_2^{-1}, R_4^{-1}, S_2^{-1}, S_4^{-1}\}$. Then, applying

Lemma 3.4, we have

$$\begin{bmatrix} \hat{\Psi}_{11} & * & * \\ \hat{\Psi}_{21} & \hat{\Psi}_{22} & * \\ \hat{\Psi}_{31} & 0 & -I \end{bmatrix} + \begin{bmatrix} \hat{\Psi}_{41} \\ 0 \\ 0 \\ 0 \end{bmatrix} \xi_i(t) \begin{bmatrix} \hat{\Psi}_{51}^{\mathrm{T}} \\ 0 \\ 0 \\ 0 \end{bmatrix} +$$

$$\begin{bmatrix} \hat{\Psi}_{41} \\ 0 \\ 0 \\ 0 \end{bmatrix} \xi_i(t) \begin{bmatrix} \hat{\Psi}_{51}^{\mathrm{T}} \\ 0 \\ 0 \\ 0 \end{bmatrix} < 0$$

If there exist scalar $\theta_i > 0$, such that LMI (3.26) hold.

Remark 3.10: The controller gain, non-fragile issue, and the complex terms are solved in Theorem 3.2. Based on Lemma 3.4, the non-fragile terms can be solved. The stability problem of the power system integrated with EVs under cyberattacks is investigated via non-fragile output sliding mode LFC. The non-fragile problem is treated the same as [21, 30, 31]. The control performance is improved by considering the non-fragile issue and proved in simulation parts.

3.4.2 Reachability Analysis

Theorem 3.3: *The trajectories of the multi-area power system integrated with EVs (3.12) under real-time frequency-triggered DoS attacks can be ensured onto the designed sliding surface in a finite time with the following SMC law*

$$\begin{aligned} u_i(t) = &-K_i x_i(t) + (G_i B_i)^{-1} K_i C_i x_i(t) \\ &-d_i (G_i B_i)^{-1} \|G_i \hat{F}_i\| \operatorname{sgn}(\sigma_i(t)) \end{aligned} \tag{3.27}$$

where $d_i > 0$, $\operatorname{sgn}(\cdot)$ is the switch function.

Proof. The Lyapunov function is chosen as

$$V_l = \frac{1}{2} \sigma_i^{\mathrm{T}}(t) \sigma_i(t) \tag{3.28}$$

Take deviation on V_l, we have

$$
\begin{aligned}
\dot{V}_l &= \sigma_i^{\mathrm{T}}(t)\dot{\sigma}_i(t) \\
&= \sigma_i^{\mathrm{T}}(t)[G_i\dot{x}_i(t) - G_i(A_i - B_iK_i)x_i(t) - K_iy_i(t)] \\
&= \sigma_i^{\mathrm{T}}(t)\{G_i[A_ix_i(t) + B_iu_i(t) + F_iw_i(t)] \\
&\quad - G_i(A_i - B_iK_i)x_i(t) - K_iy_i(t)\} \\
&= \sigma_i^{\mathrm{T}}(t)[-d_i\|G_iF_i\|\mathrm{sgn}(\sigma_i(t)) + G_iF_iw_i(t)] \\
&\leq \{-d_i\|G_iF_i\|\|\sigma_i(t)\| + \|\sigma_i(t)\|\|G_iF_i\|\|w_i(t)\|\} \\
&\leq \|G_iF_i\|\|\sigma_i(t)\|[-d_i + \|w_i(t)\|] < 0.
\end{aligned}
$$

Thus, the reaching condition is ensured by the designed control law. This completes the proof.

Remark 3.11: Based on Lyapunov theory, the trajectories of the multi-area power system under frequency-triggered DoS attacks can be ensured to reach onto the predesigned sliding mode surface in a finite time. Which proves the correctness and effectiveness of the proposed sliding mode controller.

3.5 Numerical Examples

In this part, the stability of the power system integrated with EVs under real-time frequency-triggered DoS attacks is proved by simulations with the proposed non-fragile output sliding mode LFC.

Example 3.1 *The non-fragile output sliding mode LFC technique is simulated for a three-area power system integrated with EVs under real-time frequency-triggered DoS attacks. The difference among different participation of EVs in the power system is simulated.*

System parameters of the three-area power system comes from [20] and EVs comes from [2] are outlined in Table 3.1.

The tolerable frequency deviation $\Delta f_A = 0.001$. Weighting parameters are set as $\alpha_g = 0.9$, $\alpha_e = 0.1$, $k_e = 1$. According to Theorem 3.2, the controller gain is calculated as

$K_1 = [-0.4010, -0.3313, -0.1024, -2.0974, -0.7069, 0.0020]$,
$K_2 = [-0.1293, -0.3241, -0.2596, -1.9639, -0.9447, 0.0086]$,
$K_3 = [-0.3998, -0.3303, -0.1015, -2.0947, -0.6514, 0.0002]$.

Table 3.1 Parameters of each area.

Area-1	Parameters	D_1	T_{ch1}	T_{g1}	R_1	T_{12}	T_{ev1}	M_1
	Value	0.5	0.3	0.3	0.5	0.36π	0.02	1.5
Area-2	Parameters	D_2	T_{ch2}	T_{g2}	R_2	T_{13}	T_{ev2}	M_2
	Value	0.8	0.1	0.3	0.5	0.36π	0.03	1.5
Area-3	Parameters	D_3	T_{ch3}	T_{g3}	R_3	T_{23}	T_{ev3}	M_3
	Value	0.3	0.3	0.3	0.5	0.36π	0.05	1.5

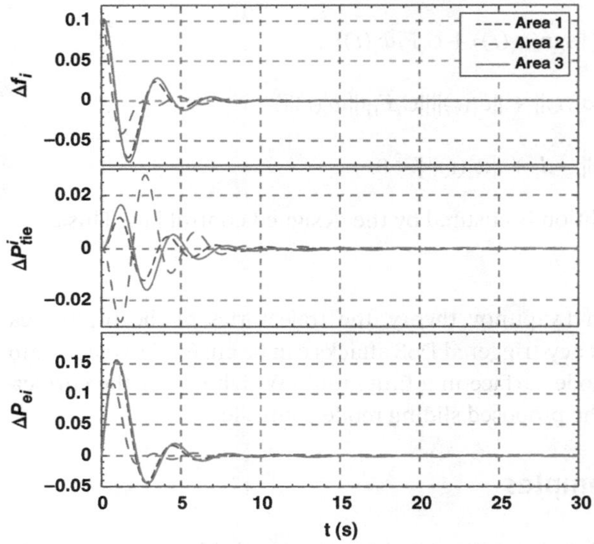

Figure 3.4 State response of three-area power system integrated with EVs.

To verify the effectiveness of the proposed non-fragile OSMLFC, the investigation is conducted from three aspects. The Δf_i, ΔP^i_{tie}, and ΔP_{ei} of the three-area power system integrated with EVs are shown in Figure 3.4. From the simulated results in Figure 3.4, one can conclude that the three-area power system integrated with EVs is ensured to be stable by the designed non-fragile OSMLFC. Therefore, the simulation results prove the usefulness of the proposed non-fragile OSMLFC to the multi-area power system integrated with EVs under real-time frequency-triggered DoS attacks.

To illustrate the influence of the suitability of EVs, their willingness to participate in frequency regulation service, and related constraints in the EVs. Those constraints are constructed as the level of participation of EVs in the power system. The simulations are tested among different participation of EVs. $\alpha_e = 0$ means no

Table 3.2 Maximum $|\Delta f_i| \times 10^{-2}$ under different participation α_e.

α_e	0	0.02	0.04	0.08	0.1	0.14	**0.15**
Value	8.821	8.855	8.804	8.828	8.838	8.776	8.734
α_e	0.16	0.18	0.2	0.25	0.5	0.7	1
Value	8.747	8.752	8.829	8.801	8.791	9.072	11.102

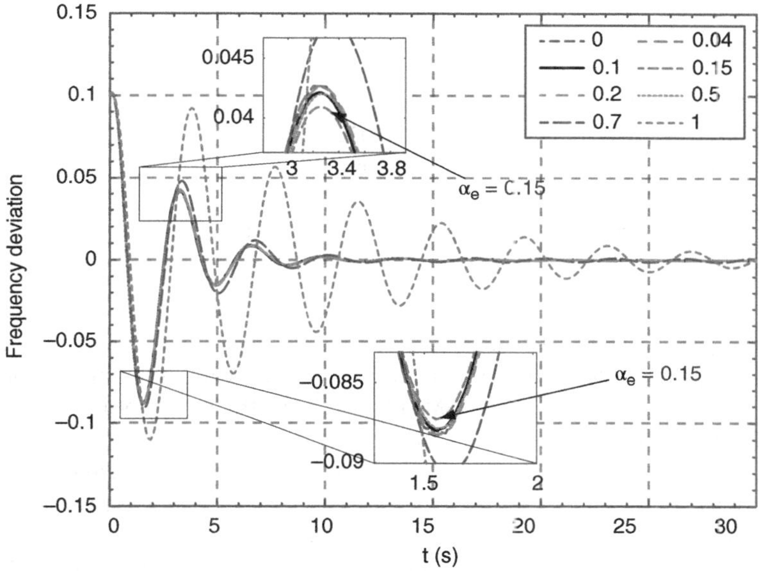

Figure 3.5 Frequency deviation of different EVs participation.

EVs are integrated, and $\alpha_e = 1$ means LFC performance are realized completely by EVs.

The results of different participation in EVs to the power system frequency are presented in Table 3.2 and Figure 3.5. The results prove that with the integration of EVs, system performance can be improved. However, an excessive or too low participation of EVs in the LFC power systems may lead to bad system performance. In the tested system, system performance is the best when the participation in EVs is at about $\alpha_e = 0.15$. Thus, it is necessary to have an appropriate proportion of EVs participate in the LFC power system.

Example 3.2 *The difference between the proposed method and the other control strategies is shown in this part. In addition, we investigate the effect of non-fragile on the controller performance.*

System parameters are the same as area-1 in Example 5.1. We made the comparisons among the proposed method, non-fragile SMC [38], PI-SMC [28], and non-fragile output.

The proposed non-fragile output sliding mode controller gain $\tilde{K}_1$ is calculated according to Theorem 3.2.

$\tilde{K}_1 = [-0.4010, -0.3313, -0.1024, -2.0974, -0.7069, -0.0020]$.

The non-fragile SMC [38] is designed as

$$\sigma_2 = G_i x_i(t) - \int_0^t G_i(A_i - B_i\tilde{K}_2)x_i(s)ds$$

and the controller gain is

$\tilde{K}_2 = [-0.0641, -0.0347, 0.0909, \ -0.2177, 0.0328, 0.0277]$.

SMC [28] controller gain through pole assignment is

$\tilde{K}_3 = [-0.5622, -0.4092, 0.8525, -3.1806, -1.6260, 7.4455]$.

The sliding mode surface with non-fragile output is designed as

$$\sigma_4 = \tilde{K}_4 P_{ei}(t)$$

and controller gain $\tilde{K}_4 = [0, 0, -0.3243, -0.8108, 0, 0]$.

Figure 3.6 is given as the changes of Δf_i, ΔP^i_{ptie}, and ΔP_{ei} with different control strategies. The proposed non-fragile OSMLFC can make the power system to be stable at about $t = 10s$, which is earlier than the others. The result between the proposed method and non-fragile SMC [38] proves that the use of system output rises the control performance. Overall, the results illustrate the superiority of the proposed method.

Next, the necessity of considering non-fragile for the controller design is proved in the following simulations.

Ignoring the non-fragile of the controller design, the output sliding mode LFC gain is calculated as

$\tilde{K}_5 = [-0.3915, -0.3208, 0.0286, -1.3704, -1.0211, 0.0017]$.

Simulated results of output sliding mode LFC and non-fragile OSMLFC are given in Figure 3.7. The changes of Δf_i, ΔP^i_{tie}, and ΔP_{ei} are given in Figure 3.7. From Figure 3.7, all two methods can assure the stability of the power system. The results show that the proposed non-fragile OSMLFC can make the power system became stable at about $t = 10s$. Besides, the non-fragile OSMLFC appears to have more excellent performance in the remaining stable state than without non-fragile controller. Figure 3.7 shows the system state to the power system with non-fragile and without non-fragile. The result proves that the non-fragile OSMLFC can assure the stability of the power system integrated with EVs while the without non-fragile controller still fluctuates. Simulated results show the necessity of considering a non-fragile controller design.

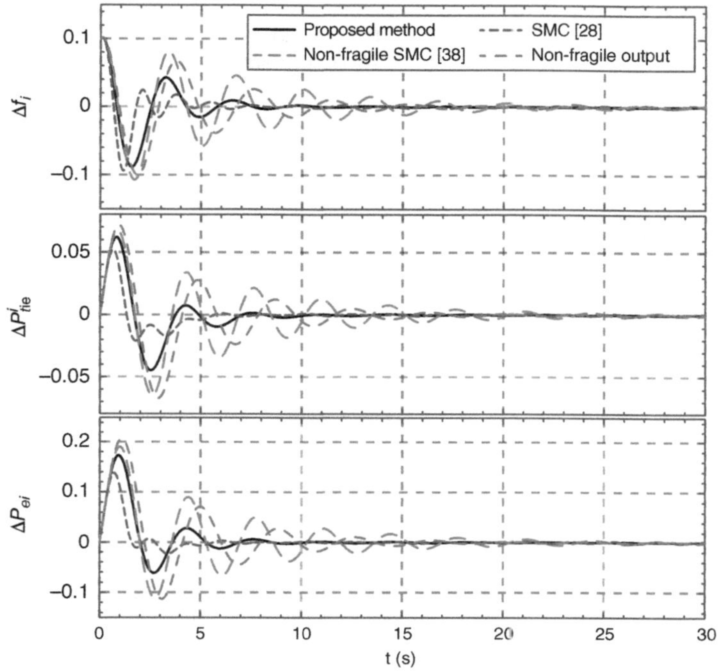

Figure 3.6 State response of power system with different control methods.

Example 3.3 *In this part, the characteristics of the considered real-time frequency-triggered DoS attacks are simulated.*

System parameters are mostly the same as area-1 in Example 5.1. To investigate the characteristics and influence of real-time frequency-triggered DoS attacks. The triggered threshold Δf_A is given as 0.002, 0.01, 0.08, and 0.1.

Figure 3.8 simulates the changes of Δf_i when the DoS attacks triggering threshold is given as $|\Delta f_A| = 0.002$. Then, the triggering threshold becomes $|\Delta f_A| = 0.01$ in Figure 3.9. When the system's real-time frequency deviation is located in the attack region, DoS attacks happen. The result shows that along with the increases in the threshold, attack frequency becomes higher. To further investigate the characteristics of frequency-triggered DoS attacks, more simulations are carried out for the cases under DoS attacks triggering threshold $|\Delta f_A| = 0.08$ and $|\Delta f_A| = 0.1$, respectively. The results are presented in Figure 3.10. The fluctuation of the system frequency increases and the stabilizing time gets longer. Simulated results reveal the influence of real-time frequency-triggered DoS attacks on the stability of the power system integrated with EVs. Furthermore, the results show that all the real-time frequency-triggered DoS attacks happened in the predefined stable region.

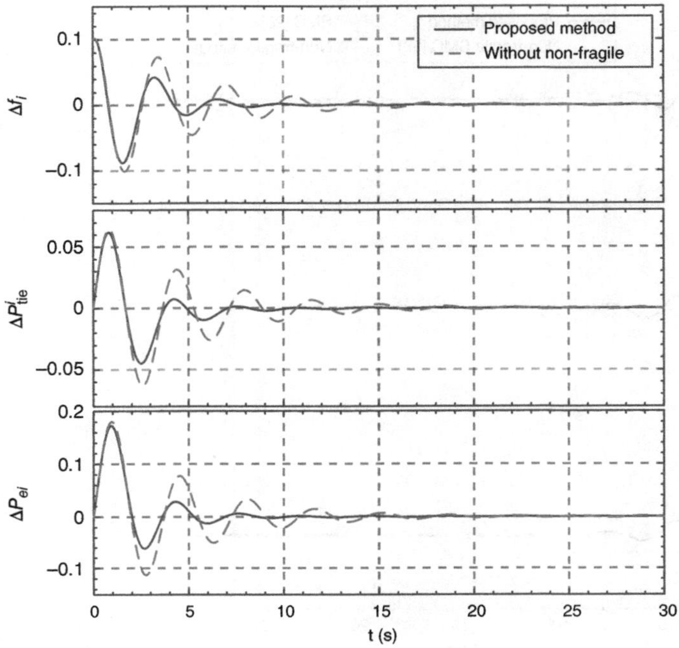

Figure 3.7 State response with non-fragile and without non-fragile.

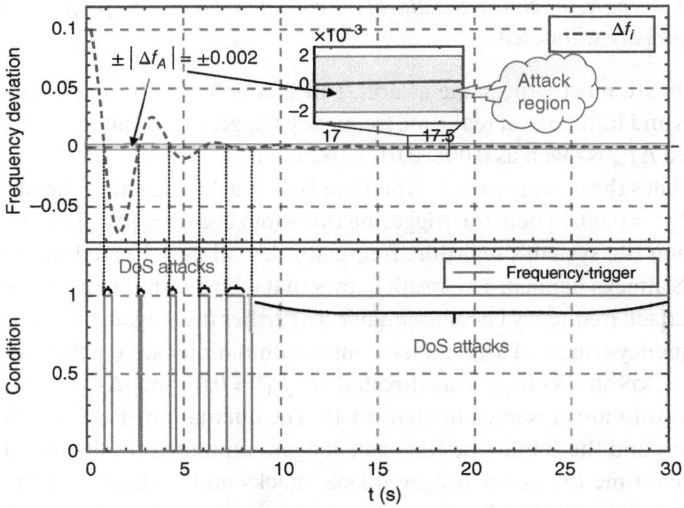

Figure 3.8 Δf_i of power system with $|\Delta f_A| = 0.002$.

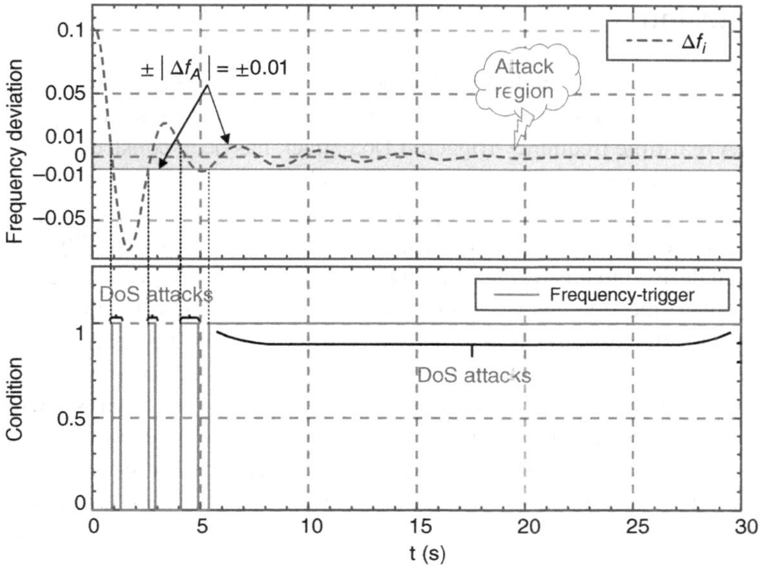

Figure 3.9 Δf_i of power system with $|\Delta f_A| = 0.01$.

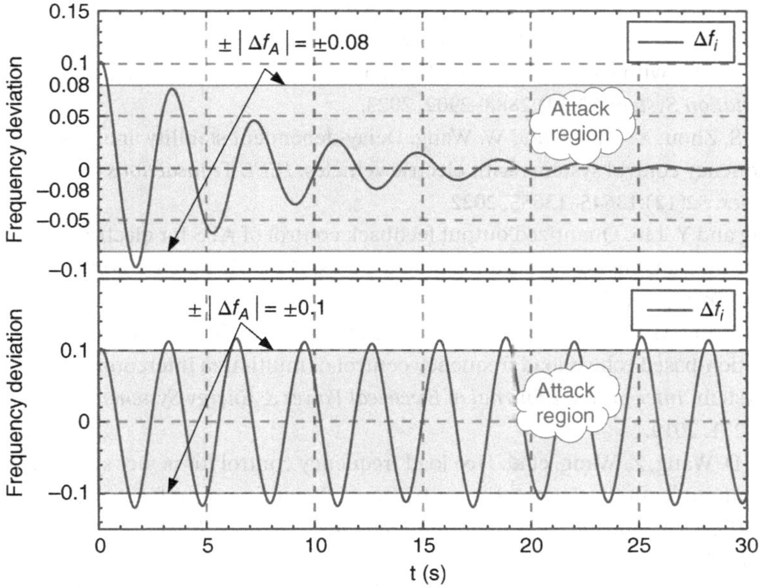

Figure 3.10 Δf_i of power system with $|\Delta f_A| = 0.08$ and 0.1.

3.6 Conclusion

A non-fragile output sliding mode LFC has been proposed for multi-area interconnected power systems integrated with EVs under real-time frequency-triggered DoS attacks. A real-time frequency-triggered DoS attack has been constructed by combining the real-time frequency, event trigger mechanism, and the attributes of DoS attacks. Considering the uncertain appearance of control gains, a non-fragile output sliding mode LFC method has been applied. Then, a sufficient stability criterion has been derived to ensure the stability of the power system, which is contaminated by real-time frequency-triggered DoS attacks. Simulated results have demonstrated the effectiveness of the proposed control scheme. The benefits and limitations of EVs in enhancing the dynamic performance of LFC power systems are discussed by considering different levels of participation of EVs. It was shown that while an appropriate participation of EVs may greatly benefit enhancing LFC power system, an excessive participation of EVs in the LFC power system may cause serious damage to system performance. Simulated results have also illustrated the characteristic of the proposed real-time frequency-triggered DoS attacks.

References

1 A. Mirheli and L. Hajibabai. Charging network design and service pricing for electric vehicles with user-equilibrium decisions. *IEEE Transactions on Intelligent Transportation Systems*, 24(3):2888–2902, 2023.

2 H. Zeng, S. Zhou, X. Zhang, and W. Wang. Delay-dependent stability analysis of load frequency control systems with electric vehicles. *IEEE Transactions on Cybernetics*, 52(12):13645–13653, 2022.

3 X. Chang and Y. Liu. Quantized output feedback control of AFS for electric vehicles with transmission delay and data dropouts. *IEEE Transactions on Intelligent Transportation Systems*, 23(9):16026–16037, 2022.

4 K. Lu, W. Zhou, G. Zeng, and Y. Zheng. Constrained population extremal optimization-based robust load frequency control of multi-area interconnected power system. *International Journal of Electrical Power & Energy Systems*, 105:249–271, 2019.

5 H. Shen, D. Wang, Z. Wang, et al. H∞ load frequency control for power systems under communication delays: An event-triggered dynamic output feedback scheme. *IEEE Transactions on Circuits and Systems II: Express Briefs*, 69(8):3495–3499, 2022.

6 R. Abolpour, K. Torabi, M. Dehghani, et al. Direct search algorithm for load frequency control of a time-delayed electric vehicle aggregator. *IEEE Transactions on Industry Applications*, 59(2):2603–2614, 2022.

7 M. Khan and H. Sun. Complete provision of MPC-based LFC by electric vehicles with inertial and droop support from DFIG-based wind farm. *IEEE Transactions on Power Delivery*, 37(2):716–726, 2021.

8 S. S. K. Madahi, A. S. Kamrani, and H. Nafisi. Overarching sustainable energy management of PV integrated EV parking lots in reconfigurable microgrids using generative adversarial networks. *IEEE Transactions on Intelligent Transportation Systems*, 23(10):19258–19271, 2022.

9 B. Khokhar and K. P. Singh Parmar. A novel adaptive intelligent MPC scheme for frequency stabilization of a microgrid considering SoC control of EVs. *Applied Energy*, 309:118423, 2022.

10 B. Gao, B. Bu, and X. Wang. A comprehensive resilient control strategy for CBTC systems through train-to-train communications under malicious attacks. *IEEE Transactions on Intelligent Transportation Systems*, 23(11):21015–21033, 2022.

11 C. Peng, H. Sun, M. Yang, and Y. Wang. A survey on security communication and control for smart grids under malicious cyber attacks. *IEEE Transactions on Systems, Man, and Cybernetics: Systems*, 49(8):1554–1569, 2019.

12 E. Mousavinejad, F. Yang, Q. Han, et al. Distributed cyber attacks detection and recovery mechanism for vehicle platooning. *IEEE Transactions on Intelligent Transportation Systems*, 21(9):3821–3834, 2019.

13 J. Liu, Y. Gu, L. Zha, et al. Event-triggered H∞ load frequency control for multi-area power systems under hybrid cyber attacks. *IEEE Transactions on Systems, Man, and Cybernetics: Systems*, 49(8):1665–1678, 2019.

14 X. Zhao, S. Zou, and Z. Ma. Decentralized resilient H∞ load frequency control for cyber-physical power systems under DoS attacks. *IEEE/CAA Journal of Automatica Sinica*, 8(11):1737–1751, 2021.

15 D. J. Wang, F. Chen, B. Meng, et al. Event-based secure h_∞ load frequency control for delayed power systems subject to deception attacks. *Applied Mathematics and Computation*, 394:125788–125801, 2021.

16 S. Yan, Z. Gu, and J. H. Park. Memory-event-triggered H∞ load frequency control of multi-area power systems with cyber-attacks and communication delays. *IEEE Transactions on Network Science and Engineering*, 8(2):1571–1583, 2021.

17 Y. Pan, Y. Wu, and H. K. Lam. Security-based fuzzy control for nonlinear networked control systems with DoS attacks via a resilient event-triggered scheme. *IEEE Transactions on Fuzzy Systems*, 30(10):4359–4368, 2022.

18 C. De Persis and P. Tesi. Input-to-state stabilizing control under denial-of-service. *IEEE Transactions on Automatic Control*, 60(11):2930–2944, 2015.

19 M. Hossain, C. Peng, Y. L. Wang, and X. Du. Handshake logic-based event-triggered load frequency control for smart grids under dos attacks. *IEEE Transactions on Industrial Informatics*, 18(6):3863–3872, 2022. https://doi.org/10.1109/TII.2021.3116165.

20 J. Yang, Q. Zhong, K. Shi, and S. Zhong. Dynamic-memory event-triggered H∞ load frequency control for reconstructed switched model of power systems under hybrid attacks. *IEEE Transactions on Cybernetics*, 53(6):3913–3925, 2022.

21 Q. Zhong, J. Yang, K. Shi, and S. Zhong. Design of observer-based discrete-type PID control for reconstructed jump model of interconnected power system under hybrid attacks. *IEEE Transactions on Smart Grid*, 14(3):1896–1906, 2022.

22 Y. Ma, Z. Nie, S. Hu, et al. Fault detection filter and controller co-design for unmanned surface vehicles under DoS attacks. *IEEE Transactions on Intelligent Transportation Systems*, 22(3):1422–1434, 2020.

23 Z. Ye, D. Zhang, Z. Wu, and H. Yan. A3C-based intelligent event-triggering control of networked nonlinear unmanned marine vehicles subject to hybrid attacks. *IEEE Transactions on Intelligent Transportation Systems*, 23(8):12921–12934, 2021.

24 C. Mu, Y. Tang, and H. He. Improved sliding mode design for load frequency control of power system integrated an adaptive learning strategy. *IEEE Transactions on Industrial Electronics*, 64(8):6742–6751, 2017.

25 P. Mani and Y. H. Joo. Fuzzy logic-based integral sliding mode control of multi-area power systems integrated with wind farms. *Information Sciences*, 545:153–169, 2021.

26 R. J. Ma, P. Shi, and L. G. Wu. Dissipativity-based sliding-mode control of cyber-physical systems under denial-of-service attacks. *IEEE Transactions on Cybernetics*, 51(5):2306–2318, 2020.

27 W. Qi, G. Zong, and H. R. Karimi. Finite-time observer-based sliding mode control for quantized semi-Markov switching systems with application. *IEEE Transactions on Industrial Informatics*, 16(2):1259–1271, 2019.

28 Y. Mi, Y. Fu, C. S. Wang, and P. Wang. Decentralized sliding mode load frequency control for multi-area power systems. *IEEE Transactions on Power Systems*, 28(4):4301–4309, 2013.

29 Q. Zhong, J. Yang, K. Shi, et al. Event-triggered H∞ load frequency control for multi-area nonlinear power systems based on non-fragile proportional integral control strategy. *IEEE Transactions on Intelligent Transportation Systems*, 23(8):12191–12201, 2021.

30 L. Liu, Z. Fu, X. Cai, and X. Song. Non-fragile sliding mode control of discrete singular systems. *Communications in Nonlinear Science and Numerical Simulation*, 18(3):735–743, 2013.

31 B. Wang, Y. Luo, Z. Zhong, and K. Li. Robust non-fragile fault tolerant control for ensuring the safety of the intended functionality of cooperative adaptive cruise control. *IEEE Transactions on Intelligent Transportation Systems*, 23(10):18746–18760, 2022.

32 S. Anand, A. Dev, M. K. Sarkar, and S. Banerjee. Non-fragile approach for frequency regulation in power system with event-triggered control and communication delays. *IEEE Transactions on Industry Applications*, 57(3):2187–2201, 2021.

33 A. Farnam and R. M. Esfanjani. Improved linear matrix inequality approach to stability analysis of linear systems with interval time-varying delays. *Journal of Computational and Applied Mathematics*, 294:49–56, 2016.

34 W. Qian, M. Yuan, L. Wang, et al. Stabilization of systems with interval time-varying delay based on delay decomposing approach. *ISA Transactions*, 70:1–6, 2017.

35 X. C. ShangGuan, Y. He, and C. K. Zhang. Switching system-based load frequency control for multi-area power system resilient to denial-of-service attacks. *Control Engineering Practice*, 107:104678–104689, 2021.

36 E. Tian and C. Peng. Memory-based event-triggering h_∞ load frequency control for power systems under deception attacks. *IEEE Transactions on Cybernetics*, 50:4610–4618, 2020.

37 M. Zhao, C. Peng, W. He, and Y. Song. Event-triggered communication for leader-following consensus of second-order multiagent systems. *IEEE Transactions on Cybernetics*, 48(6):1888–1897, 2017.

38 J. Huang, G. Yang, Z. Fang, et al. Non-fragile sliding mode control for one-sided Lipschitz chaotic systems. *ISA Transactions*, 124:311–317, 2022.

39 H. Bevrani, P. R. Daneshmand, P. Babahajyani, et al. Intelligent LFC concerning high penetration of wind power: Synthesis and real-time application. *IEEE Transactions on Sustainable Energy*, 5(2):655–662, 2013.

4

Memory-based Event-triggered Fault-tolerant Load Frequency Control of Multi-area Power Systems with Electric Vehicles

After reading this chapter, you should be able to:

- Know the impact of sensor failures and limited communication network bandwidth in power system.
- Explain the event trigger mechanism and the key threshold in triggering scheme.
- Identify the use of fault-tolerant load frequency control.

4.1 Background

The frequency stability of a power system directly impacts its operation, safety, and reliability [1]. Load frequency control (LFC) is used in multi-area power systems to maintain frequency stability at a specified level for each zone and to regulate power exchange between different regions within a multi-zone power system [2]. LFC has been proven highly effective in ensuring power system frequency stability [3–6]. In spite of this, modern power systems are becoming increasingly complex. The use of various communication devices for grid protection, monitoring, and control has become common [7]. As a result, the reliability of control loop components is becoming increasingly important. If a component fails, fault-tolerant control (FTC) ensures the system's safe operation [8]. Therefore, studying fault-tolerant LFC is essential.

On the other hand, accurate estimation of fault information is of utmost importance for FTC. The sliding mode observer (SMO) has been widely employed due to its strong robustness in regard to uncertainties and the possibility of uncertainty estimation [9]. It cannot only eliminate the influence of faults but also provide precise estimation of system states and faults [10]. In [11], the FTC based on SMO is proposed for pressurized water reactor-type nuclear power plant with actuator and sensor faults. In [12], multi-area power systems with sensor faults are examined for FTC, proposing the use of the SMO to estimate the system state and fault values. Furthermore, the limited bandwidth of the power system's communication network can result in network congestion and even system instability.

In light of this issue, the event-triggered mechanism (ETM) has attracted considerable attention [13–16]. Compared to traditional time triggers, ETM can alleviate communication burdens and conserve limited network resources [13]. In [16], introduced a dynamic ETM to solve the $H\infty$ filtering problem of network control system under multiple cyberattacks. In [17], LFC is studied in multi-area power systems, employing both the ETM and FTC. It deals with the event-triggered control problem in FTC systems with actuator faults and random parameter uncertainty in [18]. In contrast to [17, 18], it adopts a hybrid event-triggered mechanism (HETM) that avoids Zeno behavior and continuous system status measurements. However, in most ETMs, the current system states are used to construct the threshold function. In a class of Lur'e systems, [19] introduced a memory-based ETM to address the global synchronization problem, demonstrating that utilizing historical state information in designing the threshold function can reduce the number of triggers. What's more, the adaptive ETM can meet the requirements of rapidly changing systems [20]. Currently, this method has been widely applied [20–22].

The electric vehicles (EVs) have characteristics such as fast response speed and high flexibility [23]. Consequently, integrating EVs into the grid can enhance the dynamic performance of system. The advancement of Vehicle-to-Grid (V2G) technology has prompted an increasing number of researchers to investigate the participation of EVs in system frequency regulation [24–26]. A multitude of EVs can be grouped and controlled by an EV aggregator (EVA) [24]. When a signal is emitted by the control center, the EVA promptly responds and engages in frequency regulation [27]. The participation capability of EVAs in LFC was investigated in [28], revealing that EVs, via V2G, can achieve 100% LFC. The EV's state of charge (SOC) is an important feature that measures the amount of energy stored and the capacity of its batteries [29]. EVs are dependent on gain to determine their output power deviation. And the gain is closely linked to the SOC [24]. The gain of an EV should be set as a time-varying parameter when considering the SOC. The power system model incorporating parameter uncertainty is established in [29], addressing the issue of delay-dependent stability. However, there are limited researches on the LFC problem in power systems with EVs, considering both sensor faults and limited bandwidth of the communication network. This serves as the key motivation behind this study.

This chapter mainly focuses on the development of a memory-based adaptive HETM and the design of a fault-tolerant LFC for power systems with EVs. This chapter makes the following main contributions:

1. A multi-area power system model is established, taking into account the variation in SOC. In comparison with [29, 30], this chapter considers the impact of sensor failures and limited communication network bandwidth, resulting in a more comprehensive system model.
2. Different from [17], this study utilizes historical state information to construct the event trigger threshold function and introduces an adaptive factor. This memory-based adaptive HETM effectively reduces the number of triggers and conserves resources.
3. By constructing multiple Lyapunov functions and employing Lyapunov theory and matrix inequalities, the stability criteria for power systems are derived. The obtained result is less conservative than [12].

4.2 Problem Formulation

4.2.1 Simplified EV Model

EVs aggregated together can participate in LFC, as shown in Figure 4.1. The $SOC_{\exp}$, T_p, SOC_0, T_{ct} are the minimum expected battery charge to leave the grid, the dwell time, the initial SOC of the battery, and the charging time. Obviously, only if $T_{ct} < T_p$, the EVs can provide frequency regulation service.

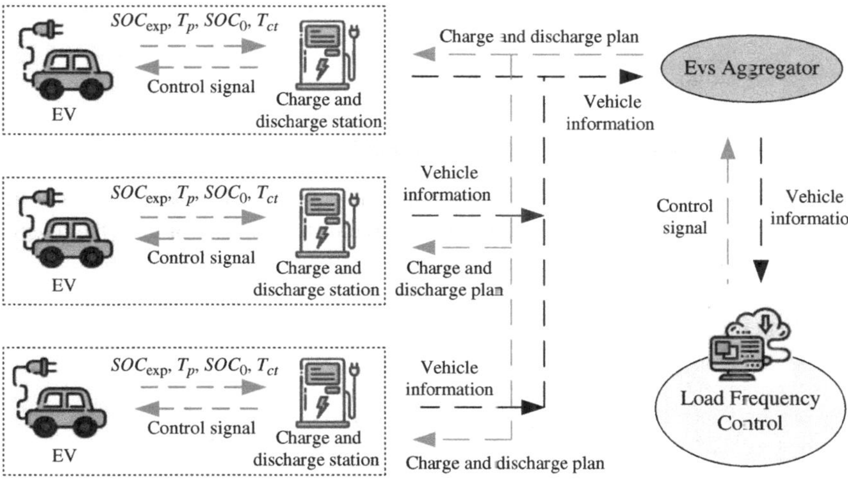

Figure 4.1 Multi-area power systems with electric vehicles.

The deviation of power output for the i_{th} EV is obtained as

$$\dot{\mathcal{P}}_{a_i}(t) = \frac{K_{a_i}}{\mathcal{T}_{a_i}} u(t) - \frac{1}{\mathcal{T}_{a_i}} \mathcal{P}_{a_i}(t) \tag{4.1}$$

The i_{th} EV gain K_a is determined by frequency and battery SOC. Therefore, the K_{a_i} is modeled as

$$K_{a_i} = K_m - \varrho_i(t) K_m \tag{4.2}$$

$$\varrho_i(t) = \left(\frac{SOC_i - SOC_{\text{low(high)},i}}{SOC_{\text{max(min)},i} - SOC_{\text{low(high)},i}} \right)^2 \tag{4.3}$$

where K_m is the maximum gain of EV, and $SOC_{\text{min},i}$ and $SOC_{\text{max},i}$, SOC_i, $SOC_{\text{low},i}$ and $SOC_{\text{high},i}$, denote the minimum and maximum battery SOC, current battery SOC, low and high battery SOC of the i_{th} EV, respectively, and $\varrho_i(t)$ is a time-varying function related to the current battery SOC.

Assume that the number of EVs participating in the LFC is $N(t)$, the $\mathcal{T}_{a_i}$ is the same for all EVs. The $u(t)$ is allocated equally among each EV as $u_i(t) = u(t)/N(t)$. The aggregated EVs gain is $K_a = \frac{\sum_{i=1}^{N(t)} K_{ai}}{N(t)} = K_m - \varrho(t) K_m$. Therefore, the deviation of power output $\mathcal{P}_a(t)$ can be determined

$$\dot{\mathcal{P}}_a(t) = \frac{K_m}{\mathcal{T}_a} u(t) - \frac{\varrho(t) K_m}{\mathcal{T}_a} u(t) - \frac{1}{\mathcal{T}_a} \mathcal{P}_a(t) \tag{4.4}$$

where $\varrho(t) = \sum_{i=1}^{N(t)} \varrho_i(t)/N(t)$.

4.2.2 Power System Model with Sensor Fault

The i_{th} area multi-area electric power system simplified model is given in Figure 4.2. Based on signal transmission, the dynamic equation of the system can be obtained as follows.

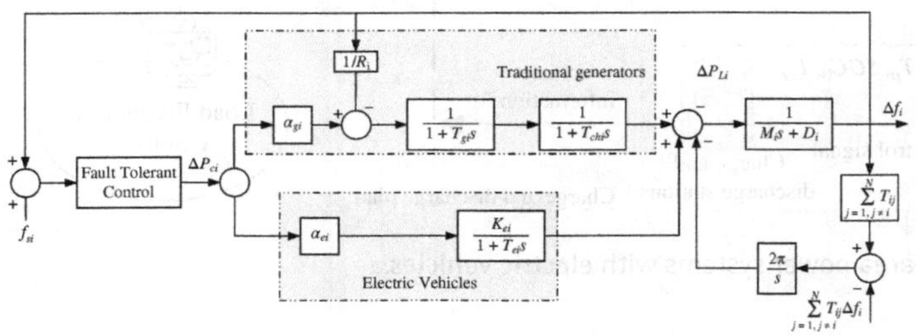

Figure 4.2 Dynamic model of the ith in multi-area power scheme.

$$\begin{cases} \Delta \dot{f}_i(t) = -\frac{D_i}{M_i}\Delta f_i(t) + \frac{1}{M_i}(\Delta \mathcal{P}_{mi}(t) + \Delta \mathcal{P}_{ei}(t) \\ \qquad -\Delta \mathcal{P}_{\text{tie}-i}(t) - \Delta \mathcal{P}_{Li}(t)) \\ \Delta \dot{\mathcal{P}}_{mi}(t) = \frac{1}{\mathcal{T}_{mi}}(\Delta \mathcal{P}_{gi}(t) - \Delta \mathcal{P}_{mi}(t)) \\ \Delta \dot{\mathcal{P}}_{gi}(t) = \frac{1}{\mathcal{T}_{gi}}(\alpha_{gi}u(t) - \Delta \mathcal{P}_{gi}(t) - \frac{1}{R_i}\Delta f_i(t)) \\ \Delta \dot{\mathcal{P}}_{ai}(t) = -\frac{1}{\mathcal{T}_{ai}}\mathcal{P}_{ai}(t) + \frac{\alpha_{ai}(1-\varrho(t))K_{mi}}{\mathcal{T}_{ai}}u(t) \\ \int ACE_i(t) = \beta_i \Delta f_i + \Delta P_{\text{tie}-i} \\ \Delta \dot{\mathcal{P}}_{\text{tie}-i}(t) = 2\pi \sum_{j=1, i\neq 1}^{N} \Delta T_{ij}\Delta f_i \end{cases} \tag{4.5}$$

where the meanings of the symbols are listed in Table 4.1.

Thus, the state space function can be represented as

$$\begin{cases} \dot{x}_i(t) = A_i x_i(t) + (B_i + \Delta B_i(t))u_i(t) + F_i w_i(t) \\ y_i(t) = C_i x_i(t) + f_{si}(t), \end{cases} \tag{4.6}$$

where $f_s(t)$ represents the sensor faults and $\|f_s(t)\| \leq \rho$,

$$x_i(t) = [\Delta f_i \ \ \Delta \mathcal{P}_{mi} \ \ \Delta \mathcal{P}_{gi} \ \ \Delta \mathcal{P}_{ei} \ \ \Delta \mathcal{P}_{tie-i} \ \ \int ACE_i]^T, B = [0 \ \ 0 \ \ \frac{\alpha_{gi}}{\mathcal{T}_{gi}} \ \ \frac{\alpha_{ei}K_{mi}}{\mathcal{T}_{ei}} \ \ 0 \ \ 0]^T$$

$$H = [0 \ \ 0 \ \ 0 \ \ -\frac{\alpha_{ei}K_{mi}}{\mathcal{T}_{ei}} \ \ 0 \ \ 0]^T, y_i(t) = [ACE_i \ \ \int_0^t ACE_i(s)ds]^T,$$

Table 4.1 Symbols of the system model.

Δf_i	Frequency deviation	$\Delta \mathcal{P}_{mi}$	Mechanical power change
$\Delta \mathcal{P}_{ei}$	EVs output power change	$\Delta \mathcal{P}_{gi}$	Governor output change
$\Delta \mathcal{P}_{Li}$	Disturbance of load	M_i	Equivalent inertia constant
D_i	Equivalent damping coefficient	$\mathcal{T}_{gi}$	Time-constant of governor
$\mathcal{T}_{mi}$	Time-constant of turbine	$\mathcal{T}_{ai}$	Time-constant of EVs
R_i	Speed droop	β_i	Frequency bias factor
$\Delta \mathcal{P}_{\text{tie}-i}$	Power deviation of tie-line	K_{mi}	The maximum gain of EVs
T_{ij}	Synchronizing coefficient of tie-line		

$$\bar{C} = \begin{bmatrix} \beta_i & 0 & 0 & 0 & 1 & 0 \\ 0 & 0 & 0 & 0 & 0 & 1 \end{bmatrix},$$

$$\Delta B_i(t) = H_i \varrho_i(t) E, E_i = 1, F_i = \begin{bmatrix} -1/M_i & 0 & 0 & 0 & 0 & 0 \\ 0 & 0 & 0 & 0 & -2\pi & 0 \end{bmatrix}^T,$$

$$A_i = \begin{bmatrix} -\frac{D_i}{M_i} & \frac{1}{M_i} & 0 & \frac{1}{M_i} & -\frac{1}{M_i} & 0 \\ 0 & -\frac{1}{\mathcal{T}_{mi}} & \frac{1}{\mathcal{T}_{mi}} & 0 & 0 & 0 \\ -\frac{1}{R_i \mathcal{T}_{gi}} & 0 & -\frac{1}{\mathcal{T}_{gi}} & 0 & 0 & 0 \\ 0 & 0 & 0 & -\frac{1}{\mathcal{T}_{ei}} & 0 & 0 \\ 2\pi \sum_{j=1, i \neq 1}^{N} T_{ij} & 0 & 0 & 0 & 0 & 0 \\ \beta_i & 0 & 0 & 0 & 1 & 0 \end{bmatrix},$$

$$w_i(t) = \begin{bmatrix} \Delta \mathcal{P}_{Li}(t) & \sum_{j=1, i \neq 1}^{N} T_{ij} \delta f_j(t) \end{bmatrix}^T.$$

4.2.3 Sliding Mode Observer Design

By using the SMO, the fault value and system state can be estimated simultaneously. To facilitate the design of the observer, we introduce augmentation matrix.

$$E_i = \begin{bmatrix} I_n & 0 \\ 0 & 0 \end{bmatrix}, \bar{A}_i = \begin{bmatrix} A_i & 0 \\ 0 & -I_p \end{bmatrix}, \bar{B}_i = \begin{bmatrix} B_i \\ 0 \end{bmatrix}, \Delta\bar{B}_i = \begin{bmatrix} \Delta B_i \\ 0 \end{bmatrix}, \bar{F}_i = \begin{bmatrix} F_i \\ 0 \end{bmatrix},$$

$$\bar{C}_i = \begin{bmatrix} C_i \\ I_p \end{bmatrix}^T, \bar{N}_i = \begin{bmatrix} 0 \\ I_{pi} \end{bmatrix}, \bar{x}_i(t) = \begin{bmatrix} x_i(t) \\ f_{si}(t) \end{bmatrix}.$$

Then, the augmented state space function as follow

$$\begin{cases} E_i \dot{\bar{x}}_i(t) = \bar{A}_i \bar{x}_i(t) + (\bar{B}_i + \Delta\bar{B}_i(t)) u_F i(t) + \bar{N}_i f_{si}(t) + \bar{F}_i w_i(t) \\ y(t) = \bar{C} \bar{x}(t) \end{cases} \tag{4.7}$$

It can be observed that the matrices satisfy the following properties

$$\text{rank} = \begin{bmatrix} E_i \\ C_i \end{bmatrix} = \text{rank} \begin{bmatrix} I_n & 0 \\ 0 & 0 \\ C & I_p \end{bmatrix} = n + p$$

There exists a matrix $\bar{L}_D = [0_{p\times n} \;\; W]^T$ such that the matrix $\bar{S}$ is nonsingular

$$\bar{S}_i = (E_i + \bar{L}_{Di}\bar{C}_i)$$

The observer as follow

$$\begin{cases} \bar{S}_i\dot{\hat{z}}_i(t) = (\bar{A}_i - \bar{L}_{Pi}\bar{C}_i)\hat{z}_i(t) + (\bar{B}_i + \Delta\bar{B}_i(t))u_{Fi}(t) - \bar{C}_{Ni}y_i(t) + \bar{L}_{si}u_{si}(t) \\ \hat{\bar{x}}_i(t) = \bar{z}_i(t) + \bar{S}_i^{-1}\bar{L}_{Di}y_i(t) \end{cases} \tag{4.8}$$

where $\bar{z}_i(t) = [z_x^T(t), z_s^T(t)]^T$ denotes the state vector of the dynamic system, and $\hat{\bar{x}}_i(t) = [\hat{x}^T(t), \hat{f}_s^T(t)]^T$ denotes the estimation of $\bar{x}(t)$, $\bar{L}_{Pi}$, $\bar{L}_{Di}$, $\bar{L}_s$ represent the observer gains. What's more, $u_{si}(t)$and $\bar{L}_{si}$ are designed to maintain stability in the event of sensor failure.

Based on the above, we can get

$$\begin{aligned} \bar{S}_i\dot{\hat{\bar{x}}}_i(t) &= (\bar{A}_i - \bar{L}_{Pi}\bar{C}_i)\hat{\bar{x}}_i(t) + (\bar{B}_i + \Delta\bar{B}_i(t))u_{Fi}(t) \\ &\quad + \bar{L}_{Pi}y_i(t) + \bar{L}_{si}u_{si}(t) + \bar{L}_{Di}\dot{y}_i(t) \end{aligned} \tag{4.9}$$

Similarly, we can get

$$\begin{aligned} \bar{S}_i\dot{\bar{x}}_i(t) &= (\bar{A}_i - \bar{L}_{Pi}\bar{C}_i)\bar{x}_i(t) + (\bar{B}_i \\ &\quad + \Delta\bar{B}_i(t))u_{Fi}(t) + \bar{L}_{Pi}y_i(t) + \bar{N}_i f_{si}(t) + \bar{F}_i w_i(t) + \bar{L}_{Di}\dot{y}_i(t) \end{aligned} \tag{4.10}$$

Furthermore, defining $\bar{e}(t) = \bar{x}(t) - \hat{\bar{x}}(t)$, we can obtain that

$$\begin{cases} \dot{\bar{e}}_i(t) = \bar{S}_i^{-1}[(\bar{A}_i - \bar{L}_{Pi}\bar{C}_i)\bar{e}_i(t) - \bar{L}_{si}u_{si}(t) + \bar{N}_i f_{si}(t) + \bar{F}_i w_i(t)] \\ e_{yi}(t) = \bar{C}_i\bar{e}_i(t) \end{cases} \tag{4.11}$$

Definition of the sliding mode surface as follow,

$$s_{ei}(t) = B_{fi}^T\bar{S}_i^{-T}P_i\bar{e}(t) \tag{4.12}$$

and for each positive P is designed to satisfy $\bar{N}_i^T \bar{S}_i^{-T} P_i = U_i \bar{C}_i$. For further,

$$u_s(t) = (-\lambda + \rho)sgn(s_e(t)) \tag{4.13}$$

where λ is a given parameter and ρ is the upper bound of the sensor fault.

Remark 4.1: The sliding mode input $u_s(t)$ is used to eliminate the influence of faults on the control system. The SMO is capable of simultaneously estimating the system state and fault signals. The non-zero estimation of fault signals indicates the occurrence of system faults, thereby obviating the necessity for the design of additional complex fault diagnosis schemes. Furthermore, the SMO exhibits advantages such as rapid response, good transient performance, and insensitivity to disturbances. Given the aforementioned description, the widespread application of SMO in addressing fault-tolerance issues is evident.

4.2.4 Memory-based Adaptive Hybrid Event-triggered

As a means of reducing the burden on communication channels, the architecture of FTC system based on the memory-based adaptive HETM is shown in Figure 4.3.

The memory-based adaptive HETM conditions are defined as follows:

$$t_{k+1} = \min\left\{ t \geq t_k + t_h \mid \epsilon_i^T(t)\Phi\epsilon(t) > \Im(t)\left(\frac{1}{\Delta t}\int_{t-\Delta t}^{t} \hat{x}_i(s)ds\right)^T \Omega_i \left(\frac{1}{\Delta t}\int_{t-\Delta t}^{t} \hat{x}(s)ds\right)\right\} \tag{4.14}$$

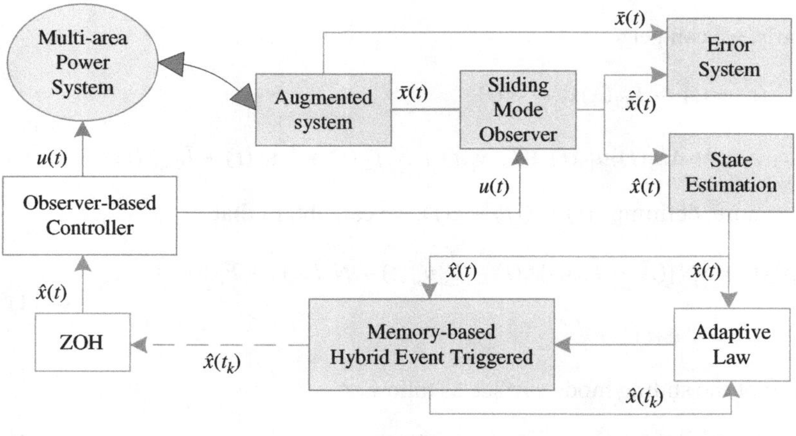

Figure 4.3 Event-triggered architecture of the FTC system.

where $\epsilon(t) = \hat{x}(t) - \hat{x}(t_k)$. Assuming that $t_h > 0$ and $\Delta t > 0$, respectively, correspond to the waiting time and the memory period. In this context, $\Phi > 0$ refers to the triggering matrix. $\mathfrak{I}(t)$ is a triggering parameter satisfying the following [31]:

$$\dot{\mathfrak{I}}_i(t) = \delta_i(t) \left(\frac{1}{\mathfrak{I}_i^2(t)} - \frac{\mathfrak{I}_0}{\mathfrak{I}_i(t)} \right) \epsilon^T(t) \Omega_i \epsilon_i(t) \tag{4.15}$$

with $\mathfrak{I}_0 > 1$, $0 < \mathfrak{I}(0) < 1$ and $\delta_i(t) = \begin{cases} 0, \text{when } t \in [t_k, t_k + t_h) \\ 1, \text{when } t \in [t_k + t_h, t_{k+1}) \end{cases}$

Remark 4.2: Combined with the proposed memory-based adaptive HETM, the triggering interval $t \in [t_k, t_{k+1})$ between two adjacent sampling moments is decomposed into two parts. And the first part $t \in [t_k, t_k + t_h)$ is the wait interval, and the second part $t \in [t_k + t_h, t_{k+1})$ is the continuous detection interval. When $t = t_k$, the system satisfies the condition, then the sampled signal $\hat{x}(t_k)$ is released immediately. And $\hat{x}(t_k)$ is transmitted to a zero-order holder (ZOH) that maintains the continuity of the control signal. Then, the condition (4.14) waits for t_h seconds. After $t_k + t_h$, the condition (4.14) is continuously checked until the next event is triggered. Obviously, there are at least t_h seconds between the two adjacent triggering moments. Thus, the triggering interval is guaranteed to be at least greater than t_h and the Zeno phenomenon is avoided.

Remark 4.3: The proposed approach in this chapter is the memory-based adaptive HETM. On the one hand, a host of ETMs typically construct threshold functions based on the instantaneous state $x(t)$ of the system. External disturbances or noise can cause frequent fluctuations in the system state. If such fluctuations occur, the ETM based on $x(t)$ may generate large data packets. However, the memory-based adaptive HETM introduces an integral term $\int_{t-\Delta t}^{t} \hat{x}(s)ds$. Compared to the HETM, it can better reduce redundant information and save network resources. On the other hand, considering that conventional event-triggering schemes use predetermined constant triggering parameters, they cannot meet the requirements of rapidly changing systems. Therefore, a varying threshold $\mathfrak{I}(t)$ is introduced in the triggering conditions, which adjusts the release frequency by changing the threshold parameters. Additionally, to avoid the Zeno phenomenon, a waiting-time period is introduced in the triggering conditions. Therefore, the proposed ETM in this chapter not only effectively reduces the number of triggering events but also better meets the requirements of rapidly changing systems and avoids the Zeno phenomenon.

4.2.5 Active Fault-tolerant Controller Design

Next, the main work is to design a fault-tolerant LFC stabilizing multi-area power system.

Based on (4.9), we can get

$$
\begin{aligned}
E\dot{\hat{x}}(t) &= (\bar{A} - \bar{L}_P\bar{C})\hat{x}(t) + (\bar{B} + \Delta\bar{B}(t))u(t) + \bar{L}_P y(t) + \bar{L}_s u_s(t) + \bar{L}_D\dot{y}(t) - \bar{L}_D\bar{C}\dot{\hat{x}}(t) \\
&= \bar{A}\hat{x}(t) - \bar{L}_P\bar{C}\bar{e}(t) + (\bar{B} + \Delta\bar{B}(t))u(t) + \bar{L}_D\bar{C}\dot{\bar{e}}(t) + \bar{L}_s u_s(t)
\end{aligned} \tag{4.16}
$$

Decompose the matrices $\bar{L}_D$, $\bar{L}_s$, u_s as follows

$$
u_s = \begin{bmatrix} u_{s1}^T & u_{s2}^T \end{bmatrix}^T, \bar{L}_D = \begin{bmatrix} L_{D1}^T & L_{D2}^T \end{bmatrix}^T, \bar{L}_P = \begin{bmatrix} L_{P1}^T & L_{P2}^T \end{bmatrix}^T, \bar{L}_s = \begin{bmatrix} L_{s1}^T & L_{s2}^T \end{bmatrix}^T.
$$

Therefore, we can obtain

$$
\dot{\hat{x}}(t) = A\hat{x}(t) + L_{P1}\bar{C}\bar{e}(t) + (B + \Delta B(t))u(t) + L_{D1}\bar{C}\dot{\bar{e}}(t) + L_{s1}u_{s1}(t) \tag{4.17}
$$

We define $\bar{L}_s = \bar{S}P^{-1}\bar{C}^T U^T$, $\bar{N}^T\bar{S}^{-T}P = U\bar{C}$ and (17) can be processed as

$$
\dot{\hat{x}}(t) = A\hat{x}(t) + L_{P1}\bar{C}\bar{e}(t) + (B + \Delta B(t))u(t) \tag{4.18}
$$

Following is the design of a PI-type controller

$$
u(t) = K_p A\hat{C}E(t) + K_I \int A\hat{C}E(t) = K\hat{y}(t) = KC\hat{x}(t) \tag{4.19}
$$

The controller $u(t)$ should be designed according to the MBET condition (4.14) in the following from

$$
u(t) = KC\hat{x}(t_k) \tag{4.20}
$$

Therefore, the multi-area power system is formulated as follows

$$
\begin{aligned}
\dot{\hat{x}}(t) = {} & A\hat{x}(t) + L_{P1}\bar{C}\bar{e}(t) + (1 - \delta(t))(B + \Delta B(t))KC\hat{x}(t - d(t)) \\
& + \delta(t)(B + \Delta B(t))KC[\hat{x}(t) + \epsilon(t)]
\end{aligned} \tag{4.21}
$$

where

$$
\delta(t) = \begin{cases} 0, \text{when } t \in [t_k, t_k + t_h) \\ 1, \text{when } t \in [t_k + t_h, t_{k+1}) \end{cases}
$$

The following definitions and lemmas are important for proving the main conclusions before moving on to the results.

Definition 4.1: For $w(t) = 0$, if there is a continuous first-order partial derivative of $V(t)$, and $V(t)$ is positively definite and $\dot{V}(t)$ is negatively definite, then the system is asymptotically stable. Besides, H_∞ performance is guaranteed under the condition of zero initial state if the inequality $||y(t)||_2 \leq \gamma ||w(t)||_2$ holds, in which $\gamma > 0$ is a pregiven constant.

Lemma 4.1: [32] *For any constant matrices* $U \in R^{n\times n}$, $G \in R^{n\times n}$, $\Pi = \begin{bmatrix} U & G \\ * & U \end{bmatrix} > 0$, *scalars* $0 \leq d_m \leq d(t) \leq d_M$, *and* $\dot{x} : [-d_m, -d_M] \to \mathbb{R}^n$, *inequality holds as follows*

$$-(d_m - d_M) \int_{t-d_M}^{t-d_m} \dot{x}^T(s) R \dot{x}(s) \leq -\wp^T(t) \Pi \wp(t)$$

where $\wp(t) = [\wp_1^T(t) \ \wp_2^T(t)]^T$, $\wp_1(t) = x(t - d_m) - x(t - d(t))$, $\wp_2(t) = x(t - d(t)) - x(t - d_M)$.

Lemma 4.2: [33] *For a continuous function* $x : [a, b] \to \mathbb{R}^n$ *and* $n \times n$*-matrix* $Q > 0$*, inequality holds as follows*

$$\int_a^b x^T(s) Q x(s) ds \geq \frac{1}{b-a} \left(\int_a^b x(s) ds \right)^T Q \left(\int_a^b x(s) ds \right)$$

Lemma 4.3: [34] *For any* $x, y \in \mathbb{R}$ *and positive scalar* ε, *we have*

$$2y^T x \leq \varepsilon x^T x + \frac{1}{\varepsilon} y^T y.$$

4.3 Main Results

4.3.1 Stability Analysis

Now, using the following theorem to give stability conditions of the system.

Theorem 4.1: *For given scalars* Δt, *if there exist real symmetric matrices* $Q_1 > 0$, $Z > 0$ *and matrix* U *with appropriate dimension, such that the following conditions hold*

$$\Gamma_1 = \begin{bmatrix} \Phi_{1i} & P_i \bar{S}_i^{-1} \bar{F} \\ * & -\gamma_i^2 I \end{bmatrix} < 0 \tag{4.22}$$

$$\begin{bmatrix} -\delta_i I & (\bar{N}_i^T \bar{S}_i^{-T} P_i - U_i \bar{C}_i)^T \\ * & -I \end{bmatrix} \leq 0 \tag{4.23}$$

where

$$\gamma_{1i} = P_i\bar{S}_i^{-1}\bar{A}_i - Y_i\bar{C}_i + \bar{A}_i^T\bar{S}_i^{-T}P_i - \bar{C}_i^TY_i^T + C_i^TC_i$$

then it follows that the power system (4.11) is asymptotically stable when H_∞ *performance index* $\xi > 0$*, the observer gains* $\bar{L}_P$ *and* $\bar{L}_s$ *are given by* $\bar{L}_s = \bar{S}P^{-1}\bar{C}^TU^T = N_i$ *and* $\bar{L}_P = \bar{S}_iP_i^{-1}Y_i$.

Proof. The Lyapunov functional is considered as

$$V_{ei}(t) = \bar{e}_i^T(t)P_i\bar{e}_i(t) \tag{4.24}$$

Taking the derivation of $V_e(t)$, one has

$$\dot{V}_{ei}(t) = 2\bar{e}^T(t)P_i\bar{S}_i^{-1}[(\bar{A}_i - \bar{L}_{Pi}\bar{C}_i)\bar{e}_i(t) + \bar{N}_if_{si}(t) - \bar{L}_{si}u_{si}(t) + \bar{F}_iw_i(t)] \tag{4.25}$$

We define $\bar{L}_{si} = \bar{S}_iP_i^{-1}\bar{C}_i^TH_i^T = N_i$, $B_{fi}^T\bar{S}_i^{-T}P_i = H_i\bar{C}_i$ and (4.25) can be processed as

$$\begin{aligned} &2\bar{e}_i^T(t)P_i\bar{S}_i^{-1}[\bar{L}_{si}u_{si}(t) + \bar{N}_if_{si}(t)] \\ &= -2\bar{e}_i^T(t)\bar{C}_i^TU_i^Tu_{si}(t) + 2\bar{e}_i^T(t)P_i\bar{S}_i^{-1}\bar{N}_if_{si}(t) \\ &\leq -2s_{ei}^T(t)u_s(t) + 2\bar{e}_i^T(t)\|\bar{C}_i^TU_i^T\|\|f_{si}(t)\| \\ &\leq -2(\lambda_i + \rho_i)sgn(s_{ei}(t)) + 2\rho_i\|s_{ei}(t)\| \\ &\leq -2\lambda_i\|s_{ei}(t)\| \end{aligned} \tag{4.26}$$

In view of Lemma 4.2, define $\jmath(t) = \int_{t-\Delta t}^{t} \bar{e}(s)ds$, one has

$$-\int_{t-\Delta t}^{t} \bar{e}_i^T(s)Q_1\bar{e}_i(s)ds \leq -\frac{1}{\Delta t}\jmath(t)^TQ_1\jmath(t) \tag{4.27}$$

Thus, combining (4.25) to (4.27), we can get

$$\dot{V}_{ei}(t) \leq -e_{yi}^T(t)e_{yi}(t) + \gamma_i^2w_i^T(t)w_i(t) \tag{4.28}$$

where $\zeta_{1i}(t) = [\bar{e}_i(t)\ \ w_i(t)]^T$.

The integration on both sides of (4.28), we can get

$$V_{ei}(+\infty) - V_{ei}(0) \leq -\int_0^\infty e_{yi}^T(t)e_{yi}(t)dt + \int_0^\infty \gamma_i^2w_i^T(t)w_i(t)dt \tag{4.29}$$

Under $V_{ei}(0) = 0$, the following inequality can be obtained

$$\int_0^{\infty} e_{yi}^T(t)e_{yi}(t)dt \leq \int_0^{\infty} \gamma^2 w_i^T(t)w_i(t)dt \tag{4.30}$$

As a result, if we apply Definition 4.1 to the error system (4.11) we can conclude that it is asymptotically stable with H_∞ performance under the disturbance attenuation level ξ as defined by Equation (4.30).

Our conditions are denoted in the following manner to facilitate their presentation,

$$\varsigma_1(t) = [\hat{x}(t)\ \dot{\hat{x}}(t)\ \hat{x}(t-t_h)\ \hat{x}(t-d(t))\ \frac{1}{\Delta t}\int_{t-\Delta t}^{t} \hat{x}(s)ds\ \overline{e}(t)]^T,$$

$$\varsigma_2(t) = [\hat{x}(t)\ \dot{\hat{x}}(t)\ \hat{x}(t-t_h)\ \epsilon(t)\ \frac{1}{\Delta t}\int_{t-\Delta t}^{t} \hat{x}(s)ds\ \overline{e}(t)]^T,$$

$$e_i = [0_{n\times(i-1)n}\ I\ 0_{n\times(7-i)n}],\ Z = \begin{bmatrix} e_0 - e_3 \\ e_3 - e_2 \end{bmatrix}.$$

Theorem 4.2: *For given scalars t_h, μ, Δt, if there exist real symmetric matrices $\mathfrak{P} > 0$, $\mathcal{R}_a > 0$, $Q_2 > 0$, $U_2 > 0$, the scalar $\varepsilon_1 > 0$ and appropriate dimension matrices M, G such that as follows*

$$\Psi_{1i} = \begin{bmatrix} \Sigma_{1i} & \Sigma_{2i} & \Theta L_{P1}\bar{C}_i \\ * & -\frac{\varepsilon_{1i}}{1+\mu_{1i}}I & 0 \\ * & * & -\xi_i^2 I \end{bmatrix} < 0 \tag{4.31}$$

$$\Psi_2 = \begin{bmatrix} \Sigma_{3i} & \Sigma_{4i} & \Theta_i L_{P1i}\bar{C}_i \\ * & \Sigma_{5i} & 0 \\ * & * & -\xi_i^2 I \end{bmatrix} < 0 \tag{4.32}$$

$$\Pi_i = \begin{bmatrix} O_i & \Upsilon_i \\ * & O_i \end{bmatrix} > 0 \tag{4.33}$$

where

$$\Sigma_{1i} = \Lambda_{1i} - \Lambda_{2i} - \Lambda_{3i} + \Lambda_{4i} + \Lambda_{5i} + E_0^T E_0,$$

$$\Lambda_{1i} = 2E_0^T \mathfrak{P}_i E_1 + E_0^T \mathcal{R}_{ai} E_0 - E_2^T \mathcal{R}_{ai} E_2 + t_h^2 E_1^T O_i E_1 + \Delta t E_0^T Q_i E_0,$$

$$\Lambda_{2i} = \Delta t E_4^T Q_i E_4,\ \Lambda_{3i} = -J_i^T \Pi_i J_i,\ \Theta_i = E_0^T M_1^T + E_1^T \mu_i M_{1i}^T,$$

$$\Lambda_{4i} = 2\Theta_i [A_i E_0 + B_i K_i C_i E_3 - E_1],$$

$$\Lambda_{5i} = \varepsilon_{1i} E_0^T M_{1i}^T H_i H_i^T M_{1i} E_0 + \varepsilon_{1i} \mu E_{1i}^T M_{1i}^T H_i H_i^T M_{1i} E_{1i},$$

$$\Sigma_{2i} = E_{3i}^T C_i^T K_i^T E_i^T,$$

$$\Sigma_{3i} = \Lambda_{1i} - \Lambda_{2i} - \Lambda_{3i} - \Lambda_{6i} + \Lambda_{7i} + \Lambda_{8i} + \Lambda_{9i} + E_0^T E_0,$$

$$\Sigma_{4i} = [E_0^T C_i^T K_i^T E_i^T \quad E_3^T C_i^T K_i^T E_i^T],$$

$$\Sigma_{5i} = \mathrm{diag}\left\{-\frac{\varepsilon_{1i}}{1+\mu_i} I, -\frac{\varepsilon_{1i}}{1+\mu_i} I\right\},\ \Lambda_{6i} = (E_0 - E_3)^T Oi (E_0 - E_3),$$

$$\Lambda_{7i} = 2\Theta_i [(A_i + B_i K_i C_i) E_0 + B_i K_i C_i E_3 + L_{p1i} \bar{C}_i E_5 - E_1],$$

$$\Lambda_{8i} = 2\varepsilon_{1i} E_0^T M_{1i}^T H_i H_i^T M_{1i} E_0 + 2\varepsilon_{1i} \mu_i E_1^T M_{1i}^T H_i H_i^T M_{1i} E_1,$$

$$\Lambda_{9i} = E_4^T \Omega_i E_4 - \mathfrak{I}_0 E_3^T \Omega E_3.$$

Then the power system (4.21) is asymptotically stable with an H_∞ performance index $\gamma > 0$.

Proof. The Lyapunov functional is considered as

$$V_{xi}(t) = V_1(t) + V_2(t) + V_3(t) + V_4(t) \tag{4.34}$$

where

$$V_1(t) = \hat{x}^T(t) \mathfrak{P}_i \hat{x}_i(t),$$

$$V_2(t) = \int_{t-t_h}^{t} \hat{x}_i^T(s) \mathcal{R}_{ai} \hat{x}_i(s) ds + t_h \int_{-t_h}^{0} \int_{t+\theta}^{t} \dot{\hat{x}}_i^T(s) O_i \dot{\hat{x}}_i(s) ds d\theta$$

$$V_3(t) = \int_{t-\Delta t}^{t} (s - t + \Delta t) \hat{x}_i^T(s) Q_i \hat{x}_i(s) ds,$$

$$V_4(t) = \frac{1}{2} \mathfrak{I}_i^2(t).$$

Remark 4.4: Combining with Theorem 4.2, we can know the given scalars $t_h > 0$, $\Delta t > 0$, $\mu > 0$ and the symmetric matrices $\mathfrak{P} > 0$, $\mathcal{R}_a > 0$, $\varepsilon_1 > 0$, $Q_2 > 0, U_2 > 0$. Considering the memory-based adaptive HETM (4.14), the triggering parameter $\mathfrak{I}(t) > 0$. Hence, the Lyapunov functional $V_{xi}(t) = V_1(t) + V_2(t) + V_3(t) + V_4(t) > 0$.

Taking the derivation of $V_x(t)$, we have

$$\dot{V}_1(t) = 2\hat{x}_i^T(t)\mathfrak{P}_i\dot{\hat{x}}_i(t),$$

$$\dot{V}_2(t) = \hat{x}_i^T(t)\mathcal{R}_{ai}\hat{x}_i(t) - \hat{x}_i^T(t-t_h)\mathcal{R}_{ai}\hat{x}_i(t-t_h)$$

$$+ t_h^2\dot{\hat{x}}_i^T(t)O_i\dot{\hat{x}}_i(t) - t_h\int_{t-t_h}^{t}\dot{\hat{x}}_i^T(s)O_i\dot{\hat{x}}_i(s)ds,$$

$$\dot{V}_3(t) = \Delta t\hat{x}_i^T(t)Q_i\hat{x}_i(t) - \int_{t-\Delta t}^{t}\hat{x}_i^T(s)Q_i\hat{x}_i(s)ds.$$

$$\dot{V}_4(t) = \mathfrak{I}_i(t)\dot{\mathfrak{I}}_i(t).$$

There are two parts to the proof. Firstly, for system (4.21) with $\delta(t) = 0$, $t \in (t_k + t_h, t_{k+1}]$:

Under condition (4.15), we can deduce that

$$\dot{V}_4(t) = \mathfrak{I}_i(t)\dot{\mathfrak{I}}_i(t) = \mathfrak{I}_i(t)\delta_i(t)\left(\frac{1}{\mathfrak{I}_i^2(t)} - \frac{\mathfrak{I}_0}{\mathfrak{I}_i(t)}\right)\epsilon_i^T(t)\Omega_i\epsilon_i(t) = 0$$

Therefore, we can get

$$\dot{V}_{xi}(t) = -\Delta t\int_{t-\Delta t}^{t}\hat{x}_i^T(s)Q_i\hat{x}_i(s)ds - t_h\int_{t-t_h}^{t}\dot{\hat{x}}_i^T(s)O_i\dot{\hat{x}}_i(s)ds + \varsigma_{1i}^T(t)\Lambda_{1i}\varsigma_{1i}(t) \tag{4.35}$$

The integral term in (4.35) can be processed as, by using Lemmas 4.1 and 4.2, we have

$$-\int_{t-\Delta t}^{t}\hat{x}_i^T(s)Q_i\hat{x}_i(s)ds \leq -\frac{1}{\Delta t}\left(\int_{t-\Delta t}^{t}\hat{x}_i(s)ds\right)^T Q_i\left(\int_{t-\Delta t}^{t}\hat{x}_i(s)ds\right)$$

$$\leq -\varsigma_{1i}^T(t)\Lambda_{2i}\varsigma_{1i}(t)$$

$$-t_h\int_{t-t_h}^{t}\dot{\hat{x}}_i^T(s)O_i\dot{\hat{x}}_i(s)ds \leq$$

$$-\begin{bmatrix} \hat{x}_i(t)-\hat{x}_i(t-d(t)) \\ \hat{x}_i(t-d(t))-\hat{x}_i(t-t_h) \end{bmatrix}^T \begin{bmatrix} O_i & \Upsilon_i \\ * & O_i \end{bmatrix} \begin{bmatrix} \hat{x}_i(t)-\hat{x}_i(t-d(t)) \\ \hat{x}_i(t-d(t))-\hat{x}_i(t-t_h) \end{bmatrix} \leq \varsigma_{1i}^T(t)\Lambda_{3i}\varsigma_{1i}(t) \tag{4.36}$$

The following equality can be processed as follows, by applying free-weighting matrix technique [34]

$$\begin{aligned} &2[\hat{x}_i^T(t)M_{1i}^T + \mu_i\dot{\hat{x}}_i^T(t)M_{1i}^T] \\ &\times [A_i\hat{x}_i(t) + L_{P1i}C_i\bar{e}_i(t) + (B_i+\Delta B_i(t))K_iC_i\hat{x}_i(t-d(t)) - \dot{\hat{x}}_i^T(t)] = 0 \end{aligned} \tag{4.37}$$

Therefore, we have

$$\varsigma_{1i}^T(t)\Lambda_{4i}\varsigma_{1i}(t) + 2\varsigma_{1i}(t)\Theta_i\delta B_i(t)K_iC_i\hat{x}_i(t-d(t)) + 2\varsigma_{1i}(t)\Theta_iL_{P1i}\bar{C}_i\bar{e}_i(t) = 0 \tag{4.38}$$

With Lemma 4.3, Equation (4.38) as follow

$$\begin{aligned} &2\varsigma_{1i}(t)\Theta_i\Delta B_i(t)K_iC_i\hat{x}_i(t-d(t)) \\ &\leq \varepsilon_{1i}\hat{x}_i^T(t)H_iH_i^TM_{1i}\hat{x}_i(t) + \frac{1}{\varepsilon_{1i}}(K_iC_i\hat{x}_i(t-d(t)))^TE_i^TE_iK_iC_i\hat{x}_i(t-d(t)) \\ &\quad + \varepsilon_{1i}\mu_i\hat{x}_i^T(t)H_iH_i^TM_{1i}\hat{x}_i(t) + \frac{\mu_i}{\varepsilon_{1i}}(K_iC_i\hat{x}_i(t-d(t)))^TE_i^TE_iK_iC_i\hat{x}_i(t-d(t)) \\ &= \Lambda_{5i} + \frac{1+\mu_i}{\varepsilon_{1i}}(K_iC_i\hat{x}_i(t-d(t)))^TE_i^TE_iK_iC_i\hat{x}_i(t-d(t)). \end{aligned} \tag{4.39}$$

Combining (4.35) to (4.39), we can get

$$\dot{V}_{xi}(t) \leq \varsigma_{1i}(t)\Psi_{1i}\varsigma_{1i}^T(t) - \hat{x}_i^T(t)\hat{x}_i(t) + \xi_i^2e_i^{-T}(t)\bar{e}_i(t) \tag{4.40}$$

Furthermore, we have

$$\dot{V}_x(t) \leq -\hat{x}_i^T(t)\hat{x}_i(t) + \xi_i^2e_i^{-T}(t)\bar{e}_i(t) \tag{4.41}$$

Integrating (4.40), we can get

$$V_{xi}(+\infty) - V_{xi}(0) \leq \int_0^\infty \xi_i^2e_i^{-T}(t)\bar{e}_i(t)dt + \int_0^\infty \hat{x}_i^T(t)\hat{x}_i(t)dt \tag{4.42}$$

Under condition $V_x(0) = 0$, we have

$$\int_0^\infty \hat{x}_i^T(t)\hat{x}_i(t)dt \leq \int_0^\infty \xi_i^2e_i^{-T}(t)\bar{e}_i(t)dt \tag{4.43}$$

As a result of Theorem 4.1, (4.43) is therefore represented as follows,

$$\dot{\hat{x}}_i(t) = A_i\hat{x}_i(t) + L_{P1i}\bar{C}_i\bar{e}_i(t) + (B_i + \Delta B_i(t))K_iC_i(\hat{x}_i(t) + \hat{o}_i(t)) \tag{4.44}$$

where $\gamma = \sqrt{\alpha\xi^2}$. Similarly to Theorem 4.1, equation (4.44) corresponds to Definition 4.1. Furthermore, when $w(t) = 0$, (4.44) can guarantee $\dot{V}_{xi}(t) < 0$. Hence, the closed-loop system (4.21) is asymptotically stable with H_∞ performance under disturbance attenuation level γ.

Next, we consider the system (4.21) with $\delta(t) = 1$, $t \in (t_k + t_h, t_{k+1}]$:

With Lemma 4.1, Equation (4.35) needs to be modified as

$$-t_h \int_{t-t_h}^{t} \dot{\hat{x}}_i^T(s)O_i\dot{\hat{x}}(s)ds \leq -\varsigma_{2i}^T(t)\Lambda_{6i}\varsigma_{2i}(t) \tag{4.45}$$

The following equality can be processed as follows, by applying free-weighting matrix technique [30]

$$\begin{aligned} &2[\hat{x}_i^T(t)M_{1i}^T + \mu_i\dot{\hat{x}}_i^T(t)M_{1i}^T] \times [A_i\hat{x}_i(t) + L_{P1i}C_ie_i(t) \\ &+ (B_i + \Delta B_i)K_iC_i[\hat{x}_i(t) + \epsilon_i(t)] - \dot{\hat{x}}_i^T(t)] = 0 \end{aligned} \tag{4.46}$$

Hence, Equation (4.46) as follow

$$\varsigma_{2i}^T(t)\Lambda_{7i}\varsigma_{2i}(t) + 2\varsigma_{2i}(t)\Theta_i\Delta B_i(t)K_iC_i(\hat{x}_i(t) + \hat{o}_i(t)) + 2\varsigma_{2i}(t)\Theta_iL_{P1i}\bar{C}_i\bar{e}_i(t) = 0 \tag{4.47}$$

With Lemma 4.3, Equation (4.47) as follow

$$\begin{aligned} &2\varsigma_{2i}(t)\Theta_i\Delta B_i(t)K_iC_i(\hat{x}_i(t) + \hat{o}_i(t)) \\ &\leq 2\varepsilon_{1i}\hat{x}_i^TM_{1i}\hat{x}_i(t) + \frac{1}{\varepsilon_{1i}}(K_iC_i\hat{x}_i(t))^TE_i^TE_iK_iC_i\hat{x}_i(t) \\ &+ \frac{1}{\varepsilon_{1i}}(K_iC_i\epsilon_i(t))^TE_i^TE_iK_iC_i\hat{x}_i(t) + 2\varepsilon_{1i}\mu_i\dot{\hat{x}}_i^T(t)M_{1i}^TH_iH_i^TM_{1i}\hat{x}_i(t) \\ &+ \frac{\mu_i}{\varepsilon_{1i}}(K_iC_i\hat{x}_i(t))^TE_i^TE_iK_iC_i\hat{x}_i(t) + \frac{\mu_i}{\varepsilon_{1i}}(K_iC_i\epsilon_i(t))^TE_i^TE_iK_iC_i\hat{x}_i(t) \\ &= \varsigma_{2i}^T(t)\lambda_{8i}\varepsilon_{2i}(t) + \frac{1+\mu}{\varepsilon_1}(e_0 + e_3)^TC^TK^TE^T \times EKC(e_0 + e_3). \end{aligned} \tag{4.48}$$

Under the memory-based adaptive HETM (4.14) and (4.15), we can deduce that

$$\begin{aligned} \dot{V}_4(t) &= \mathfrak{J}_i(t)\delta_i(t)\left(\frac{1}{\mathfrak{J}_i^2(t)} - \frac{\mathfrak{J}_0}{\mathfrak{J}_i(t)}\right)\epsilon_i^T(t)\Omega_i\epsilon_i(t) \\ &= \varsigma_{2i}^T(t)\Lambda_{9i}\varsigma_{2i}(t) \end{aligned} \tag{4.49}$$

Combining (4.45) to (4.49), we can get

$$\dot{V}_{xi}(t) \leq \varsigma_{2i}^T(t)\Psi_{2i}\varsigma_{2i}(t) - y^T(t)y(t) + \xi_i e^T(t)e(t) \tag{4.50}$$

According to the same procedure in $\delta_i(t) = 1$, Equation (4.50) is the same as Definition 4.1. Based on Definition 4.1, closed-loop system (4.21) is asymptotically stable with H_∞ performance under disturbance attenuation level γ. This end the proof of Theorem 4.2.

Remark 4.5: This chapter investigates the frequency stability issue in multi-area power systems considering EVs. In the modeling process, the SOC is taken into account, leading to a parameter-uncertain model for the power system. Meanwhile, in order to alleviate the heavy burden on communication channels, a varying threshold $\mathfrak{I}(t)$ and an integral term $\int_{t-\Delta t}^{t} \hat{x}(s)ds$ are introduced in the triggering condition. Therefore, the design of a suitable Lyapunov function to establish feasible LMI-based conditions is a challenge in this study. To tackle this challenge, Lemma 4.3 is introduced to transform the uncertainty parameter terms into inequality terms. Additionally, $V_3(t)$ and $V_4(t)$ are constructed in the Lyapunov function. By taking derivatives to obtain $\mathfrak{I}(t)$ and an integral term $\frac{1}{\Delta t}\int_{t-\Delta t}^{t} \hat{x}(s)ds$, the feasibility of the resulting LMI can be ensured.

Theorem 4.3: *For given scalars $t_h > 0$, $\mu_i > 0$, $\Delta t > 0$, if there exist real symmetric matrices $\bar{\mathfrak{P}}_i > 0$, $\bar{Q}_i > 0$, $\bar{\mathcal{R}}_{ai} > 0$, $\bar{O}_i > 0\Omega_i > 0$, the scalar $\varepsilon_1 > 0$ and matrices $\mathcal{X}$,G such that as follows*

$$\bar{\Psi}_{1i} = \begin{bmatrix} \bar{\Sigma}_{1i} & \bar{\Sigma}_{2i} & E_0^T X_i^T & \bar{\Theta}_i L_{P1i} C_i^T \\ * & \frac{\varepsilon_{1i}}{1+\mu_i} I_1 & 0 & 0 \\ * & 0 & -I_3 & 0 \\ * & 0 & 0 & -\xi_i^2 I \end{bmatrix} < 0 \tag{4.51}$$

$$\bar{\Psi}_{2i} = \begin{bmatrix} \bar{\Sigma}_{3i} & \bar{\Sigma}_{4i} & E_0^T X_i^T & \bar{\Theta}_i L_{P1i} C_i \\ * & \bar{\Sigma}_{5i} & 0 & 0 \\ * & 0 & -I_3 & 0 \\ * & 0 & 0 & -\xi_i^2 I \end{bmatrix} < 0 \tag{4.52}$$

$$\begin{bmatrix} -\varsigma_i I & (V_i C_i - C_i \mathcal{X}_i)^T \\ * & -I \end{bmatrix} < 0 \tag{4.53}$$

$$\bar{\Pi}_i = \begin{bmatrix} \bar{O}_i & \bar{Z}_i \\ * & \bar{O}_i \end{bmatrix} > 0 \tag{4.54}$$

where

$$\bar{\Sigma}_{1i} = \bar{\Lambda}_{1i} - \bar{\Lambda}_{2i} - \bar{\Lambda}_{3i} + \bar{\Lambda}_{4i} + \bar{\Lambda}_{5i},$$

$$\bar{\Lambda}_{1i} = 2E_0^T \bar{\mathfrak{P}}_i E_1 + E_0^T \bar{\mathcal{R}}_{ai} E_0 - E_2^T \bar{\mathcal{R}}_{ai} E_2 + t_h^2 E_1^T \bar{H}_i E_1 + \Delta t E_0^T \bar{Q}_i E_0,$$

$$\bar{\Lambda}_{2i} = \Delta t E_4^T \bar{Q}_i E_4,\ \Lambda_{3i} = -J_i^T \bar{\Pi} J_i,\ \bar{\Theta} = E_0^T + E_1^T \mu_i,$$

$$\bar{\Lambda}_{4i} = 2\bar{\Theta}_i [A_i \mathcal{X}_i E_0 + B_i W_i C_i E_3 - \mathcal{X}_i E_1],\ \bar{\Lambda}_{5i} = \varepsilon_1 E_0^T H_i H_i^T E_0 + \varepsilon_{1i} \mu_i E_1^T H_i H_i^T E_1$$

$$\bar{\Sigma}_{2i} = E_3^T C_i^T W_i^T E_i^T,\ \bar{\Sigma}_{3i} = \bar{\Lambda}_{1i} - \bar{\Lambda}_{2i} - \bar{\Lambda}_{3i} - \bar{\Lambda}_{6i} + \bar{\Lambda}_{7i} + \bar{\Lambda}_{8i} + \bar{\Lambda}_{9i},$$

$$\bar{\Sigma}_{4i} = [E_0^T C_i^T W_i^T E_i^T \ \ E_3^T C_i^T W_i^T E_i^T],\ \Sigma_{5i} = diag - \frac{\varepsilon_{1i}}{1+\mu_i} I_1, -\frac{\varepsilon_{1i}}{1+\mu_i} I_2,$$

$$\bar{\Lambda}_{6i} = (E_0 - E_3)^T \bar{H}_i (E_0 - E_3),\ \bar{\Lambda}_{7i} = 2\bar{\Theta}[(AX + BWC)E_0 + BWCE_3 - XE_1],$$

$$\bar{\Lambda}_{8i} = 2\varepsilon_{1i} E_0^T H_i H_i^T E_0 + 2\varepsilon_{1i} \mu_i E_1^T H_i H_i^T E_1,\ \bar{\Lambda}_{9i} = E_4^T \bar{\Omega}_i E_4 - \mathfrak{J}_0 E_3^T \bar{\Omega}_i E_3.$$

Then the power system (4.21) is asymptotically stable with an H_∞ performance index $\gamma > 0$, the controller gain $K_i = W_i V_i^{-1}$.

Proof. By defining $\mathcal{X}_i = M_{1i}^{-1}$, $\bar{\mathfrak{P}}_i = \mathcal{X}_i \mathfrak{P}_i \mathcal{X}_i$, $\bar{O}_i = \mathcal{X}_i O_i \mathcal{X}_i$, $\bar{Z}_i = \mathcal{X}_i Z_i \mathcal{X}_i$, $\bar{Q}_i = \mathcal{X}_i Q_i \mathcal{X}_i$, $\bar{\mathcal{R}}_{ai} = \mathcal{X}_i \mathcal{R}_{ai} \mathcal{X}_i$, $\bar{\Omega}_i = \mathcal{X}_i \Omega_i \mathcal{X}_i$ and pre- and post-multiplying both sides of (4.31) with diag$\{\mathcal{X}_i, \mathcal{X}_i, \cdots, \mathcal{X}_i, \mathcal{X}_i\}$, we can obtain (4.51).

What's more, inversion of matrix C makes it difficult to find gain K. Following the method in [35], it is possible to transform the above problem into a minimization problem. Defining $W_i C_i = K_i C_i \mathcal{X}_i$, $V_i C_i = C_i \mathcal{X}_i$, we can get

$$(V_i C_i - C_i \mathcal{X}_i)^T (V_i C_i - C_i \mathcal{X}_i) = 0$$

It can be further converted into (4.53). Hence, the gain K is $K_i = W_i V_i^{-1}$ This completes the proof.

4.3.2 Reachability of the Sliding Mode Surface

Sliding surface is proved to be reachable in the error estimation space by presenting a theorem in this section.

Theorem 4.4: *Assuming real symmetric matrices $P_i > 0$ and matrix H_i have appropriate dimensions such that the constraints in Theorem 4.1 hold, it is guaranteed by sliding mode law (4.13) that sliding motion is achieved on the sliding surface $s_{ei} = 0$.*

Proof. The Lyapunov functional as follow

$$V_{si}(t) = 0.5 s_{ei}^T \left(\bar{N}_i \bar{S}_i^{-T} P_i \bar{S}_i^{-1} \bar{N}_i^T\right)^{-1} s_{ei} \tag{4.55}$$

Defining $Z_i = \bar{N}_i^T \bar{S}_i^{-T}$, taking the derivation of the constructed Lyapunov function, we can get

$$\dot{V}_{si}(t) = s_{ei}^T (Z_i P_i Z_i)^{-1} Z_i P_i \bar{S}_i^{-1} [(\bar{A}_i - \bar{L}_{Pi} \bar{C}_i)\bar{e}_i(t) - \bar{N}_i f_{si}(t) + \bar{L}_{si} u_{si}(t) + \bar{F}_i w_i(t)] \tag{4.56}$$

Combining the $\bar{N}_i^T \bar{S}_i^{-T} P = U_i \bar{C}_i$, we have the following equality

$$\begin{aligned} \dot{V}(t) &= s_{ei}^T (Z_i P_i Z_i)^{-1} F_i P_i \bar{S}_i^{-1} [(\bar{A}_i - \bar{L}_{Pi} \bar{C}_i)\bar{e}_i(t) - \bar{N}_i f_{si}(t) + \bar{L}_{si} u_{si}(t)] \\ &\quad + s_{ei}^T (Z_i P_i Z_i)^{-1} U_i \bar{C}_i \bar{S}_i^{-1} \bar{F}_i w_i(t) \end{aligned} \tag{4.57}$$

It is observed that $\bar{C}_i \bar{S}_i^{-1} \bar{F}_i = 0$, therefore

$$\dot{V}(t) = s_{ei}^T (Z_i P_i Z_i)^{-1} Z_i P_i \bar{S}_i^{-1} [(\bar{A}_i - \bar{L}_{Pi} \bar{C}_i)\bar{e}_i(t) + \bar{N}_i f_{si}(t) - \bar{L}_{si} u_{si}(t)] \tag{4.58}$$

Using $\bar{L}_{si} = \bar{S}_i \bar{P}_i^{-1} \bar{C}_i^T U_i^T = \bar{N}_i$ and $\bar{N}_i \bar{S}_i^{-T} P_i = U_i \bar{C}_i$, we can get

$$\begin{aligned} & s_{ei}^T (Z_i P_i Z_i)^{-1} Z_i P_i \bar{S}_i^{-1} [\bar{L}_{si} u_{si}(t) - \bar{N}_i f_{si}(t)] \\ & \leq s_{ei}^T [u_{si}(t) - f_{si}(t)] \leq -\lambda \| s_e \| \end{aligned} \tag{4.59}$$

We design

$$\varphi_i = \|(Z_i P_i Z_i)^{-1} Z_i P_i \bar{S}_i^{-1} (\bar{A}_i - \bar{L}_{Pi} \bar{C}_i)\| \tag{4.60}$$

Combining (4.55) to (4.60), we can get

$$\dot{V}_{si}(t) \leq -\|s_{ei}(t)\| \{\lambda_i - \varphi_i \|\bar{e}_i(t)\|\} \tag{4.61}$$

Then, we define

$$\Omega_i \triangleq \bigcap_{i=1}^{s} \Omega(\varphi_i, \nu_i) \tag{4.62}$$

$$\Omega_i(\varphi_i, \nu_i) \triangleq \{\lambda_i - \varphi_i \|\bar{e}_i(t)\| > 0\} \tag{4.63}$$

From (4.62) and (4.63), it is noticed that $\dot{V}_{si}(t) < 0$ is in the domain Ω_i. This means that the trajectory of $e_i(t)$ will enter Ω_i and remain in finite time after the stabilization of (4.11) is proved in Theorem 4.1. We complete the proof.

4.4 Numerical Examples

The multi-region system under sensor faults is analyzed in this section to verify the effectiveness of the proposed fault-tolerant LFC. And a three-area model with EVs is established on the Matlab/Simiulink environment.

Choosing $t_h = 0.01$, $\Delta t = 0.04$, defined the $T_{12} = 0.03$, $T_{23} = 0.02$, $T_{13} = 0.035$. According to Table 4.2, the system parameters are as follows.

The number of EVs that meet the conditions to participate in the system frequency modulation $N(t)$ is shown in Figure 4.4. N changes over time, as can be seen. The SOC parameters of EVs are $SOC_{min,i} = 10\%$, $SOC_{high,i} = 85\%$. Therefore, Figure 4.4 illustrates the aggregated EV gains.

Table 4.2 Parameters of each area.

Parameter	Area 1	Area 2	Area 3
M_i	1.5	1.5	1.5
D_i	2.5	2	3
R_i	1.5	1.5	1.5
$\mathcal{T}_{mi}$	0.5	0.5	0.4
$\mathcal{T}_{gi}$	0.25	0.2	0.3
$\mathcal{T}_{ai}$	1	1	1
K_{mi}	1	1	1
ε_i	0.1	0.1	0.1

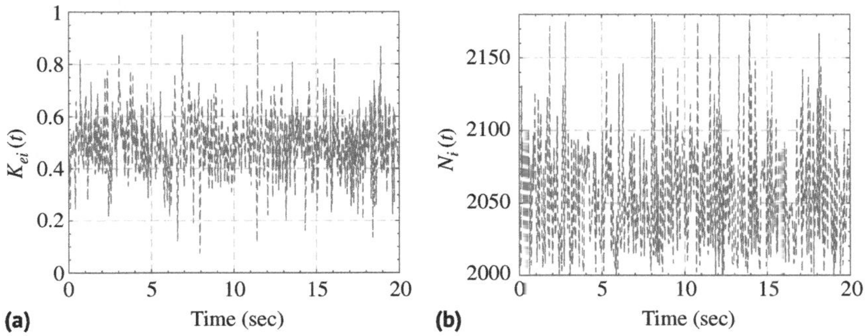

Figure 4.4 (a) Aggregated EVs gain $K_{ei}(t)$; (b) Number of EVs $N(t)$.

The effectiveness of the proposed method is verified by the following three simulation cases.

4.4.1 Unstable Case with Sensor Fault

The sensor faults are defined as

$$f_{si}(t) = \begin{cases} \eta_i \sin(1.8t) & t > 10 \\ 0 & \text{otherwise} \end{cases} \tag{4.64}$$

where $\eta_1 = 0.1, \eta_2 = 0.3, \eta_2 = 0.5$.

Figure 4.5 shows the state comparison under sensor faults. Obviously, the stability of a system without FTC will be lost when it is under the sensor faults. Moreover, it is observed that the larger the value of the faults, the more severe the resulting impact.

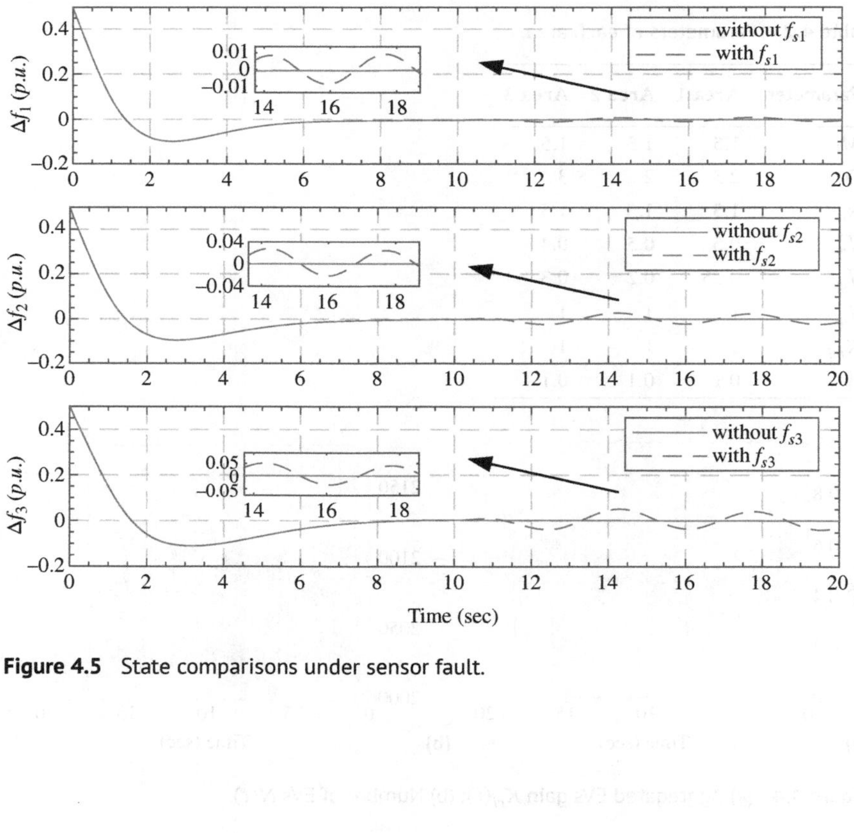

Figure 4.5 State comparisons under sensor fault.

4.4.2 Stable Case with Sensor Fault

In order to deal with sensor faults, we propose fault-tolerant LFC. The observer gain L_D is designed as

$$L_{D1} = L_{D2} = L_{D3} = \begin{bmatrix} 0 & 0 & 0 & 0 & 0 & 0 & 1 & 0 & 1 \\ 0 & 0 & 0 & 0 & 0 & 0 & 0 & 1 & 1 \end{bmatrix}.$$

Based on Theorems 4.1 and 4.3, the observer gain L_P and controller gain K can be obtained as

$$L_{P1} = \begin{bmatrix} -0.03 & 1.06 & -1.96 & 0.55 & 0.21 & 1.75 & 0.95 & -0.01 \\ -0.91 & 1.22 & -1.55 & 0.66 & 0.03 & 1.58 & 0.59 & 0.42 \end{bmatrix}^T,$$
$$L_{P2} = \begin{bmatrix} -0.30 & 0.64 & -1.69 & 0.17 & 0.31 & 0.57 & 0.79 & -0.33 \\ 0.62 & -0.50 & 0.81 & -0.02 & -0.05 & 0.15 & -0.29 & -0.33 \end{bmatrix}^T,$$
$$L_{P3} = \begin{bmatrix} -0.72 & 0.41 & -1.04 & 0.29 & 0.25 & 0.40 & 1.23 & 0.58 \\ 1.11 & -0.49 & 0.78 & -0.07 & -0.04 & -0.09 & -0.58 & -0.29 \end{bmatrix}^T$$

$$K_1 = \begin{bmatrix} -0.46 \\ -0.61 \end{bmatrix}^T, K_2 = \begin{bmatrix} -0.47 \\ -0.65 \end{bmatrix}^T, K_3 = \begin{bmatrix} -0.41 \\ -0.53 \end{bmatrix}^T.$$

Figure 4.6 shows the simulation results using the above parameters and control gain. It can be observed that when sensor faults occur, multi-area power systems can be operated more reliably with the proposed fault-tolerant LFC.

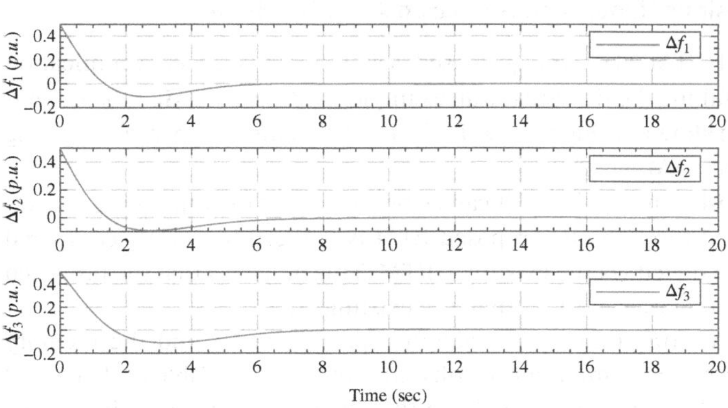

Figure 4.6 The frequency deviation Δf_i.

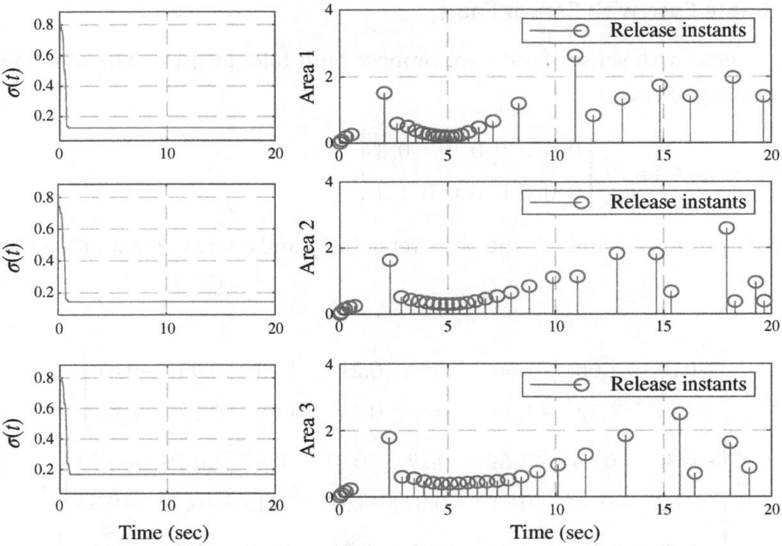

Figure 4.7 Release instants of three area.

And Figure 4.7 depicts release instants and intervals under the memory-based adaptive HETM [18].

To demonstrate the effectiveness of the proposed SMO, we considered three different types of fault signals. Taking the first area as an example, the simulation results are shown in Figure 4.8. It is evident that the proposed observer can effectively handle different types of fault signals.

4.4.3 Transmissions Under Memory-based Adaptive HETM

Compared with HETM [18], the proposed scheme can save communication resources better. Taking the first area as an example, set $\Delta t = 0.08$, $\Im_0 = 8$, the simulation results are shown in Figure 4.9. Table 4.3 shows the number of triggering events for different ETM.

Based on Figure 4.9 and Table 4.3, it can be observed that the number of triggering events obtained through our proposed ETM is reduced by 39.13% compared with the results produced via the existing HETM. This indicates that the proposed ETM saves more network resources than the existing HETM.

By varying the waiting time t_h, the adaptive parameter $\Im_0$ and the memory period parameter Δt, the simulation results are presented in Tables 4.4–4.6 and Figure 4.10. From the results, we can obtain that the number of triggering events increases with the higher value of the adaptive parameter $\Im_0$. And it can be

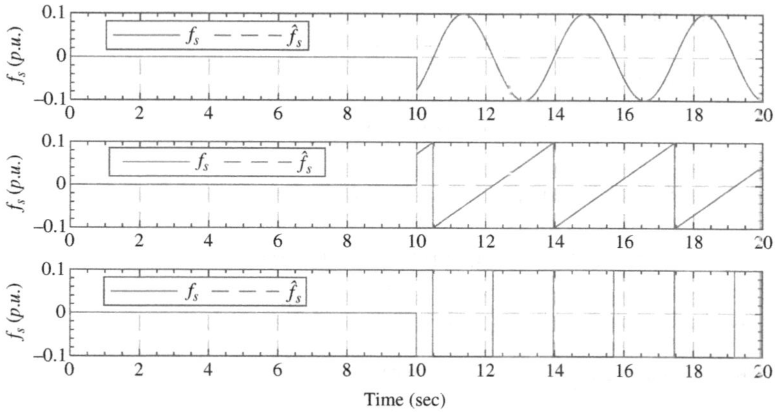

Figure 4.8 State comparisons under different types of fault signals.

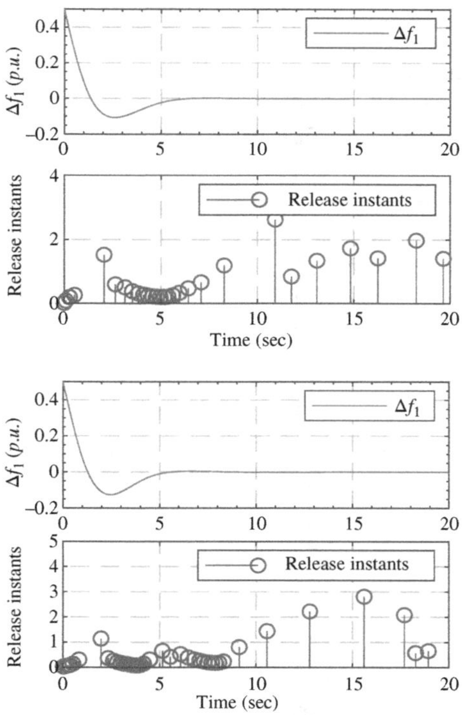

Figure 4.9 Δf_1 and release instants under different ETMs.

Table 4.3 The number of triggering events.

	Area 1		Area 1
Our method	28	Method in [18]	46

Table 4.4 The number of triggering with $\Delta t = 0.08$, $\mathfrak{I}_0 = 8$.

t_h	$t_h = 0.001$	$t_h = 0.01$	$t_h = 0.1$	$t_h = 0.5$
$\mathcal{N}$	30	28	27	20

Table 4.5 The number of triggering with $t_h = 0.01$, $\mathfrak{I}_0 = 8$.

Δt	$\Delta t = 0.03$	$\Delta t = 0.05$	$\Delta t = 0.08$	$\Delta t = 0.1$
$\mathcal{N}$	29	27	28	30

Table 4.6 The number of triggering with $\Delta t = 0.08$, $t_h = 0.01$.

$\mathfrak{I}_0$	$\mathfrak{I}_0 = 6$	$\mathfrak{I}_0 = 8$	$\mathfrak{I}_0 = 9$	$\mathfrak{I}_0 = 10$
$\mathcal{N}$	26	28	31	32

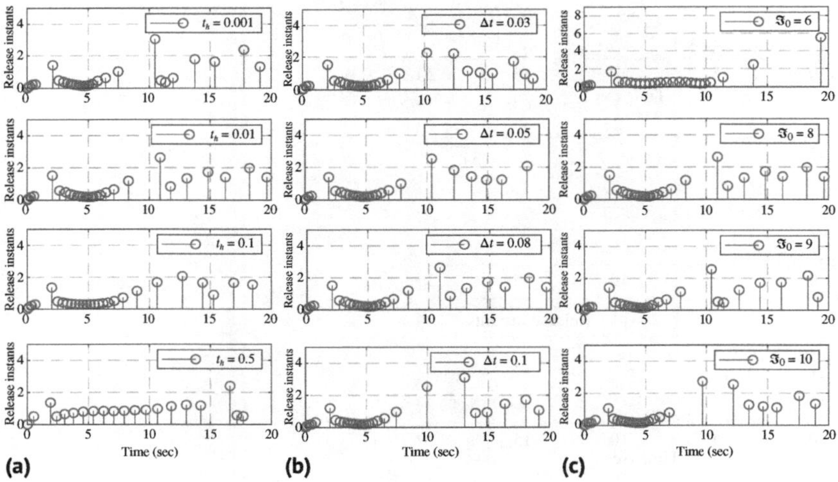

Figure 4.10 The comparisons of release instants under different parameter.

observed that as the memory parameter Δt increases, the number of triggering times decreases first and then increases. For t_h, when $t_h < 0.1$, the value of t_h has little influence on the number of triggering times. When the t_h exceeds a certain range, the number of triggering times decreases sharply. Therefore, it is possible to select appropriate event-triggering parameters t_h, $\Im_0$ and Δt based on the real-time system requirements to effectively save network resources.

4.5 Conclusion

Under sensor faults in multi-area power systems with EVs, this chapter investigates event-triggered fault-tolerant LFC. It has been proposed to incorporate EVs into a multi-area power system model, which accounts for the uncertainty of battery state. Additionally, a memory-based adaptive hybrid event-triggered scheme has been designed to efficiently utilize limited network resources. Furthermore, considering the impact of sensor faults, a fault-tolerant LFC method has been proposed. Based on simulations of a three-area power system, the method was demonstrated to be effective. In the future, research will focus on fault-tolerant LFC under network attacks. The aim is to ensure accurate and efficient control even in the face of system faults and deliberate attacks.

References

1 P. Mani and Y. H. Joo. Fuzzy logic-based integral sliding mode control of multi-area power systems integrated with wind farms. *Information Sciences*, 545:153–169, 2021.

2 S. Yan, Z. Gu, and J. H. Park. Memory-event-triggered H∞ load frequency control of multi-area power systems with cyber-attacks and communication delays. *IEEE Transactions on Network Science and Engineering*, 8(2):1571–1583, 2021.

3 Q. Zhong, J. Yang, K. Shi, et al. Event-triggered H∞ load frequency control for multi-area nonlinear power systems based on non-fragile proportional integral control strategy. *IEEE Transactions on Intelligent Transportation Systems*, 23(8):12191–12201, 2021.

4 D. J. Wang, F. Chen, B. Meng, et al. Event-based secure h_∞ load frequency control for delayed power systems subject to deception attacks. *Applied Mathematics and Computation*, 394:125788–125801, 2021.

5 B. Li, S. Hu, Q. Zhong, et al. Dynamic memory event-triggered proportional-integral-based load frequency control for multi-area wind power systems. *Applied Mathematics and Computation*, 435:128070–128083, 2023.

6 J. Ye and X. Yu. Detection and estimation of false data injection attacks for load frequency control systems. *Journal of Modern Power Systems and Clean Energy*, 10:861–870, 2022.

7 A. Ameli, A. Hooshyar, E. F. El-Saadany, and A. M. Youssef. Attack detection and identification for automatic generation control systems. *IEEE Transactions on Power Systems*, 33:4760–4774, 2018.

8 S. Arunagirinathan, P. Muthukumar, and Y. H. Joo. Robust reliable control design for networked control systems with nonlinear actuator faults and randomly missing measurements. *International Journal of Robust and Nonlinear Control*, 29:4168–4190, 2018.

9 R. J. Ma, P. Shi, and L. G. Wu. Sparse false injection attacks reconstruction via descriptor sliding mode observers. *International Journal of Robust and Nonlinear Control*, 66:5369–5376, 2020.

10 H. Y. Yang, Y. C. Jiang, and S. Yin. Fault-tolerant control of time-delay Markov jump systems with stochastic process and output disturbance based on sliding mode observer. *IEEE Transactions on Industrial Informatics*, 14:5299–5307, 2018.

11 P. V. Surjagade, J. Deng, S. R. Shimjith, and A. John Arul. Descriptor sliding mode observer based fault tolerant control for nuclear power plant with actuator and sensor faults. *Progress in Nuclear Energy*, 162:10744–10755, 2023.

12 X. Su, X. Liu, and Y. D. Song. Fault-tolerant control of multiarea power systems via a sliding-mode observer technique. *IEEE/ASME Transactions on Mechatronics*, 23:38–47, 2018.

13 Y. Wang and J. Zhao. Periodic event-triggered sliding mode control for switched uncertain t-s fuzzy systems with a logistic adaptive event-triggering scheme. *IEEE/ASME Transactions on Mechatronics*, 30:4115–4126, 2022.

14 L. Xie, J. Cheng, Y. Zou, et al. A dynamic-memory event-triggered protocol to multiarea power systems with semi-Markov jumping parameter. *IEEE Transactions on Cybernetics*, 53:6577–6587, 2023.

15 X. Wang, J. H. Park, Z. Liu, and H. Yang. Dynamic event-triggered control for GSES of memristive neural networks under multiple cyber-attacks. *IEEE Transactions on Neural Networks and Learning Systems*, 35(6):7602–7611, 2022. https://doi.org/10.1109/TNNLS.2022.3217461.

16 X. Wang and G. Feng. Dynamic event-triggered filtering for NCSs under multiple cyber-attacks. *IEEE Transactions on Cybernetics*, 53:4705–4714, 2023.

17 H. Shen, Y. Xia, J. Wang, and J. H. Park. Fault-tolerant event-triggered load frequency control for multiarea power systems with communication delay. *IEEE Systems Journal*, 16:6624–6634, 2022.

18 C. M. Kaneba, X. Mu, X. Li, and X. Wu. Event triggered control for fault tolerant control system with actuator failure and randomly occurring parameter uncertainty. *IEEE Transactions on Cybernetics*, 415:126714–126725, 2022.

19 Y. Ni, Z. Wang, Y. Fan, et al. Memory-based event-triggered control for global synchronization of chaotic Lur'e systems and its application. *IEEE Transactions on Systems, Man, and Cybernetics: Systems*, 53:1920–1931, 2023.

20 J. Liu, Y. Wang, J. Cao, et al. Secure adaptive-event-triggered filter design with input constraint and hybrid cyber attack. *IEEE Transactions on Cybernetics*, 51:4000–4010, 2021.

21 L. Liu, Y. Cui, Y. J. Liu, and S. Tong. Adaptive event-triggered output feedback control for nonlinear switched systems based on full state constraints. *IEEE Transactions on Circuits and Systems II: Express Briefs*, 53:2370–2381, 2023.

22 L. Zhang, H. Liang, Y. Sun, and C. K. Ahn. Adaptive event-triggered fault detection scheme for semi-Markovian jump systems with output quantization. *IEEE Transactions on Systems, Man, and Cybernetics: Systems*, 51:3779–3783, 2023.

23 K. S. Ko and D. K. Sung. The effect of EV aggregators with time-varying delays on the stability of a load frequency control system. *IEEE Transactions on Power Systems*, 33:669–680, 2018.

24 M. Khan, H. Sun, Y. Xiang, and D. Shi. Electric vehicles participation in load frequency control based on mixed. *IEEE Transactions on Cybernetics*, 125:106420–106430, 2021.

25 A. Naveed, Ş. Sönmez, and S. Ayasun. Impact of electric vehicle aggregator with communication time delay on stability regions and stability delay margins in load frequency control system. *Journal of Modern Power Systems and Clean Energy*, 9:595–601, 2021.

26 R. Abolpour, K. Torabi, M. Dehghani, et al. Direct search algorithm for load frequency control of a time-delayed electric vehicle aggregator. *IEEE Transactions on Industry Applications*, 59:106420–106430, 2023.

27 Y. Ota, H. Taniguchi, T. Nakajima, et al. Autonomous distributed V2G (vehicle-to-grid) satisfying scheduled charging. *IEEE Transactions on Cybernetics*, 3:559–564, 2012.

28 M. Khan and H. Sun. Complete provision of MPC-based LFC by electric vehicles with inertial and droop support from DFIG-based wind farm. *IEEE Transactions on Power Delivery*, 37(2):716–726, 2021.

29 H.-B Zeng, S.-J Zhou, X.-M Zhang, and W. Wang. Delay-dependent stability analysis of load frequency control systems with electric vehicles. *IEEE Transactions on Cybernetics*, 52(12):13645–13653, 2022.

30 T. N. Pham, S. Nahavandi, L. V. Hien, et al. Static output feedback frequency stabilization of time-delay power systems with coordinated electric vehicles state of charge control. *IEEE Transactions on Cybernetics*, 32:3862–3874, 2017.

31 Z. Gu, P. Shi, D. Yue, and Z. T. Ding. Decentralized adaptive event-triggered filtering for a class of networked nonlinear interconnected systems. *IEEE Transactions on Cybernetics*, 49:1570–1579, 2018.

32 C. Peng, E. Tian, J. Zhang, and D. Du. Decentralized event-triggering communication scheme for large-scale systems under network environments. *Information Sciences*, 380:132–144, 2017.

33 B. Yang, M. Hao, J. Cao, and X. Zhao. Delay-dependent global exponential stability for neural networks with time-varying delay. *Neurocomputing*, 338:172–180, 2019.

34 X. Zhang, M. Li, M. Wu, and J. She. Further results on stability and stabilisation of linear systems with state and input delays. *IEEE Transactions on Cybernetics*, 40:1–10, 2009.

35 Y. Yang, Y. Niu, and Z. Zhang. Dynamic event-triggered sliding mode control for interval type-2 fuzzy systems with fading channels. *ISA Transactions*, 110:53–62, 2021.

5

Active Fault-tolerant LFC for Power Systems with EVs Under Deception Attacks

After reading this chapter, you should be able to:

- Study the way of improving the usability of EVs in participating the LFC power system.
- Explain the harm of deception attacks in LFC power system aggregated with EVs.
- Identify the difference between the fault-tolerant control and active fault-tolerant control.

5.1 Background

Frequency deviation from nominal values directly affects the operation, safety, and reliability of the system [1]. Load frequency control (LFC) is an important part of automatic power generation control (AGC). It has been proven to be very effective in ensuring the stable performance of power systems [2–6]. In recent years, a large number of electric vehicles have been injected into the grid due to environmental pollution and the depletion of fossil fuels. However, the random and uncoordinated charging of electric vehicles will increase the demand in the power system [7]. With the development of vehicle-to-grid (V2G) technology, the EVs are used to participate in the frequency regulation of power systems because the battery of EVs have the characteristics of faster response and adjustment speed than traditional generator sets [8]. In [9], the authors studied the participation of EVs in smart grid frequency regulation as controllable loads. The results show that EVs have the sufficient frequency regulation capability. Due to the limited capacity of a single EVs, a large number of EV participating system frequency modulation should be aggregated by an EV Aggregator (EVA) [10]. In [11], the authors proposed an integrated EVs management method developed by EVAs, which can manage the charge/discharge operations of EVs fleets and provide frequency re-services. The battery's state of charge (SOC) is the ratio between stored energy and its capacity, which is an important feature of EVs [12]. In fact, when EVs take part in frequency regulation of the power system, we should also consider the requirements of the owners. Therefore, the SOC value of EVs should be controlled within a reasonable range. SOC is closely related to EVs gain, which determines EVs output power deviation [13]. Besides, EVs gain should vary within the expected range. In related studies, EVs gain is simply treated as constant [14–16]. A robust frequency control scheme was proposed in [17] for the power system model in which the uncertainty caused by the change of battery SOC was considered. However, modern power networks are often in harsh environments such as high voltage and high magnetic field, leading to the failure of related actuators, sensors, and controller modules. This may cause the system to lose the expected performance index and cause huge losses, but such problems are not considered in the above literature.

Fault-tolerant control can ensure the safe operation of the controlled system when faults occur [18]. Among related researches, fault-tolerant control was used in different areas, such as islanded microgrids [19], satellite attitude control systems [20], wind turbines [21], and battery storage systems [22]. Therefore, we propose fault-tolerant LFC to improve the safety and reliability of power system. In [23], the fault-tolerant LFC scheme with known and unknown actuator faults was designed for multi-area wind power generation system. In [24], the authors constructed a novel actuator fault model that contains linear and nonlinear terms, and a corresponding fault-tolerant control scheme was proposed to analyze the performance of the novel model. However, [23] and [24] adopt passive fault-tolerant control (PFTC), which is highly limited and only effective for known faults. On the contrary, after detecting the fault information of the controlled system in the fault detection and identification link, the active fault-tolerant control (AFTC) adjusts the structure and parameters of the controlled system online to achieve the desired performance index. It can achieve better robustness and fault acceptability than PFTC [20]. In [23], the AFTC problems of the multi-area power system under sensor fault were studied. In [25], AFTC based on interval observer was designed to solve the problem of LFC and sensor fault tolerance control in power grid. However, these methods require a fault diagnosis scheme that is sensitive to changes in model parameters and operating conditions. This chapter mainly introduces the integrated design of fault estimation (FE) and AFTC, which can directly estimate the fault component of the system. When the fault estimate value is non-zero, the fault diagnosis result is automatically indicated, and there is no need for additional complex fault diagnosis scheme.

Modern power systems usually use open communication networks. Therefore, it is more vulnerable to communication channel interference, false data injection, load change of power system and other malicious attacks [26]. Deception attacks destroy data integrity by replacing the normally transmitted data in the communication channel with fake data [27]. It poses a great threat to the safe operation of power system. At present, there has been increasing attention to the study of deception attacks [28–30]. The LFC problem of networked power system with random deception attacks and limited bandwidth was studied [28]. In [29], the problem of LFC under mixed network attacks was studied. The security control of large-scale systems under joint attacks was studied in [30]. The considered attacks could happen in both sensor-to-observer and controller-to-actuator. It is worth noting that the LFC problem of multi-area power systems with electric vehicles that involve sensor faults and deception attacks has not been well studied, which is the motivation for this chapter.

1. The random and uncoordinated charging of electric vehicles will increase the demand in the power system [7]. V2G technology has been proposed to improve the usability of EVs in participating the frequency regulation of power systems [8].
2. Due to the limited capacity of a single EVs, a large number of EV participating system frequency modulation should be aggregated by an EVA [10]. In [11], the authors proposed an integrated EVs management method developed by EVAs, which can manage the charge–discharge operations of EVs fleets and provide frequency re-services.
3. A multiple Lyapunov function is constructed, and then the stability criterion of the multi-area power system is established by Lyapunov theory and matrix inequalities, where the obtained result is less conservative than that of [31].

5.2 Problem Formulation

In recent years, due to the gradual depletion of fossil fuels, EVs have been widely used. Aggregated electric vehicles can be used to participate in frequency regulation of the power system. When the grid load is too high, the EVs are discharged to the grid. When the grid load is too low, the EVs are charged by the grid to ensure the stability of the system frequency. Because the power provided by one EV is too low, the EVA is used to aggregate and control a large number of EVs. Firstly, the EVA collects information about the EVs from the charging–discharging stations and sends it to the LFC control center. Then the control center processes the information to make a decision and sends the command to the EVA. Finally, the EVA controls the orderly charging and discharging of the EVs according to

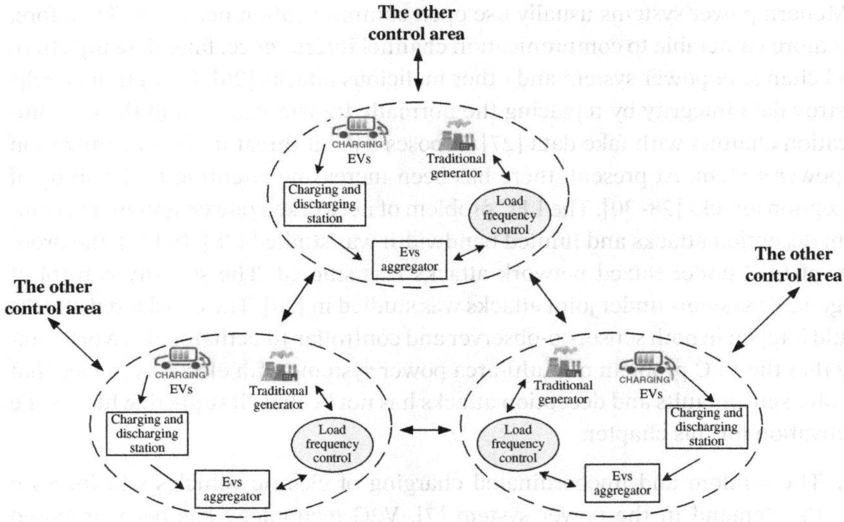

Figure 5.1 Multi-area power systems integrated with electric vehicles.

the adjustment command. Thereby achieving the purpose of frequency regulation. A simple structure for the multi-area power systems integrated with EVs is shown in Figure 5.1.

5.2.1 Electric Vehicle Model

The information contains the minimum expected battery charge to leave the grid $\mathrm{SOC}_{\mathrm{exp}}$, the dwell time T_p, and the initial SOC of the battery SOC_0. The charging time of an electric vehicle from SOC_0 to $\mathrm{SOC}_{\mathrm{exp}}$ can be expressed as

$$T_{ct} = \frac{Q_{ev}(\mathrm{SOC}_{\mathrm{exp}} - \mathrm{SOC}_0)}{\eta_c P_{\mathrm{lim}}} \tag{5.1}$$

where Q_{ev} is the battery capacity of EVs, η_c is the charging efficiency, P_{lim} is the limitation of EVs charging and discharging power. Obviously, only if $T_{ct} < T_p$, the EVs can provide frequency regulation service.

The deviation of power output for the i_{th} EV battery system is described by

$$\dot{P}_{ei}(t) = -\frac{1}{T_{ei}}P_{ei}(t) + \frac{K_{ei}}{T_{ei}}u_i(t) \tag{5.2}$$

where we assume that all EVs have the same time constant T_{ei}, K_{ei} is the EV gain, and the system input $u(t)$ is reallocated equally to each EV as $u_i(t) = u(t)/N$. And N is the total number of EVs. Accordingly, the aggregated EVs can be modeled as a first-order dynamic model. Hence, the deviation of power output for the aggregated EVs is obtained as

$$\dot{P}_e(t) = -\frac{1}{T_e}P_e(t) + \frac{K_e}{T_e}u(t) \tag{5.3}$$

where aggregated EVs gain is $K_e = \frac{\sum_{i=1}^{N} K_{ei}}{N}$.

Individual EV gain K_{ei} is determined by frequency and battery SOC. To prevent overcharging (over discharging), if the SOC is almost full (almost empty), no high power charging (discharging) should be done. Therefore, EV gain K_{ei} is modeled as

$$K_{ei} = K_m \left(1 - \left(\frac{\mathrm{SOC}_i - \mathrm{SOC}_{\mathrm{low(high)},i}}{\mathrm{SOC}_{\mathrm{max(min)},i} - \mathrm{SOC}_{\mathrm{low(high)},i}}\right)^2\right) \tag{5.4}$$

where K_m is the maximum gain of EV, and SOC_i, $\mathrm{SOC}_{\mathrm{low},i}$, $\mathrm{SOC}_{\mathrm{high},i}$, $\mathrm{SOC}_{\mathrm{min},i}$, and $\mathrm{SOC}_{\mathrm{max},i}$ denote the current, known low and high, and known minimum and maximum battery SOC of the i_{th} EV, respectively.

Therefore, K_{ei} is

$$K_{ei} = K_m - K_m g_i(t) \tag{5.5}$$

Assuming that at any time t, there are N EVs participating in the multi-area power system, the aggregated EVs gain can be obtained as

$$K_e = \frac{\sum_{i=1}^{N} K_{ei}}{N} = K_m - g(t)K_m \tag{5.6}$$

where $g_i(t)$ is a time-varying function related to the current SOC value.

$$g_i(t) = \left(\frac{\mathrm{SOC}_i - \mathrm{SOC}_{\mathrm{low(high)},i}}{\mathrm{SOC}_{\mathrm{max(min)},i} - \mathrm{SOC}_{\mathrm{low(high)},i}}\right)^2 \tag{5.7}$$

and $g(t) = \sum_{i=1}^{N} g_i(t)/N$ is the average of N different values $g_i(t)$.

Therefore, the power output deviation of the aggregated EVs can be obtained

$$\dot{P}_e(t) = -\frac{1}{T_e}P_e(t) + \frac{K_m}{T_e}u(t) - \frac{g(t)K_m}{T_e}u(t) \tag{5.8}$$

5.2.2 System Model with Delay and Sensor Fault

The i_{th} area multi-area electric power system simplified model is given in Figure 5.2. Based on signal transmission, the dynamic equation of the system can be obtained as follows

$$\begin{cases} \Delta\dot{f}_i(t) = \dfrac{1}{M_i}(\Delta P_{mi}(t) + \Delta P_{ei}(t) - \Delta P_{tie-i}(t) \\ \qquad -\Delta P_{Li}(t)) - \dfrac{D_i}{M_i}\Delta f_i(t) \\ \Delta\dot{P}_{mi}(t) = \dfrac{1}{T_{mi}}(\Delta P_{gi}(t) - \Delta P_{mi}(t)) \\ \Delta\dot{P}_{gi}(t) = \dfrac{1}{T_{gi}}(u(t) - \Delta P_{gi}(t) - \dfrac{1}{R_i}\Delta f_i(t) \\ \qquad - \Delta ACE_i(t - h(t))) \\ \Delta\dot{P}_{ei}(t) = -\dfrac{1}{T_{ei}}P_{ei}(t) + \dfrac{K_{mi}}{T_{ei}}u(t) - \dfrac{g(t)K_{mi}}{T_{ei}}u(t) \\ \int ACE_i(t) = k_i\beta_i\Delta f_i + k_i\Delta P_{tie-i} \\ \Delta\dot{P}_{\text{tie}-i}(t) = 2\pi \sum\limits_{j=1,i\neq 1}^{N} \Delta T_{ij}\Delta f_i \end{cases} \tag{5.9}$$

Remark 5.1: The multi-area power system consists of the conventional generator and the EVA. The traditional generator consists of generator, turbine, and governor. According to Figure 5.2, the transfer function of the power system can be obtained. By applying the Laplace transform of the transfer function, the above equation can be obtained. The treatment method can be referred to [3, 4].

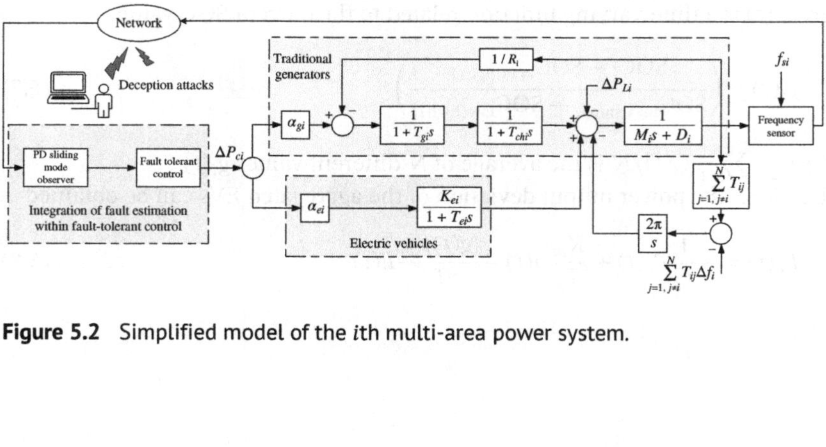

Figure 5.2 Simplified model of the ith multi-area power system.

Thus, the state space function can be represented as

$$\begin{cases} \dot{x}(t) = Ax(t) + A_d x(t - h(t)) + (B + \Delta B(t))u(t) + Dw(t) \\ y(t) = Cx(t), \end{cases} \tag{5.10}$$

where

$$x_i(t) = \left[\triangle f_i \;\; \triangle P_{mi} \;\; \triangle P_{gi} \;\; \triangle P_{ei} \;\; \triangle P_{tie-i} \;\; \int ACE_i\right]^T,$$

$$y_i(t) = [ACE_i \quad \int ACE_i]^T, \omega_i(t) = \Delta P_{Li},$$

$$A = \begin{bmatrix} -\frac{D}{M} & 0 & \frac{1}{M} & \frac{1}{M} & 0 & 0 \\ -\frac{1}{R_g T_g} & -\frac{1}{T_g} & 0 & 0 & 0 & 0 \\ 0 & \frac{1}{T_g} & -\frac{1}{T_g} & 0 & 0 & 0 \\ 0 & 0 & 0 & -\frac{1}{T_e} & 0 & 0 \\ 2\pi \sum_{j=1, i \neq 1}^{N} \Delta T_{ij} & 0 & 0 & 0 & 0 & 0 \\ \beta & 0 & 0 & 0 & 1 & 0 \end{bmatrix},$$

$$A_d = \begin{bmatrix} 0_{2x5} & 0_{2x1} \\ 0_{1x5} & -\frac{1}{T_{gi}} \\ 0_{3x5} & 0_{3x1} \end{bmatrix}, B = \left[0 \;\; \frac{\alpha_{gi}}{T_{gi}} \;\; 0 \;\; \frac{\alpha_{ei} K_{mi}}{T_{ei}} \;\; 0 \;\; 0\right]^T,$$

$$\Delta B(t) = Hg(t)E, H = \left[0 \;\; 0 \;\; 0 \;\; -\frac{K_{mi}}{T_e} \;\; 0 \;\; 0\right]^T,$$

$$E = [1 \;\; 0 \;\; 0 \;\; 0 \;\; 0 \;\; 0], D = [-1/M \;\; 0 \;\; 0 \;\; 0 \;\; 0 \;\; 0]^T,$$

$$\bar{C} = \begin{bmatrix} \beta & 0 & 0 & 0 & 1 & 0 \\ 0 & 0 & 0 & 0 & 0 & 1 \end{bmatrix}.$$

Considering the existence of delay in actual transmission, $h(t)$ is defined to represent the time delay. And $h(t)$ is a time-varying differentiable function as follows

$$0 \leq h(t) \leq \bar{h}, \; 0 \leq \dot{h}(t) \leq \phi \leq 1$$

Considering the influence of sensor failure, the system state equation can be rewritten as

$$\begin{cases} \dot{x}(t) = Ax(t) + A_d x(t - h(t)) + (B + \Delta B(t))u(t) + Dw(t) \\ y(t) = Cx(t) + f_s(t) \end{cases} \tag{5.11}$$

where $f_s(t)$ represents the sensor fault, which satisfies $\|f_s(t)\| \leq \alpha$ and α is a known constant [31].

5.2.3 Deception Attacks Model

Compared to traditional, modern deregulated power system LFC schemes use open communication facilities. The output signal under deception attacks can be described as

$$y(t) = Cx(t) + f_s(t) + f_{ag}(t) \tag{5.12}$$

where $f_{ag}(t)$ stands for the deception attack launched by the adversary.

From the attacker's point of view, they are more willing to attack when the system is stable than when the system state close to stable situation. Therefore, the system remains unstable for a long time, the attack status is as follows

$$f_{ag}(t) = \begin{cases} -aCx(t) + \varphi(t) & |\Delta f_i| < v \\ 0 & \text{otherwise} \end{cases} \tag{5.13}$$

where $\varphi(t)$ is an unknown-but-bounded injected signal satisfying $\|\varphi(t)\| \leq \sigma$. a and v are given positive scalars. When the system frequency satisfies $|\Delta f_i| < v$, the system is considered stable and the adversary sends an attack signal.

Therefore, the system state space function can be expressed as

$$\begin{cases} \dot{x}(t) = Ax(t) + A_d x(t - h(t)) + (B + \Delta B(t))u(t) + D\omega(t) \\ y(t) = Cx(t) + f_s(t) + f_{ag}(t) \end{cases} \tag{5.14}$$

Before proceeding with the main results, we present the following definition and lemmas, which play an important role in the proof of the main result.

Definition 5.1 For $w(t) = 0$, if there is a continuous first-order partial derivative of $V(t)$, and $V(t)$ is positively definite and $\dot{V}(t)$ is negatively definite, then the system (5.14) is asymptotically stable. Besides, H_∞ performance is guaranteed under the condition of zero initial state if the inequality $\|y(t)\|_2 \leq \gamma \|w(t)\|_2$ holds, in which $\gamma > 0$ is a pregiven constant.

Lemma 5.1 [32] *For a positive definite matrix $R > 0$, any continuously differentiable function $x : [a, b] \to \mathbb{R}^n$, the following inequality holds*

$$\int_a^b \dot{x}^T(s)R\dot{x}(s) \geq \frac{1}{b-a}\delta\tilde{R}\delta$$

where $\tilde{R} = diag\{R, 3R, 5R\}$ and

$$\delta = \text{col}\left\{x(b) - x(a), x(b) + x(a) - \frac{2}{b-a}\int_a^b x(s)\text{d}s, x(b) - x(a) - \frac{6}{b-a}\int_a^b x(s)\text{d}s\right\}.$$

Lemma 5.2 [33] *For any $x, y \in \mathbb{R}$ and positive scalar ε, we have*

$$2y^T x \leq \varepsilon x^T x + \frac{1}{\varepsilon}y^T y.$$

5.3 Main Results

5.3.1 Sliding Mode Observer Design

An efficient observer is proposed for the extended system (5.14), which can estimate both the original state and the sensor fault value.

In order to facilitate further analysis, the following matrix is constructed

$$E = \begin{bmatrix} I_n & 0 & 0 \\ 0 & I_p & 0 \\ 0 & 0 & 0 \end{bmatrix}, \bar{A} = \begin{bmatrix} A & 0 & 0 \\ 0 & -I_p & 0 \\ 0 & 0 & -I_q \end{bmatrix}, \bar{B} = \begin{bmatrix} B \\ 0 \\ 0 \end{bmatrix}, \Delta\bar{B} = \begin{bmatrix} \Delta B \\ 0 \\ 0 \end{bmatrix}, \bar{D} = \begin{bmatrix} D \\ 0 \\ 0 \end{bmatrix},$$

$$\bar{C} = \begin{bmatrix} C \\ I_p \\ I_q \end{bmatrix}^T, N = \begin{bmatrix} 0 & 0 \\ I_p & 0 \\ 0 & I_q \end{bmatrix}, G_1 = \begin{bmatrix} 1 \\ 0 \end{bmatrix}^T, G_2 = \begin{bmatrix} 0 \\ 1 \end{bmatrix}^T, \bar{x}(t) = \begin{bmatrix} x(t) \\ f_s(t) \\ f_{ag}(y(t)) \end{bmatrix}.$$

Then, an augmented state space function based on (5.14) can be obtained

$$\begin{cases} E\dot{\bar{x}}(t) = \bar{A}\bar{x}(t) + A_d x(t-h(t)) + (\bar{B} + \Delta\bar{B}(t))u(t) \\ \qquad + NG_1\bar{f}(t) + NG_2 f_{ag}(t) + \bar{D}w(t) \\ y(t) = \bar{C}\bar{x}(t) \end{cases} \tag{5.15}$$

where $\bar{f} = f_s + \dot{f}_s$.

It can be observed that the matrices satisfy the following properties

$$\text{rank}\begin{bmatrix} E \\ \bar{C} \end{bmatrix} = \text{rank}\begin{bmatrix} I_n & 0 \\ 0 & 0 \\ C & I_p \end{bmatrix} = n + p$$

There exists a matrix $\bar{L}_D = [0_{p\times n} \;\; W]^T$ such that the matrix $\bar{S}$ is nonsingular

$$\bar{S} = (E + \bar{L}_D\bar{C})$$

Now, the PD sliding mode observer-based power system model can be given as

$$\begin{cases} \bar{S}\dot{\bar{z}}(t) = (\bar{A} - \bar{L}_P\bar{C})\hat{\bar{x}}(t) + (\bar{A}_d - \bar{L}_H\bar{C})\hat{\bar{x}}(t-h(t)) \\ \qquad + (\bar{B} + \Delta\bar{B}(t))u(t) + \bar{L}_P y(t) \\ \qquad + \bar{L}_H y(t-h(t)) + \bar{L}_s u_s(t) + \bar{D}w(t) \\ \hat{\bar{x}}(t) = \bar{z}(t) + \bar{S}^{-1}\bar{L}_D y(t) \end{cases} \tag{5.16}$$

where $\bar{z}(t) = [z_x^T(t), z_s^T(t)]^T$ denotes the intermediate variable and $\hat{\bar{x}}(t) = [\hat{x}^T(t), \hat{f}_s^T(t)]^T$ denotes the estimation of $\bar{x}(t)$, $\bar{L}_P$, $\bar{L}_D$, $\bar{L}_s$ represent the observer gains. What's more, $u_s(t)$and $\bar{L}_s$ are designed to eliminate the effects of sensor failure and keep the system stable.

Remark 5.2: The fault diagnosis is sensitive to the change of model parameters and operating conditions, but the system state and fault signal can be estimated simultaneously by PD sliding mode observer. When the fault estimate is non-zero, it indicates that the system has failed. Therefore, there is no need to design additional complex fault diagnosis scheme. In addition, PD sliding mode observer has the advantages of fast response, good transient performance, and insensitivity to interference. Based on the above description, PD sliding mode observer is widely used to solve fault tolerance problems [31, 33].

Based on (5.16), we have

$$\begin{aligned}\bar{S}\dot{\hat{x}}(t) &= (\bar{A} - \bar{L}_P\bar{C})\hat{x}(t) + (\bar{A}_d - \bar{L}_H\bar{C})\hat{x}(t-h(t)) \\ &+ (\bar{B} + \Delta\bar{B}(t))u(t) + \bar{L}_P y(t) + \bar{L}_s u_s(t) \\ &+ \bar{L}_H y(t-h(t)) + \bar{L}_D \dot{y}(t) + \bar{D}w(t)\end{aligned} \tag{5.17}$$

Similarly, adding $\bar{L}_D\dot{y}(t)$ on both sides of (5.15), we can get

$$\begin{aligned}\bar{S}\dot{\bar{x}}(t) &= (\bar{A} - \bar{L}_P\bar{C})\bar{x}(t) + (\bar{A}_d - \bar{L}_H\bar{C})\hat{x}(t-h(t)) \\ &+ (\bar{B} + \Delta\bar{B}(t))u(t) + \bar{L}_P y(t) + NG_1\bar{f}(t) \\ &+ \bar{L}_H y(t-h(t)) + NG_2 f_{ag}(t) + \bar{L}_D\dot{y}(t) + \bar{D}w(t)\end{aligned} \tag{5.18}$$

Furthermore, defining $\bar{e}(t) = \hat{x}(t) - \bar{x}(t)$ and subtracting (5.17) to (5.18), we can obtain that

$$\begin{aligned}\bar{S}\dot{\bar{e}}(t) &= \bar{S}\dot{\hat{x}}(t) - \bar{S}\dot{\bar{x}}(t) \\ &= (\bar{A} - \bar{L}_P\bar{C})\bar{e}(t) + (\bar{A} - \bar{L}_H\bar{C})\bar{e}(t-h(t)) \\ &- NG_1\bar{f}(t) - NG_2 f_{ag}(t) + \bar{L}_s u_s(t)\end{aligned} \tag{5.19}$$

The sliding mode surface is first defined as follows

$$s_e = \bar{N}^T\bar{S}^{-T}P\bar{e}(t) \tag{5.20}$$

and for each positive P is designed to satisfy $\bar{N}^T\bar{S}^{-T}P = H\bar{C}$. For further, $u_s(t)$ is designed as

$$u_s(t) = (-\lambda + G_1\alpha + G_2 v)sgn(s_e(t)) \tag{5.21}$$

5.3.2 Active Fault-tolerant Controller Design

A proportional differential sliding mode observer is designed, which can simultaneously estimate system status and fault value. Next, the main work is to design a fault-tolerant controller to stabilize the system frequency.

The PI-type controller is designed as follows

$$u(t) = K_p A\hat{C}E(t) + K_I \int A\hat{C}E(t) = K\hat{y}(t) = KC\hat{x}(t) \tag{5.22}$$

Therefore, the system state function (5.14) can be rewritten as

$$\begin{aligned}
\dot{x}(t) &= Ax(t) + A_d x(t-\tau(t)) + (B+\Delta B(t))u(t) + Dw(t) \\
&= Ax(t) + A_d x(t-\tau(t)) + (B+\Delta B(t))(KC\hat{x}(t)) + Dw(t) \\
&= Ax(t) + A_d x(t-\tau(t)) + (B+\Delta B(t))KCx(t) \\
&\quad - (B+\Delta B(t))KCx(t) + (B+\Delta B(t))KC\hat{x}(t) + Dw(t) \\
&= (A + (B+\Delta B(t))KC)x(t) + A_d x(t-\tau(t)) \\
&\quad + (B+\Delta B(t))KCe_x(t) + Dw(t) \\
&= (A + (B+\Delta B(t))KC)x(t) + A_d x(t-\tau(t)) \\
&\quad + (B+\Delta B(t))KCUe(t) + Dw(t)
\end{aligned} \tag{5.23}$$

Then, the overall closed-loop multi-area power system is formulated in the following form

$$\begin{cases}
\dot{x}(t) = (A + (B+\Delta B(t))KC)x(t) + A_d x(t-\tau(t)) \\
\qquad + (B+\Delta B(t))KCU\bar{e}(t) + Dw(t) \\
\bar{S}\dot{\bar{e}}(t) = (\bar{A} - \bar{L}_P\bar{C})\bar{e}(t) + (\bar{A} - \bar{L}_H\bar{C})\bar{e}(t-\tau(t)) \\
\qquad - NG_1\bar{f}(t) - NG_2 f_{ag}(t) + \bar{L}_s u_s(t)
\end{cases} \tag{5.24}$$

5.3.3 Stability Analysis

To facilitate in presenting our conditions, we denote

$$\begin{aligned}
\eta(t) &= \left[x(t) \ \int_{t-h}^{t} x(s)ds \ \int_{t-h}^{t}\int_{s}^{t} x(u)duds\right]^T, \\
\varsigma(t) &= [x(t) \ \dot{x}(t) \ x(t-h(t)) \ x(t-h) \\
&\quad \frac{1}{h}\int_{t-h}^{t} x(s)ds \ \bar{e}(t) \ \bar{e}(t-h) \ \bar{e}(t-h(t))]^T, \\
e_i &= \left[0_{n\times(i-1)n} \ \ I \ \ 0_{n\times(9-i)n}\right], \\
\Gamma_1 &= \left[e_1 \ he_5 \ \frac{h^2}{2}e_6\right]^T, \Gamma_2 = [e_2 \ e_1 - e_4 \ h(e_1 - e_5)]^T,
\end{aligned}$$

$$G_0 = \begin{bmatrix} I & 0 & 0 & -I & 0 & 0 & 0 & 0 \\ I & 0 & 0 & I & -2I & 0 & 0 & 0 \\ I & 0 & 0 & -I & -6I & 0 & 0 & 0 \end{bmatrix}.$$

Theorem 5.1 *For given scalars $\bar{h}$, μ, ε_1, power system (5.24) is asymptotically stable with an H_∞ performance index $\gamma > 0$ if there exist real symmetric matrix $Z > 0$, $P > 0$, $Q_1 > 0$, $Q_2 > 0$, $Q_3 > 0$, $Q_4 > 0$, $R > 0$ and appropriate dimension matrix M, U such that as follows*

$$\Gamma = \begin{bmatrix} \Sigma_1 & \Sigma_2 & e_0^T C^T & \Theta D \\ 0 & -\frac{\varepsilon_1}{1+\mu} I_1 & 0 & 0 \\ 0 & 0 & -I_2 & 0 \\ 0 & 0 & 0 & 0 \\ 0 & 0 & 0 & -\gamma^2 I_3 \end{bmatrix} < 0 \tag{5.25}$$

$$N^T S^{-T} P = UC \tag{5.26}$$

where

$$\Sigma_1 = \Lambda_1 - \Lambda_2 + \Lambda_3 + \Lambda_5 + \Lambda_6 + \Lambda_7 + \Lambda_8,$$

$$\Lambda_1 = 2\Gamma_1^T Z \Gamma_2 + e_1^T (Q_1 + Q_2) e_1 - e_3^T Q_1 e_3 - e_4^T Q_2 e_4 + h^2 e_2^T R e_2,$$

$$\Lambda_2 = G_0^T \tilde{R} G_0, \tilde{R} = \mathrm{diag}\{R, 3R, 5R\},$$

$$\Lambda_3 = 2\Theta[(A + BKC)e_1 + A_d e_3 - e_2],$$

$$\Lambda_5 = \varepsilon_1 e_1^T M_1^T H H^T M_1 e_1 + \varepsilon_1 \mu e_2^T M_1^T H H^T M_1 e_2, -e_8^T Q_3 e_8 - e_9^T Q_4 e_9,$$

$$\Lambda_7 = 2\Theta B U e_7, \Lambda_8 = \frac{1+\mu}{\varepsilon_1} e_7 C^T K^T U^T E^T E U K C e_7, \Sigma_2 = e_1^T C^T K^T E^T,$$

$$\Theta = e_1^T M_1^T + e_2^T \mu M_1^T,$$

$$A_h = (\bar{A} - \bar{L}_P \bar{C}) e_7 + (\bar{A} - \bar{L}_H \bar{C}) e_8.$$

Proof. The Lyapunov-functional is considered as

$$V(t) = V_x(t) + V_e(t) \tag{5.27}$$

where

$$\begin{aligned} V_x(t) &= \eta^T(t)Z\eta(t) + \int_{t-h(t)}^{t} x(s)^T Q_1 x(s)ds + \int_{t-\bar{h}}^{t} x(s)^T Q_2 x(s)ds \\ &\quad + \bar{h}\int_{t-\bar{h}}^{t}\int_{s}^{t} \dot{x}(s)^T R\dot{x}(s)dsdt \end{aligned} \tag{5.28}$$

$$\begin{aligned} V_e(t) &= 2e^T(t)P\dot{e}(t) + e^T(t)(Q_1 + Q_2)e(t) - e^T(t-h(t))Q_1 e(t-h(t)) \\ &\quad - e^T(t-\bar{h})Q_2 e(t-\bar{h}) \end{aligned} \tag{5.29}$$

Taking the derivation of $V(t)$, we have

$$\begin{aligned} \dot{V}_x(t) &= \varsigma^T(t)\Lambda_1\varsigma(t) - h^2\int_{t-\bar{h}}^{t} x^T(s)Rx(s)ds \\ \dot{V}_e(t) &= \varsigma^T(t)\Lambda_6\varsigma(t) + 2e_7^T PS^{-1}[L_s u_s(t) - NG_1 f(t) \\ &\quad - NG_2\varphi(t)] \end{aligned} \tag{5.30}$$

By Lemma 5.1, the integral term in (5.30) can be processed as

$$-h\int_{t-\bar{h}}^{t} x^T(s)Rx(s)ds \leq -\varsigma^T(t)\Lambda_2\varsigma(t) \tag{5.31}$$

By applying free-weighting matrix technique [32], we have the following equality

$$\begin{aligned} &2[x^T(t)M_1^T + \dot{x}^T(t)\mu M_1^T] \times [(A + (B + \Delta B(t))KC)x(t) + A_d x(t-h(t)) \\ &+ (B + \Delta B(t))KCU\bar{e}(t) + Dw(t) - \dot{x}^T(t)] = 0 \end{aligned} \tag{5.32}$$

Therefore, Equation (5.32) can be expressed as

$$\varsigma^T(t)(\Lambda_3 + \Lambda_4 + \Lambda_7)\varsigma(t) + 2\varsigma^T(t)\Theta Dw(t) = 0 \tag{5.33}$$

By applying Lemma 5.2, we can obtain that

$$
\begin{aligned}
\Lambda_4 &= 2[e_1^T M_1^T + e_2^T \mu M_1^T] \times [\Delta B(t) K C e_1 + \Delta B(t) N e_7] \\
&\leq 2\varepsilon_1 e_1^T M_1^T H g(t) g(t) H^T M_1 e_1 \\
&\quad + \frac{1}{\varepsilon_1}(KCe_1 + KCUe_7)^T E^T E(KCe_1 + KCUe_7) \\
&\quad + 2\varepsilon_1 \mu e_2^T M_1^T H g(t) g(t) H^T M_1 e_2 \\
&\quad + \frac{\mu}{\varepsilon_1}(KCe_1 + KCUe_7)^T E^T E(KCe_1 + KCUe_7) \\
&= \Lambda_5 + \Lambda_8 + \frac{1+\mu}{\varepsilon_1} e_1^T C^T K^T E^T E K C e_1.
\end{aligned}
\tag{5.34}
$$

We define $L_s = \bar{S}P^{-1}C^T U^T$, $N^T S^{-T} P = UC$ and (5.31) can be processed as

$$
\begin{aligned}
& 2e_7^T P\bar{S}^{-1}[\bar{L}_s u_s(t) - NG_1 \bar{f}(t) - NG_2 \varphi(t)] \\
&\quad = 2e_7^T C^T U^T u_s(t) - 2e_7^T P\bar{S}^{-1} N G_1 \bar{f}(t) \\
&\quad - 2e_7^T P\bar{S}^{-1} N G_2 \varphi(t) \\
&\quad \leq -2\lambda \|s_e(t)\|
\end{aligned}
\tag{5.35}
$$

Combining (5.27) to (5.35), we can get

$$
\dot{V} \leq \varsigma(t)\Gamma\varsigma^T(t) - y^T(t)y(t) + \gamma^2 w^T(t)w(t) \tag{5.36}
$$

If the matrix inequality (5.25) holds, the (5.36) can be processed as

$$
\dot{V}(t) \leq -y^T(t)y(t) + \gamma^2 w^T(t)w(t) \tag{5.37}
$$

The integration of both sides of (5.37), we can get

$$
V(+\infty) - V(0) \leq \int_0^\infty \gamma^2 w^T(t)w(t)dt - \int_0^\infty y^T(t)y(t)dt \tag{5.38}
$$

Under initial condition $V(0) = 0$, the following inequality can be obtained

$$
\int_0^{+\infty} y^T(t)y(t)dt \leq \int_0^\infty \gamma^2 w^T(t)w(t)dt \tag{5.39}
$$

Hence, we have $\|y(t)\|_2 \leq \gamma\|w(t)\|_2$ for $w(t) \neq 0$. Based on Definition 5.1, there exists a scalar $\epsilon > 0$ such that $\dot{V}(x(t)) \leq -\epsilon\|\bar{x}(t)\|^2$. Therefore, the proposed active fault-tolerant LFC can guarantee the multi-area power system asymptotically stable. This completes the proof.

Theorem 5.2 *For given scalars $\bar{h}$, μ, ε_1, if there exist real symmetric matrix $\bar{Z} > 0$, $P > 0, \bar{Q}_1 > 0, \bar{Q}_2 > 0, Q_3 > 0, Q_4 > 0, R > 0$ and matrix X, W, V, U, Y, Y_H such that as follows*

$$\bar{\Gamma} = \begin{bmatrix} \bar{\Sigma}_1 & \bar{\Sigma}_2 & e_1^T X^T C^T & \Theta D \\ 0 & -\frac{\varepsilon_1}{1+\mu} I_1 & 0 & 0 \\ 0 & 0 & -I_2 & 0 \\ 0 & 0 & 0 & 0 \\ 0 & 0 & 0 & -\gamma^2 I_3 \end{bmatrix} < 0 \tag{5.40}$$

$$\begin{bmatrix} -\sigma_1 I & (N^T S^{-T} P - UC)^T \\ 0 & -I \end{bmatrix} < 0 \tag{5.41}$$

$$\begin{bmatrix} -\sigma_2 I & (VC - CX)^T \\ 0 & -I \end{bmatrix} < 0 \tag{5.42}$$

where

$$\bar{\Sigma}_1 = \bar{\Lambda}_1 - \bar{\Lambda}_2 + \bar{\Lambda}_3 + \bar{\Lambda}_5 + \bar{\Lambda}_6 + \bar{\Lambda}_7 + \bar{\Lambda}_8,$$

$$\bar{\Lambda}_1 = 2\Gamma_1^T \bar{Z} \Gamma_2 + e_1^T(\bar{Q}_1 + \bar{Q}_2)e_1 - e_2^T \bar{Q}_1 e_2 - e_3^T \bar{Q}_2 e_3 + h^2 e_2^T \bar{R} e_2,$$

$$\bar{\Lambda}_2 = G_0^T \hat{R} G_0, \hat{R} = \text{diag}\{\bar{R}, 3\bar{R}, 5\bar{R}\},$$

$$\bar{\Lambda}_3 = 2\bar{\Theta}[(A + BKCX)e_1 + A_d X e_3 - X e_2],$$

$$\bar{\Lambda}_5 = \varepsilon_1 e_1^T H H^T e_1 + \varepsilon_1 \mu e_2^T H H^T e_2,$$

$$\Lambda_6 = 2e_7^T P S^{-1}[A_h + 0.5NG_2Ce_1] + e_7^T(Q_3 + Q_4)e_7 - e_8^T Q_3 e_8 - e_9^T Q_4 e_9,$$

$$\bar{\Lambda}_7 = 2\bar{\Theta} B U e_7, \Lambda_8 = \frac{1+\mu}{\varepsilon_1} e_7 U^T E^T E U e_7,$$

$$\Sigma_2 = e_1^T X^T C^T K^T E^T,$$

$$\bar{\Theta} = e_1^T + e_2^T \mu, A_h = (\bar{A} - \bar{L}_P \bar{C})e_7 + (\bar{A} - \bar{L}_H \bar{C})e_8.$$

Then the power system (5.24) is asymptotically stable with an H_∞ performance index $\gamma > 0$, the controller gain is $K = WV^{-1}$ and the observer gain is $\bar{L}_P = \bar{S}P^{-1}Y$, $\bar{L}_H = \bar{S}P^{-1}Y_H$, $\bar{L}_S = \bar{S}P^{-1}C^TU^T$.

Proof. By defining $X = M^{-1}$, $\bar{Z} = XZX$, $\bar{Q}_1 = XQ_1X$, $\bar{Q}_2 = XQ_2X$, $\bar{R} = XRX$, and pre- and post-multiplying both sides of (5.25) with diag$\{X, X, \cdots, I, I, I, I\}$, we can obtain (5.40). What's more, for $BKCX$, $EKCX$, it is difficult to find observer gain K because the matrix C is not invertible. Following the method in [20], above problem can be transformed into a minimization issue. By defining $WC = KCX$, $VC = CX$, we can get

$$(VC - CX)^T(VC - CX) = 0$$

It can be further converted into (5.42) by using Lemma 5.1 and the same can be found in (5.41). Hence, the controller gain matrix K can be described as $K = WV^{-1}$. This completes the proof.

Remark 5.3: This chapter mainly proposes LMI-based LFC. Compared with the pole assignment method, we don't require a set of expected poles to calculate the controller gain. The controller gain $K = WV^{-1}$, observer gain $\bar{L}_P = \bar{S}P^{-1}Y$, $\bar{L}_H = \bar{S}P^{-1}Y_H$, and $\bar{L}_S = \bar{S}P^{-1}C^TU^T$ are obtained by solving the LMI matrices in Theorem 5.2, which can be found in [23, 31].

5.3.4 Reachability of the Sliding Mode Surface

In this subsection, in order to ensure the accessibility of the sliding surface in the error estimation space, the following theorem is proposed.

Theorem 5.3 *If there exist parameter matrix U and matrices P with appropriate dimension such that the constraints hold, the sliding mode input $u_s(t)$ guarantees the sliding motion to be driven to the sliding mode surfaces $s_e = 0$ in a finite time.*

Proof. Defining $F = NS^{-T}$, the Lyapunov-functional can be considered as

$$V(t) = 0.5s_e^T\left(FPF^T\right)^{-1}s_e \tag{5.43}$$

Taking the derivation of the constructed Lyapunov function, we can get

$$\begin{aligned}\dot{V}(t) = {} & s_e^T(FPF)^{-1}FPS^{-1} \times [(\bar{A} - \bar{L}_P\bar{C})\bar{e}(t) \\ & + (\bar{A} - \bar{L}_H\bar{C})\bar{e}(t - \tau(t)) - NG_1\bar{f}(t) \\ & + aNG_2Cx(t) - NG_2\varphi(t) + \bar{L}_s u_s(t)]\end{aligned} \tag{5.44}$$

Using $L_s = \bar{S}P^{-1}C^TU^T$ and $N^TS^{-T}P = UC$, we can get

$$\begin{aligned} & s_e^T(FPF)^{-1}FPS^{-1}[\bar{L}_s u_s(t) - NG_1\bar{f}(t) - NG_2\varphi(t)] \\ & \quad \leq s_e^T[u_s(t) - G_1\bar{f}(t) - G_2\varphi(t)] \\ & \quad \leq -\lambda\|s_e\| \end{aligned} \tag{5.45}$$

Combining (5.44) and (5.45), we can get

$$\begin{aligned} \dot{V}(t) \leq & -\|s_e(t)\|\{\lambda - \|(FPF)^{-1}FP\bar{S}^{-1}(\bar{A} - \bar{L}_p\bar{C})\| \\ & \times \|\bar{e}(t)\| - \|(FPF)^{-1}FP\bar{S}^{-1}aNG_2\|y(t))\| \\ & - \|(FPF)^{-1}FP\bar{S}^{-1}(\bar{A} - \bar{L}_h\bar{C})\|\|\bar{e}(t - h(t))\|\} \end{aligned} \tag{5.46}$$

Using Theorem 5.1, we can get $\|y(t)\| \leq \gamma\|w(t)\|$, then we define

$$\begin{aligned} \varphi &= \|(FPF)^{-1}FP\bar{S}^{-1}(\bar{A} - \bar{L}_p\bar{C})\| \\ \psi &= \|(FPF)^{-1}FP\bar{S}^{-1}(\bar{A} - \bar{L}_h\bar{C})\| \\ \xi &= \|(FPF)^{-1}FP\bar{S}^{-1}aNG_2\| \end{aligned} \tag{5.47}$$

From (5.45) to (5.47), we can obtain

$$\dot{V}_s(t) \leq -\|s_e(t)\|\{\lambda - \xi\gamma\|w(t)\| - \varphi\|\bar{e}(t)\| - \psi\|\bar{e}(t - h(t))\|\} \tag{5.48}$$

Then, we define the following domain as

$$\Omega \triangleq \bigcap_{i=1}^{s} \Omega(\xi, \varphi, \psi) \tag{5.49}$$

$$\begin{aligned} \Omega(\xi, \varphi, \eta) \triangleq \{ & \lambda - \xi\gamma\|w(t)\| - \varphi\|\bar{e}(t)\| \\ & - \psi\|\bar{e}(t - h(t))\| > 0\} \end{aligned} \tag{5.50}$$

From (5.49) and (5.50), it is noticed that $\dot{V}(t) < 0$ is in the domain Ω. Furthermore, we have proved the stabilization of (5.25) in Theorem 5.1. This implies that the trajectories of $\bar{e}(t)$ will enter Ω in finite time and remains thereafter. Until now, we have completed the proof.

5.4 Numerical Examples

In this section, four cases are provided to verify the effectiveness of the proposed active fault-tolerant LFC for an IEEE 39-bus 10-generator power system with sensor faults and deception attacks. A three-area power system model with EVs is established on the Matlab/Simulink environment.

The first one illustrates that the proposed active fault-tolerant LFC can guarantee the stability of the power system under deception attacks and sensor faults. The second one studies the sensitivity analysis of the proposed active fault-tolerant LFC with respect to the parameters. The third one studies that the proposed active fault-tolerant LFC strategy has been compared with the other LFC methods subject to the same condition. We test the proposed active fault-tolerant LFC on an IEEE 39-bus system (Figure 5.3).

The parameters of the three-area power system are shown in Table 5.1 and are similar to [31, 34]. What's more, defined the $T_{12} = 0.03$, $T_{23} = 0.02$, $T_{13} = 0.035$, the upper bound of time-varying delay $\bar{h} = 0.01s$.

The number of EVs $N(t)$ participating in LFC is shown in Figure 5.4. It can be seen that N change with time. The SOC parameters of EVs are $SOC_{min,i} = 10\%$, $SOC_{high,i} = 85\%$. Therefore, combining (5.7), the aggregated EVs gain is shown in Figure 5.4.

Example 5.1 *We design the active fault-tolerant LFC to ensure the stability of the closed loop system under deception attack and sensor fault.*

Assuming that the fault time exists from $t = 5s$ to $t = 12s$ and the sensor fault is shown in Figure 5.5.

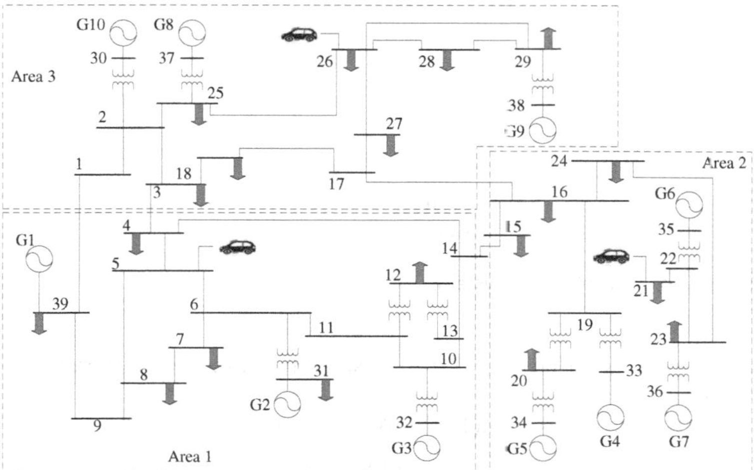

Figure 5.3 Single-line diagram of IEEE 39-bus system.

Table 5.1 Parameters of each area.

Parameter	Area 1	Area 2	Area 3
M_i	0.12	0.1	0.08
D_i	2	1.5	2
R_i	2	1	1.8
$\mathcal{T}_{mi}$	0.5	0.5	0.55
$\mathcal{T}_{gi}$	0.15	0.25	0.2
$\mathcal{T}_{ei}$	1	1	1
K_{mi}	1	1	1
ε_i	0.1	0.11	0.1

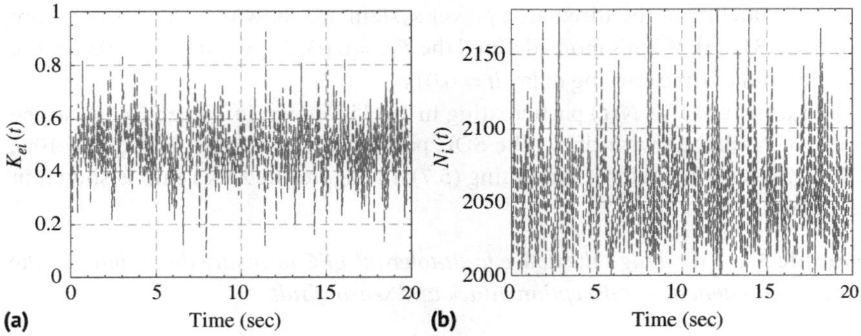

Figure 5.4 (a) Aggregated EVs gain $K_{ei}(t)$; (b) number of EVs $N(t)$.

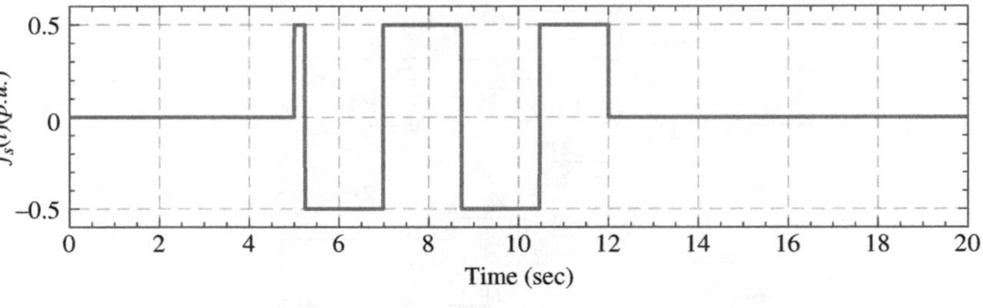

Figure 5.5 The sensor fault of power system.

The deception attack is defined as $\varphi(t) = 0.01\sin(1.8t)$ and $\nu = 0.01$, therefore

$$f_{ag}(t) = \begin{cases} -0.5Cx(t) + 0.01\sin(1.8t) & |\Delta f_i| < 0.01 \\ 0 & \text{otherwise} \end{cases}$$

We choose $\alpha_1 = \alpha_2 = \alpha_3 = 0.6$, $\sigma_1 = \sigma_2 = \sigma_3 = 0.02$, the adjustable parameters $\lambda_1 = \lambda_2 = \lambda_3 = 1$. The sliding mode input $u_s(t)$ is designed as

$$u_{s1}(t) = u_{s2}(t) = u_{s3}(t) = (-0.5 + G_1 0.6 + G_2 0.02) sgn(s_e(t))$$

The observer gain L_D is designed as

$$L_{D1} = L_{D2} = L_{D3} = \begin{bmatrix} 0 & 0 & 0 & 0 & 0 & 0 & 1 & 0 & 1 \\ 0 & 0 & 0 & 0 & 0 & 0 & 0 & 1 & 1 \end{bmatrix}.$$

Based on Theorem 5.2, the observer gain L_P, L_H and controller gain K can be obtained by the MATLAB toolbox.

$$L_{P1} = \begin{bmatrix} -3.23 & 4.06 & -12.81 & 0.32 & -0.11 & -0.72 & -278.43 & -25.89 & -2.47 \\ 4.34 & 32.40 & 62.90 & 1.04 & 1.81 & 21.41 & 517.20 & 44.62 & 23.14 \end{bmatrix},$$

$$L_{P2} = \begin{bmatrix} -3.27 & 5.66 & -12.16 & 0.29 & -0.12 & -0.67 & -280.65 & -25.58 & -2.13 \\ 4.01 & 31.80 & 59.39 & 1.09 & 1.79 & 20.69 & 496.50 & 41.70 & 21.45 \end{bmatrix},$$

$$L_{P3} = \begin{bmatrix} -2.95 & 9.67 & -12.22 & 0.24 & -0.16 & -1.40 & -304.16 & -27.04 & -2.54 \\ 3.87 & 40.67 & 65.41 & 1.48 & 2.10 & 25.19 & 569.37 & 45.82 & 25.00 \end{bmatrix},$$

$$L_{H1} = \begin{bmatrix} 0_{2\times1} & 0_{2\times1} \\ 0 & -1.43 \\ 0_{6\times1} & 0_{6\times1} \end{bmatrix}^T, L_{H2} = \begin{bmatrix} 0_{2\times1} & 0_{2\times1} \\ 0 & -1.42 \\ 0_{6\times1} & 0_{6\times1} \end{bmatrix}^T, L_{H3} = \begin{bmatrix} 0_{2\times1} & 0_{2\times1} \\ 0 & -1.47 \\ 0_{6\times1} & 0_{6\times1} \end{bmatrix}^T.$$

The system estimates obtained by the observer were used for controller design, and the system controller gain as

$$K_1 = \begin{bmatrix} -1.02 \\ -9.23 \end{bmatrix}^T, K_2 = \begin{bmatrix} -1.08 \\ -9.34 \end{bmatrix}^T, K_3 = \begin{bmatrix} -1.15 \\ -8.91 \end{bmatrix}^T.$$

We take the frequency deviation and the deviation of the tie-line power as examples. The simulative results of the three-area power system are shown in Figure 5.6. Therefore, it can be seen that the proposed active fault-tolerant LFC can effectively ensure the stability of the system.

We take the first area as an example. Figure 5.7 shows the true and the estimated system state. In Figure 5.7, the proposed observer can quickly and accurately estimate the system state.

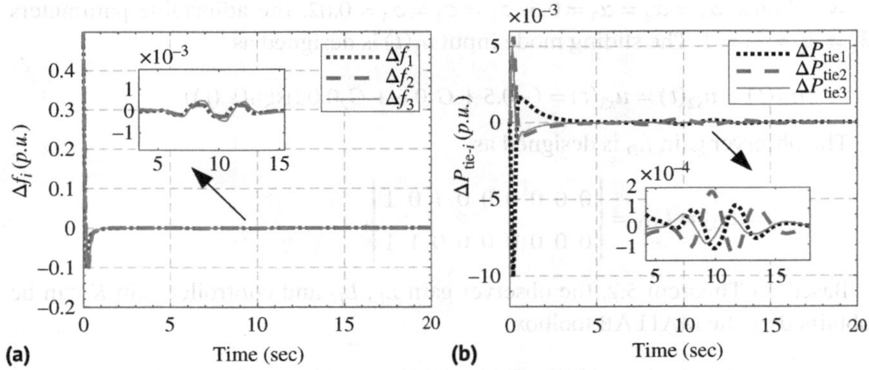

Figure 5.6 (a) The frequency deviation Δf_i; and (b) the tie-line power $\Delta P_{\text{tie}-i}$.

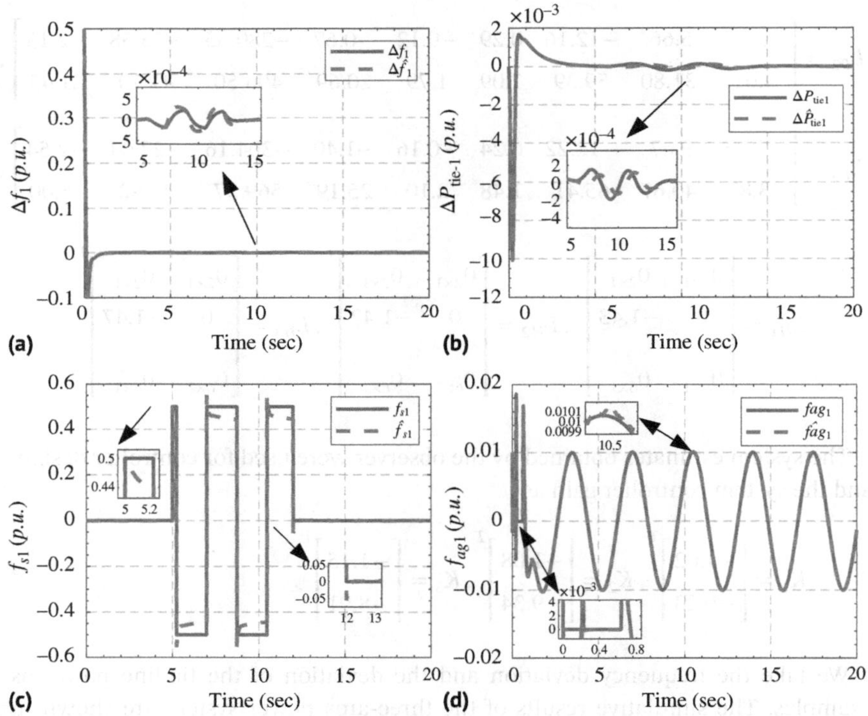

Figure 5.7 The true and the estimated system state in the first area: (a) the frequency deviation Δf_1; (b) the tie-line power $\Delta P_{\text{tie}1}$; (c) the sensor fault f_{s1}; (d) the deception attack fag_1.

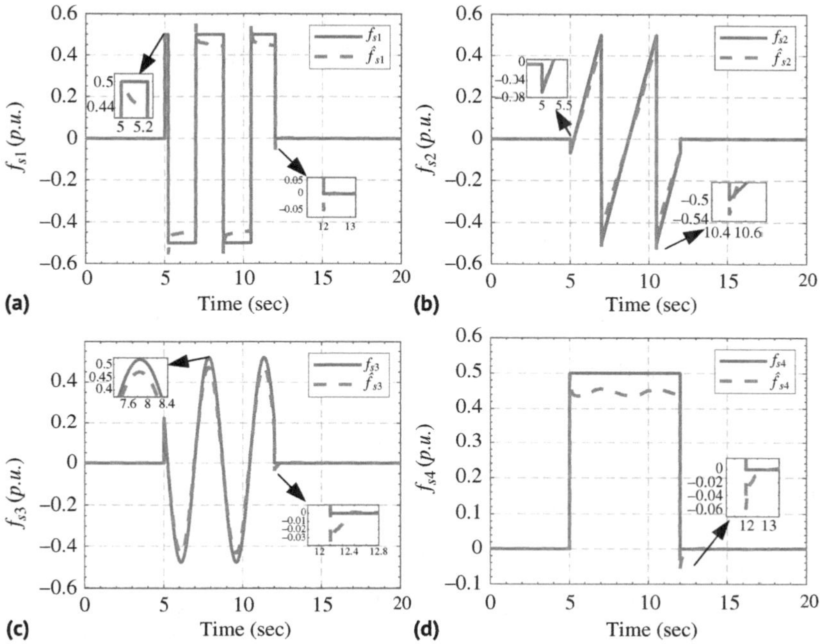

Figure 5.8 The true and the estimated sensor faults in the first area: (a) a square wave fault signal; (b) a triangular wave fault signal; (c) a sinusoidal fault signal; (d) a rectangular wave fault signal.

To further illustrate the effectiveness of the proposed PD sliding mode observer, we consider four different fault signals occurring in the system. We take the first area as an example. The true and the estimated sensor faults are shown in Figure 5.8. It is obvious that the proposed observer can effectively deal with different types of fault signals.

Example 5.2 *We study the sensitivity analysis of the proposed active fault-tolerant LFC with respect to the some parameters.*

The parameters of the system are identical to Example 5.1. Firstly, the observer gain $\bar{L}_D$ can be expressed as:

$$\bar{L}_D = \begin{bmatrix} 0 & 0 & 0 & 0 & 0 & 0 & a & 0 & a \\ 0 & 0 & 0 & 0 & 0 & 0 & 0 & a & a \end{bmatrix}.$$

The comparison of observer performance with different parameters is shown in Figure 5.9.

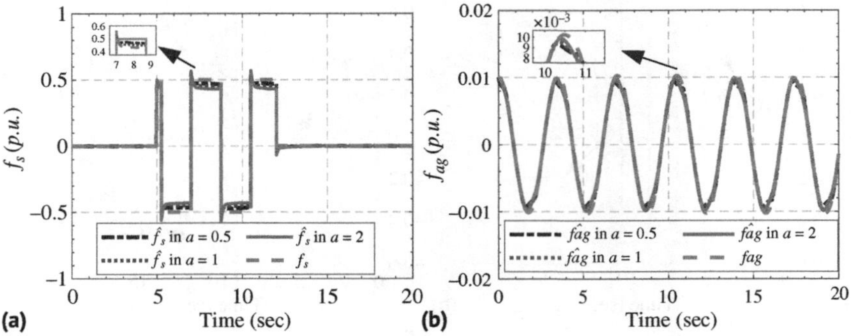

Figure 5.9 Observer performance of different α.

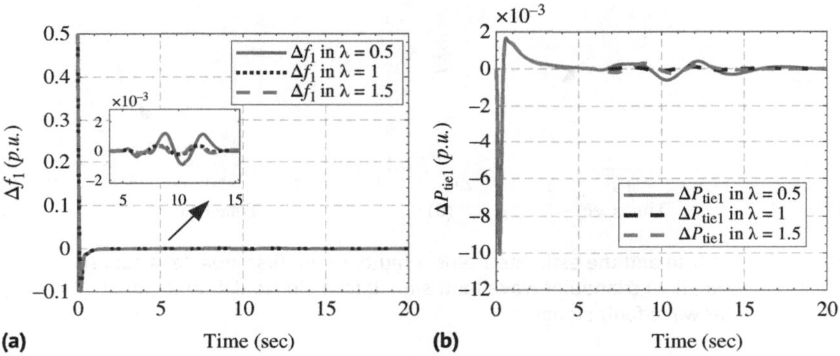

Figure 5.10 State response of different λ: (a) Δf_i; (b) $\Delta P_{\text{tie}-i}$.

Then we study the effect of different parameter λ_i on the power system. The comparison of Δf_i and $\Delta P_{\text{tie}-i}$ with different parameters λ_i is shown in Figure 5.10.

Example 5.3 *The proposed active fault-tolerant LFC scheme is compared with the LFC used in* [17] *and the traditional LFC without EVs.*

The parameters of the system are identical to Example 5.1. The simulated results are shown in Figure 5.11. It can be seen that the method in [17] is difficult to avoid the interference of sensor fault and deception attack. Therefore, it makes the frequency deviation and the deviation of the tie-line power have a large deviation. However, the proposed active fault-tolerant LFC scheme can guarantee the stability of the power system. In addition, the existence of EVs can effectively reduce the power oscillations.

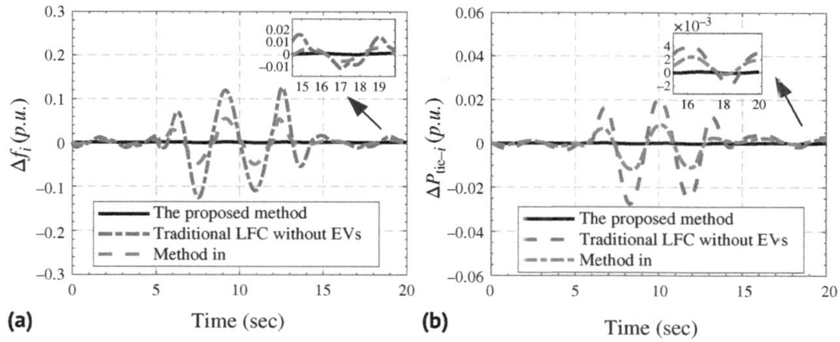

Figure 5.11 (a) The frequency deviation Δf_i; (b) The tie-line power $\Delta P_{\text{tie}-i}$.

Table 5.2 Parameters of each area.

Parameter	Area 1	Area 2	Area 3
M_i	10	10	11
D_i	4	3.5	4
R_i	3	3.5	4
T_{mi}	0.05	0.04	0.03
T_{gi}	0.03	0.03	0.03
T_{ei}	1	1	1
K_{mi}	1	1	1
ε_i	0.1	0.11	0.1
Area 1	$\alpha_{11g} = 0.3, \alpha_{12g} = \alpha_{13g} = 0.2, \alpha_{11e} = 0.3$		
Area 2	$\alpha_{21g} = 0.3, \alpha_{22g} = \alpha_{23g} = 0.2, \alpha_{21e} = 0.3$		
Area 3	$\alpha_{31g} = \alpha_{32g} = \alpha_{33g} = \alpha_{34g} 0.2, \alpha_{31e} = 0.2$		

Example 5.4 *We test the proposed scheme on IEEE 39-bus system. The IEEE 39-bus test system is shown in Figure 5.3. The system consists of 10 generators, 19 loads, 34 transmission lines, and 12 transformers* [35]. *The power system is divided into three areas. Furthermore, three EVAs are introduced in buses 5, 21, and 26, respectively.*

Here, the parameters, including the participation factor of every governor, are given in Table 5.2 and can be found in [36]. The values of deception attacks and sensor faults are identical to Example 5.1.

We take frequency deviation and the deviation of the tie-line power as examples. The simulated results of the IEEE 39-bus system are shown in Figure 5.12. It can be seen that the proposed active fault-tolerant LFC scheme can effectively guarantee the stability of the IEEE 39-bus system.

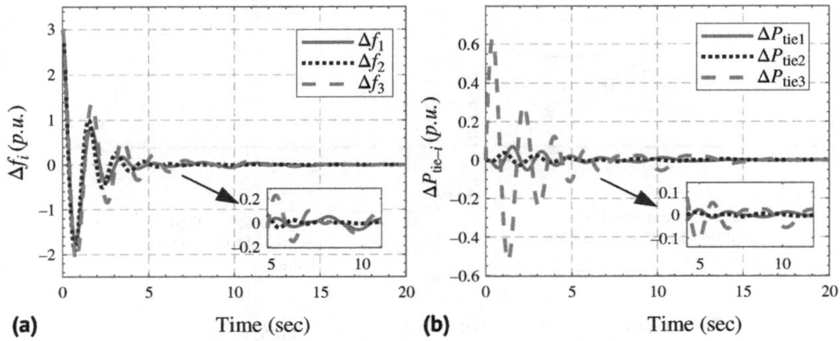

Figure 5.12 (a) The frequency deviation Δf_i and (b) the tie-line power ΔP_{tie-i}.

5.5 Conclusion

A multi-area power system model has been proposed under sensor faults and deception attacks with the uncertainty of electric vehicle battery status. An active fault-tolerant LFC scheme has been presented to ensure the stability of the multi-area power systems. The simulation with respect to a three-area power system has verified the effectiveness of the proposed active fault-tolerant LFC in this chapter.

References

1 P. Mani and Y. H. Joo. Fuzzy logic-based integral sliding mode control of multi-area power systems integrated with wind farms. *Information Sciences*, 545:153–169, 2021.

2 Y. Yang, Y. Niu, and Z. Zhang. Dynamic event-triggered sliding mode control for interval type-2 fuzzy systems with fading channels. *ISA Transactions*, 110:53–62, 2021.

3 P. C. Chen, D. Zhang, L. Yu, and H. C. Yan. Dynamic event-triggered output feedback control for load frequency control in power systems with multiple cyber attacks. *IEEE Transactions on Systems, Man, and Cybernetics: Systems*, 52(10):6246–6258, 2022.

4 M. Hossain, C. Peng, Y. L. Wang, and X. Du. Handshake logic-based event-triggered load frequency control for smart grids under dos attacks. *IEEE Transactions on Industrial Informatics*, 18(6):3863–3872, 2022. https://doi.org/10.1109/TII.2021.3116165.

5 X. X. Lv, S. W. Zhai, Y. H. Sun, et al. Load frequency control for multi-area power system based on Markov model. *Journal of the Franklin Institute*, 358(16): 8377–8395, 2021.

6 M. V. Mahendran and V. Vijayan . Model-predictive control-based hybrid optimized load frequency control of multi-area power systems. *IET Generation, Transmission & Distribution*, 15(9):1521–1537, 2021.

7 M. Khan, H. S. Sun, Y. M. Xiang, and D. Shi. Electric vehicles participation in load frequency control based on mixed H_2/H_∞. *International Journal of Electrical Power & Energy Systems*, 125:106420, 2021.

8 S. Vachirasricirikul and I. Ngamroo. Robust LFC in a smart grid with wind power penetration by coordinated V2G control and frequency controller. *IEEE Transactions on Smart Grid*, 5(1):371–380, 2014.

9 C. Lu, T. S. Bi, M. Ilic, et al. Guest editorial: New trends in wide-area monitoring and control of power systems with large scale renewables. *IET Generation, Transmission & Distribution*, 11(18):4403–4405, 2017.

10 K. S. Ko and D. K. Sung. The effect of EV aggregators with time-varying delays on the stability of a load frequency control system. *IEEE Transactions on Power Systems*, 33:669–680, 2018.

11 E. L. Karfopoulos, K. A. Panourgias, and N. D. Hatziargyriou. Distributed coordination of electric vehicles providing V2G regulation services. *IEEE Transactions on Power Systems*, 31(4):2834–2846, 2015.

12 H. Zeng, S. Zhou, X. Zhang, and W. Wang. Delay-dependent stability analysis of load frequency control systems with electric vehicles. *IEEE Transactions on Cybernetics*, 52(12):13645–13653, 2022.

13 T. N. Pham, A. M. T. Oo, and H. Trinh. Event-triggered mechanism for multiple frequency services of electric vehicles in smart grids. *IEEE Transactions on Power Systems*, 37(2):967–981, 2021.

14 D. Aravindh, R. Sakthivel, B. Kaviarasan, et al. Design of observer-based non-fragile load frequency control for power systems with electric vehicles. *ISA Transactions*, 91:21–31, 2019.

15 Z. H. Wang, Y. H. Liu, Z. H. Yang, and W. H. Yang. Load frequency control of multi-region interconnected power systems with wind power and electric vehicles based on sliding mode control. *Energies*, 14(8):2288, 2021.

16 S. Tripathi, V. P. Singh, N. Kishor, and AS. Pandey. Load frequency control of power system considering electric vehicles' aggregator with communication delay. *International Journal of Electrical Power & Energy Systems*, 145:108697, 2023.

17 T. N. Pham, S. Nahavandi, L. V. Hien, et al. Static output feedback frequency stabilization of time-delay power systems with coordinated electric vehicles state of charge control. *IEEE Transactions on Cybernetics*, 32:3862–3874, 2017.

18 S. Arunagirinathan, P. Muthukumar, and Y. H. Joo. Robust reliable control design for networked control systems with nonlinear actuator faults and randomly missing measurements. *International Journal of Robust and Nonlinear Control*, 29:4168–4190, 2018.

19 M. A. Shahab, B. Mozafari, S. Soleymani, et al. Distributed consensus-based fault tolerant control of islanded microgrids. *IEEE Transactions on Smart Grid*, 11(1):37–47, 2019.

20 X. H. Liang, Q. Wang, C. H. Hu, and C. Y. Dong. Observer-based H_∞ fault-tolerant attitude control for satellite with actuator and sensor faults. *Aerospace Science and Technology*, 95:105424, 2019.

21 V. Sharmila, R. Rakkiyappan, and Y. H. Joo. Fuzzy sampled-data control for DFIG-based wind turbine with stochastic actuator failures. *IEEE Transactions on Systems, Man, and Cybernetics: Systems*, 51(4):2199–2211, 2019.

22 S. Dey, S. Mohon, P. Pisu, and B. Ayalew. Sensor fault detection, isolation, and estimation in lithium-ion batteries. *IEEE Transactions on Control Systems Technology*, 24(6):2141–2149, 2016.

23 S. Kuppusamy, Y. H. Joo, and H. S. Kim. Fault-tolerant load frequency control for DFIG-based interconnected wind power systems. *Information Sciences*, 582:73–88, 2022.

24 B. Kaviarasan, R. Sakthivel, and OM. Kwon. Robust fault-tolerant control for power systems against mixed actuator failures. *Nonlinear Analysis: Hybrid Systems*, 22:249–261, 2016.

25 X. C. Zhou, Z. L. Dou, W. M. Zhang, et al. Distributed sliding mode fault-tolerant LFC for multiarea interconnected power systems under sensor fault. *Complexity*, 2022(1):6271232, 2022.

26 A. M. Mohan, N. Meskin, and H. Mehrjerdi. A comprehensive review of the cyber-attacks and cyber-security on load frequency control of power systems. *Energies*, 13(15):3860, 2020.

27 X. Zhang, F. Yang, and X. Sun. Resilient adaptive event-triggered load frequency control of network-based power systems against deception attacks. *Sensors*, 21(21):7047, 2021.

28 E. Tian and C. Peng. Memory-based event-triggering h_∞ load frequency control for power systems under deception attacks. *IEEE Transactions on Cybernetics*, 50:4610–4618, 2020.

29 S. P. Wen, X. H. Yu, Z. G. Zeng, and J. J. Wang. Event-triggering load frequency control for multiarea power systems with communication delays. *IEEE Transactions on Industrial Electronics*, 63(2):1308–1317, 2015.

30 X. L. Wang, D. R. Ding, H. L. Dong, and X. J. Yi. Pi-based security control against joint sensor and controller attacks and applications in load frequency control. *IEEE Transactions on Systems, Man, and Cybernetics: Systems*, 53(2):970–980, 2022.

31 X. Su, X. Liu, and Y. D. Song. Fault-tolerant control of multiarea power systems via a sliding-mode observer technique. *IEEE/ASME Transactions on Mechatronics*, 23:38–47, 2018.

32 X. Zhang, M. Li, M. Wu, and J. She. Further results on stability and stabilisation of linear systems with state and input delays. *IEEE Transactions on Cybernetics*, 40:1–10, 2009.

33 S. Yin, H. Y. Yang, and O. Kaynak. Sliding mode observer-based FTC for Markovian jump systems with actuator and sensor faults. *IEEE Transactions on Automatic Control*, 62(7):3551–3558, 2017.

34 C. Peng, J. Zhang, and H. C. Yan. Adaptive event-triggering H_∞ load frequency control for network-based power systems. *IEEE Transactions on Industrial Electronics*, 65(2):1685–1694, 2017.

35 R. A. Ramos, R. Kuiava, T. C. C. Fernandes, et al. IEEE pes task force on benchmark systems for stability controls. *Pacific Northwest National Laboratory, Technical Report*, pages 1–49, 2012.

36 C. H. Yan, B. Liu, P. Xiao, and C. Y. Zhang. Stabilization of load frequency control system via event-triggered intermittent control. *IEEE Transactions on Circuits and Systems II: Express Briefs*, 69(12):4934–4938, 2022.

6

Reliable AMD Event-triggered Control of Wind Power Systems Under Frequency-based Deception Attacks

After reading this chapter, you should be able to:

- Recognize the impact of renewable energy on the LFC power system.
- Know the effective design of triggering mechanism reduces the communication resources.
- Identify the happening of the frequency-based deception attacks.

6.1 Background

The generation and effective utilization of clean and sustainable energy is emerging as a research hotspot [1–3]. Renewable energies are expected to fulfill the growing demands for energy while at the same time protecting the environment. Popular renewable energy resources typically include wind, solar, and tidal energies, etc [4]. Wind power, with its positive impacts on alleviating power supply, however, may be easily affected by various environmental factors. Specifically, the frequency and strength of winds have major effects on the wind power generation. Frequent changes in the wind power generation may impose a threat to the normal operation of the power systems, generating some serious stability and balance problems [5]. Proper control methods have to be developed to handle such challenges.

To achieve the maximum environmental benefits, renewable energies are supposed to be integrated into interconnected power systems. Therefore, multi-area wind power systems have come into being. In such systems, load frequency control (LFC) is widely used in keeping the balance between load and power variations [6–8]. Sliding mode control (SMC) is well used in resolving disturbances or uncertainties. As a kind of robust control method, SMC is to stabilize the system with prespecified performance. To achieve the equilibrium of frequency and load, sliding mode load frequency control (SMLFC) method has been applied to lighten or remove the negative influences of uncertainties coming from the wind turbines. In [9], the stable problem of power systems connected with wind farms based on the fuzzy-based integral SMC was investigated. A decentralized SMLFC controller for tackling the unmatched uncertainties for interconnected power systems was given in [10]. Utilization of event-triggered device is also introduced in this study for lightening the burdens of the communication channels.

The event-triggered schemes have been widely adopted to reduce resource utilization of communication channels [11]. Traditional event-triggered strategies typically note the dissimilarity between the new signals and the latest transmitted signals [12]. Such strategies, however, may miss sending some important signals when the current sampling signals, important as they are, are not significantly different from the last released signals. For such cases, if the historical records of earlier released signals can be properly utilized, better control performance may be achieved. Such observations have stimulated the research related to the memory event-triggered (MET) schemes [13–15]. Stability of interconnected delayed power systems adopting a MET LFC when contaminated by cyberattacks was analyzed in [13]. In [14], a MET device is applied to regulate the output of a networked switched system. Adaptive memory-event-trigger (AMET) control was proposed to solve the asynchronous premise constraints problems for fuzzy systems [15].

We develop an adaptive-memory-derivative (AMD)-based event-triggered strategy to further improve control performance and save communication channel resources. Compared with the AMET mechanism or MET mechanism, the AMD event-triggered strategy adds a state-based derivative term into the event-triggered condition. Benefiting from the newly introduced derivative term, the proposed triggering rule works under more relaxed conditions [13–15], which allows more data to be transmitted. Moreover, even when event trigger devices are adopted to screen the sampling data, the number of transmitted signals may still be too large, imposing pressure on the power system communication networks. Signal quantization [16, 17] provides a possible solution for tackling this challenge. To reduce the communication rate, a hysteresis quantizer is applied for signal quantization in this chapter.

As the information transfer of wind power systems are typically utilizing the open network, it is vulnerable to cyberattacks. The cyberattacks may lead to major influences, including publicity of confidential information, equipment damage, and so on [18–20]. Unreliable communication networks lead to unexpected damage to the system. Therefore, a reliable control is proposed to ensure that the system under cyberattack would operate normally, which is of great significance. By using the LFC method, a MET strategy was proposed for a wind power system under deception attacks in [21]. The authors of [22] combined the sliding mode observers and adaptive dual-layer super-twisting method together to solve the contamination of communication attacks in power networks. The stability problem of Takagi–Sugeno fuzzy networked control systems under deception attacks via a reliable sampling mechanism was investigated in [23]. However, in most of the existing studies, deception attacks are modeled as having a Bernoulli distribution [24–26]. In such models, the occurrence of deception attacks follows a given probability and is not directly related to the wind power system.

Therefore, we develop a frequency-based deception attack model which is based on the power system real-time frequency. The attacker is assumed to be able to get the real-time frequency of the power system by a certain sensing device. It is wished to the attacker to let a deception attack be with destructive force and last for a long time. As carrying out an attack consumes certain resources, and it is reasonable to assume that the attacker has only limited resources. Inspired by the sleeping areas of the denial-of-service (DoS) attacks [27, 28], we introduce sleeping areas for deception attacks that may allow the attacks to last longer. When the system sampling time is in the sleeping area, it will not be attacked. In reality, power system frequency can clearly reflect the performance of the system. A smart attacker may also note the importance of frequency to the normal operation of the power system. In this chapter, the real-time frequency of the power system is used to divide the sleeping areas and the attack areas. The real-time frequency also reveals the stability of the power system. By using the frequency deviation, the overall state of the wind power system can be divided into stable and unstable regions. In a stable region, the frequency deviation is smaller than a certain threshold value; otherwise it is an unstable region. We set the limitation that the deception attacks only happen in stable regions, thus stable regions are equivalent to the deception attack areas. Considering that the wind power system cannot be normally operating in unstable regions, it makes sense to take unstable regions as sleeping areas where no deception attacks would happen. The main purpose of the frequency-based deception attacks can be set as destroying the stable state or preventing the wind power system from approaching the stable state. The motivations for considering the frequency-based deception attack model can be summarized as follows.

- The frequency-based deception attack model helps us to study the harm of cyberattacks in the context of the power system. In most cases, the destructive effect of a cyberattack is only considered on the communication network, without any further connection to the target system. Due to the use of power system frequency information, the proposed frequency-based deception attack model has a deep relationship with the power system.

- Different kinds of cyberattack models have been built to describe the actual attack behaviors or to maximize the damage of the attacks [29]. Such studies motivate us to propose the frequency-based deception attack model to describe the actual attack behavior.
- The purpose of malicious cyberattacks is to disrupt the stable operation of the system. However, there is an implicit assumption adopted in most existing work that the cyberattack launches attacks indiscriminately [29]. Differently, the frequency-based deception attack model only occurs when the power system state is in the predefined stable region. The proposed frequency-based deception attack model maximizes the damage of the attacks.

In this chapter, we propose different conditions to realize the reliable control of the wind power systems under frequency-based deception attacks. Our main contributions may be summarized in the following.

1. An adaptive-memory-derivative event-triggered scheme is developed to reduce the use of communication resources while improving control performance. Different from the normal event-trigger schemes [30–32], the AMD event-triggered device uses the historic signals to design the triggering situation such that the consumption of the communication resources can be significantly further reduced.
2. A frequency-based deception attack model is proposed to formulate the communication disturbances caused by the network contamination. Different from the widely studied deception attacks, which happen with the Bernoulli distribution [24–26], the proposed frequency-based deception attacks are carried out based on the system frequency.
3. By constructing a multiple Lyapunov function, a reliable condition for a wind power system under the AMD event-triggered QSMLFC to be stable is derived as a set of matrix inequalities. The cross terms appearing in the inequalities are well solved by introducing new complex item processing methods [33].

6.2 Problem Formulation

For convenience, we mainly study on a subsystem i with a turbine, a power generator, and a governor. The subsystems are connected through tie lines. Then, the $i-th$ wind power system model is given in Figure 6.1, where a renewable energy is represented as the wind generator.

6.2.1 Wind Power System Model

Mismatched uncertainties in the wind power system include the operation point variation, the renewable source fluctuation, and the load. The model is chosen as

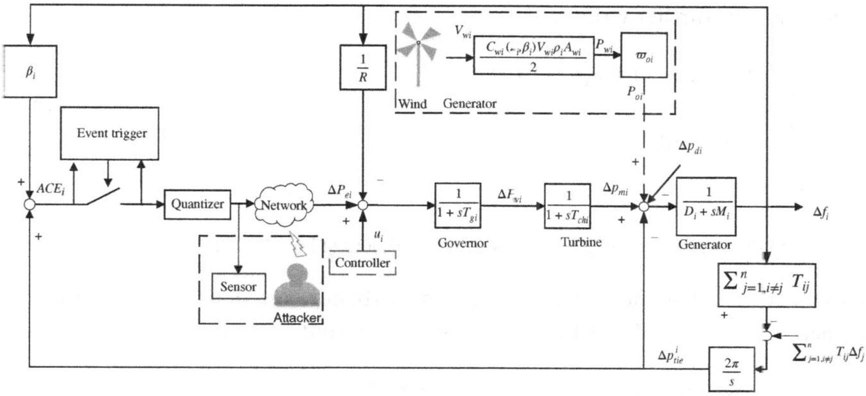

Figure 6.1 Wind power system model under frequency-based deception attacks.

the $i-th$ area wind power system which consists of a generator, a governor, and a turbine, and so on.

Based on information transfer between devices in Figure 6.1, the dynamic equations of the wind power system can be obtained as follows

$$\begin{cases} \Delta\dot{P}_{mi}(t) = -\dfrac{\Delta P_{mi}(t)}{T_{\text{chi}}} + \dfrac{\Delta P_{vi}(t)}{T_{chi}}, \\ \Delta\dot{P}^{i}_{\text{tie}}(t) = 2\pi \sum\limits_{j=1,i\neq j}^{n} T_{ij}(\Delta f_i(t) - \Delta f_j(t)), \\ \Delta\dot{P}_{vi}(t) = -\dfrac{\Delta f_i(t)}{R_i T_{gi}} - \dfrac{\Delta P_{vi}(t)}{T_{gi}} - \dfrac{\Delta P_{ei}(t)}{T_{gi}}, \\ \Delta\dot{f}_i(t) = \dfrac{\Delta P_{mi}(t)}{M_i} - \dfrac{D_i \Delta f_i(t)}{M_i} - \dfrac{\Delta P^{i}_{\text{tie}}(t)}{M_i} - \dfrac{\Delta P_{di}(t) + P_{oi}(t)}{M_i}, \\ ACE_i = \Delta P_{\text{tie}} + \beta_i \Delta f_i. \end{cases}$$

Due to unpredictable wind strength and the modeling mismatches between wind power system model and the practical system. There exist parameter uncertainties in the $i-th$ area wind power system, which can be considered as ΔA_i, ΔB_i, and ΔF_i. We have the following state space function with uncertainties

$$\begin{cases} \dot{x}_i(t) = (\tilde{A}_i + \Delta A_i)x_i(t) + (\tilde{B}_i + \Delta B_i)u_i(t) \\ \qquad + (\tilde{F}_i + \Delta F_i)(w_i(t) + P_{oi}(t)), \\ y_i(t) = C_i x_i(t), \end{cases} \tag{6.1}$$

which may be further written as

$$\begin{cases} \dot{x}_i(t) = \tilde{A}_i x_i(t) + \tilde{B}_i u_i(t) + \tilde{\omega}_i(t), \\ y_i(t) = C_i x_i(t), \end{cases} \tag{6.2}$$

where

$$\tilde{\omega}_i(t) = \Delta A_i x_i(t) + \Delta B_i u_i(t) + (\tilde{F}_i + \Delta F_i)(w_i(t) + P_{oi}(t)).$$

$w_i(t) = \Delta P_{di} < d_i$ is the system bounded disturbance, and d_i defined as a positive constant. ACE_i signal is defined as the system output. Define

$$x_i(t) = [\Delta P_{mi} \ \Delta P_{vi} \ \Delta P^i_{tie} \ \Delta f_i \ \int ACE_i]^{\mathrm{T}},$$

$$\tilde{A}_i = \begin{bmatrix} -\frac{1}{T_{\text{chi}}} & \frac{1}{T_{\text{chi}}} & 0 & 0 & 0 \\ 0 & -\frac{1}{T_{gi}} & 0 & -\frac{1}{R_i T_{gi}} & 0 \\ 0 & 0 & 0 & 2\pi \sum_{j=1, j\neq i}^{n} T_{ij} & 0 \\ \frac{1}{M_i} & 0 & -\frac{1}{M_i} & -\frac{D_i}{M_i} & 0 \\ 0 & 0 & 0 & 1 & \beta_i \end{bmatrix},$$

$$\tilde{B} = \left[0, \frac{1}{T_{gi}}, 0, 0, 0\right]^{\mathrm{T}}, \tilde{F} = \left[0, 0, 0, -\frac{1}{M_i}, 0\right]^{\mathrm{T}}.$$

6.2.2 Model of Renewable Energy

In this chapter, renewable energy is represented by a wind power generator. We adopt the wind power generator model proposed in [34], where windmill is connected to the generator by using the gearbox.

The wind power generator can be described by a few parameters. Below we introduce these parameters and discuss on the relations between them. The relation between output power P_{wi} of the windmill and some of its main parameters can be presented as follows

$$P_{wi} = \frac{C_{wi}(\lambda, \beta) V_{wi}^3 \rho A_{wi}}{2}, \tag{6.3}$$

where V_{wi} is the wind speed, ρ and A_{wi} are the air density and rotor cross-section, respectively, $C_{wi}(\lambda, \beta)$ is defined as the power coefficient, $\lambda = R\varpi/V_w$ is the tip speed ratio, β is the pitch angle, R is radius of a windmill, ϖ is the speed of rotor angular and it equals

$$\varpi^2 = \int \frac{2}{P_j}(P_{wi} - P_{oi})dt, \tag{6.4}$$

where P_j is the windmill inertia moment. Further, the output power of generator P_{oi} is given as

$$P_{oi} = \frac{-3V_P^2 s(1+s)R_2}{(R_2 - sR_1)^2 + (sX_1 + sX_2)^2}, \tag{6.5}$$

where V_P is the phase voltage, $s = (\varpi_0 - \varpi)/\varpi_0$ is the slip of generator, ϖ is the synchronous angular speed, R_1 and R_2 are the stator resistance and rotor resistance, and X_1 and X_2 are the stator reactance and rotor reactance, respectively. A graphic representation of the relations between the main parameters of a wind power generator is presented in Figure 6.2.

6.2.3 Selection of Quantizer

As the network transmission capacity is limited, the transmitted signals are subject to quantization. In this chapter, a hysteresis quantizer is used to help signal quantization. Due to the characteristics of the hysteresis quantizer [35], the system chattering comes from the function sgn() can be naturally avoided. $Q(x_i(t))$ is the hysteretic quantizer and modeled as

$$Q(x_i(t)) = \begin{cases} \hat{x}_i(t)\mathrm{sgn}(x_i(t)), & \frac{\hat{x}_i(t)}{1+\theta} \le |x_i(t)| \le \hat{x}_i(t),\ \dot{x}_i(t) < 0 \\ (1+\theta)\hat{x}_i(t)\mathrm{sgn}(x_i(t)), & \hat{x}_i(t) < |x_i(t)| \le \frac{\hat{x}_i(t)}{1-\theta},\ \dot{x}_i(t) < 0 \\ 0, & 0 \le |x_i(t)| < \frac{x_{\text{dead}}}{1+\theta},\ \dot{x}_i(t) < 0 \\ \hat{x}_i(t)\mathrm{sgn}(x_i(t)), & \hat{x}_i(t) < |x_i(t)| \le \frac{\hat{x}_i(t)}{1-\theta},\ \dot{x}_i(t) > 0 \\ (1+\theta)\hat{x}_i(t)\mathrm{sgn}(x_i(t)), & \frac{\hat{x}_i(t)}{1-\theta} < |x_i(t)| \le \frac{\hat{x}_i(t)}{\psi},\ \dot{x}_i(t) > 0 \\ 0, & \frac{x_{\text{dead}}}{1+\theta} \le |x_i(t)| \le x_{\text{dead}},\ \dot{x}_i(t) > 0 \\ Q(x_i(t^-)), & \dot{x}_i(t) = 0. \end{cases} \tag{6.6}$$

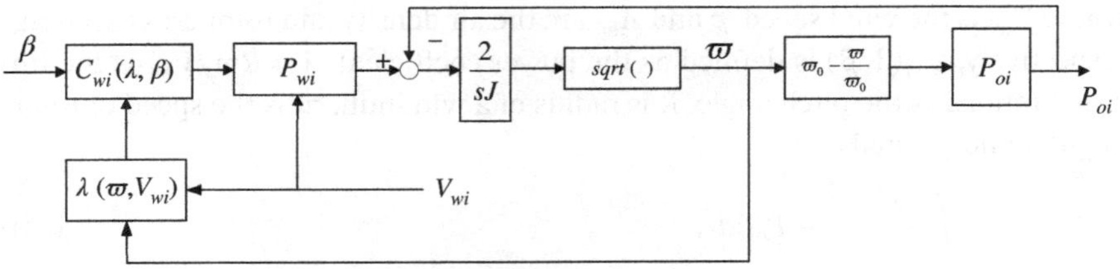

Figure 6.2 The relations between the main parameters of a wind power generator.

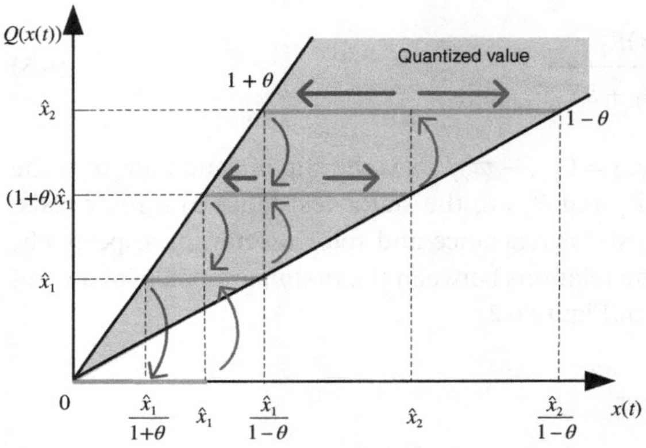

Figure 6.3 The map of hysteresis quantizer.

where $\hat{x}_{\hat{i}}(t) = \psi^{(1-\hat{i})} x_{\text{dead}}$, $x_{\text{dead}} > 0$ is the dead zone in the adjacency of zero, $\hat{i}$ is the level of quantization and $\psi = \dfrac{1-\theta}{1+\theta}$, $\psi \in (0, 1)$ is given as the density of quantization, $\theta \in (0, 1)$.

The map of hysteresis quantizer for $x_i(t) > 0$ is illustrated in Figure 6.3, and we have $(1+\theta)x_i(t) \geq Q(x_i(t)) \geq (1-\theta)x_i(t)$. We have that $(1+\theta)x_{\hat{i}+1} \geq Q(x_i(t))x_i(t) \geq (1-\theta)\hat{x}_{\hat{i}+1}$ for $x_{\text{dead}} \leq |x_i(t)|$. Thus, the quantized value belongs to the sector area that is marked in Figure 6.3. The quantized value $Q(x_i(t))$ belongs to $\{\pm(1+\theta)\hat{x}_{\hat{i}}(t), \pm\hat{x}_{\hat{i}}(t), 0\}$.

Remark 6.1: Hysteresis quantizer has two different quantization functions with the same density of quantization. But, there is only one quantizer working at each moment. When the quantizer is switched, the quantized value changes. In the uniform quantizer, the logarithmic quantizer will have a chattering problem. The hysteresis quantizer handles this problem well with two different quantization functions.

After quantification, there always exist some deviations between the input signals and the output signals. $Q(x_i(t)) - x_i(t_{k-p+1}h) = \Delta_i x_i(t_{k-p+1}h)$ is defined as the quantization error. The signal after quantification is denoted as

$$x_Q(t_{k-p+1}h) = \Delta_i x_i(t_{k-p+1}h) + x_i(t_{k-p+1}h), \tag{6.7}$$

where $-\psi < \Delta_i \leq \psi$ and $\Delta_i = \text{diag}\{\Delta_1(t) \cdots \Delta_i(t)\}$, $x_i(t_{k-p+1}h)$ is the transmitted signal supported by AMD event-triggered device and will be given later.

6.2.4 AMD Event-triggered Scheme

In the existing event-triggered mechanism, the event generator only uses the recently released signal or the latest sampled data. The error between the recently released signal and the new sampled signal may be small, making some important newly sampled data easy to be dropped. To handle such problems, in this subsection, we developed an AMD-based event-triggered scheme

$$t_{k+1}h = t_k h + \min_{l\in\mathbb{N}}\{lh|\alpha e_{di}(t) + \sum_{p=1}^{m} \varepsilon e_{ti}^T(t)\Phi e_{ti}(t) < \rho(t_k h)\hat{x}_i^T(t_k h)\Phi\hat{x}_i(t_k h)\}, \tag{6.8}$$

where $\alpha \in (0,1)$, P_i is defined as the Lyapunov variable matrix, $e_{di}(t) = 2x_i^{\mathrm{T}}(t)P_i\dot{x}_i(t)$, ε is the weighting coefficient satisfying $\sum_{p=1}^{m} \varepsilon = 1$, Φ is an appropriate positive definite matrix, $e_{ti}(t) = x_i(t_{k-p+1}h) - x_i(t_k h + lh)$, $l \in \mathbb{N}$, $\hat{x}_i(t_k h) = \frac{1}{m}\sum_{p=1}^{m} x_i(t_{k-p+1}h)$, p is the number of historical information and the details of the weighting matrix Φ will be given later. The adaptive term $\rho(t_k h)$ is set as the following rule

$$\rho(t_{k+1}h) = \max\{\rho(0), \gamma\rho(t_k h)\}, \tag{6.9}$$

where $\rho(0)$ is the initial value of $\rho(t_k h)$, parameter γ is defined as

$$\gamma = \begin{cases} 0, & \text{for} ||x_i(t_k h)|| \leq ||x_i(t_{k+1}h)||, \\ \kappa\left(1 - \frac{2}{\pi}\arctan\left(\frac{||x_i(t_{k+1}h)|| - ||x_i(t_k h)||}{||x_i(t_{k+1}h)||}\right)\right), & \text{otherwise}, \end{cases}$$

where κ is the given parameter.

In real-life control systems, there always exist communication delays between the power plant and controller [21], which are represented as τ_{t_k}. The control signal is remained in $[t_k h + \tau_{t_k}, t_k h + \tau_{t_{k+1}}] \triangleq \mathfrak{T}$. Using the virtual partition method, we obtain $\mathfrak{T} = \cup_{l=0}^{l=t_{k+1}-t_k-1}\mathfrak{T}_l$, $\mathfrak{T}_l = [t_kh + lh + \delta_{t_k}, t_kh + (l+1)h + \delta_{t_{k+1}})$, where

$$\delta_{t_k} = \begin{cases} \tau_{t_k}, l = 0, 1, \cdots, t_{k+1} - t_k - 2, \\ \tau_{t_{k+1}}, l = t_{k+1} - t_k - 1, \end{cases}$$

$0 \leq \tau_{t_k} \leq \tau(t) \leq h + \bar{\tau} = \tau_M$, where τ_M is the upper bound of the time-varying delay [36]. The system stops working when the delay is τ_M.

Remark 6.2: Compared with some existing normal event-trigger conditions, which typically only make use of the current system state [12], the presented AMD-based event-triggered scheme makes use of the historic data, which provides a more reasonable triggering condition, helping to lower the consumption of network resources [13].

Remark 6.3: In the AMD-based event-triggered scheme, m denotes the number of the historically transmitted packets. It can be seen that when $m = 1$, the proposed scheme reduces to the traditional adaptive-derivative-based event-trigger. Different from the AMET scheme [15], the event-triggered condition is designed using a Lyapunov-based derivative term $V(t) = x_i^\top(t)P_ix(t)$. This newly added Lyapunov function ensures the stable state of the event-triggered control system. Besides, the proposed triggering rule (6.8) is more relaxed compared to the triggering rule in [15].

6.2.5 Frequency-based Deception Attacks

As a case study on the effects of contamination in the shared network channels, deception attacks are assumed to happen in interconnected wind power systems. We developed a frequency-based deception attack model, which contained the frequency deviation of the multi-area wind power system.

We assume that the system state $x_Q(t_{k-p+1}h)$ can be caught by the attacker in a certain way and be replaced by the attack signal $x_a(t_{k-p+1}h)$. Based on **Assumption 4,** the attack signal consists of the system state $x_Q(t_{k-p+1}h)$ and interference signal $f_a(x_Q(t_{k-p+1}h))$, and is supposed to fulfill

$$\begin{aligned} x_a(t_{k-p+1}h) &= f_a(x_Q(t_{k-p+1}h)) + x_Q(t_{k-p+1}h), \\ ||f_a(x_Q(t_{k-p+1}h))||_2 &\leq ||G_ix_i(t)||_2, \end{aligned} \tag{6.10}$$

where G_i is a constant matrix.

The occurring status of the deception attacks is assumed to be as follows

$$\hat{x}_a(t_{k-p+1}h) = \begin{cases} x_a(t_{k-p+1}h), & |\Delta\bar{f}_i| < \Delta f_A, \\ x_Q(t_{k-p+1}h), & \text{otherwise}, \end{cases} \tag{6.11}$$

where $\Delta f_A > 0$ is a given constant, and $\Delta\bar{f}_i$ is the deviation between real-time frequency f_s and normal operation frequency f_i.

In Figure 6.1, the sensor is used by the attackers to get the frequency of the wind power system. Before the frequency-based deception attacks contaminate the wind power system, the frequency f_i at normal operation was got. Then the attacker can patch the deception attack signal into the wind power system. When deviation $|\Delta\bar{f}_i| < \Delta f_A$ is satisfied, the attack happens, and system state $x_i(t)$ is replaced by attack signal $x_a(t)$. An example is provided in Figure 6.4 to show the characteristics of frequency-based deception attacks.

Based on the occurrence condition of frequency-based deception attacks $|\Delta\bar{f}_i| < \Delta f_A$, the system frequency can be divided into two areas: one is close to the stable state, the other is far away from the stable state. The closer area is defined as the stable region from the perspective of the attacker rather than the definition of a stable state in control performance. Based on our current understanding of cyberattacks, deception attacks is aiming at destroying the stable performance of power systems. From the perspective of the attacker, the stable region will be the target. Therefore, the frequency-based deception attacks region is equal to the stable region which is defined based on the threshold Δf_A.

Remark 6.4: In this chapter, there exist two definitions for frequency deviation. From the perspective of attackers, attack frequency deviation $\Delta\bar{f}_i$ is used to design the frequency-based deception attack model. From the perspective of the control strategy, frequency deviation Δf_i is applied to the LFC power system stability analysis.

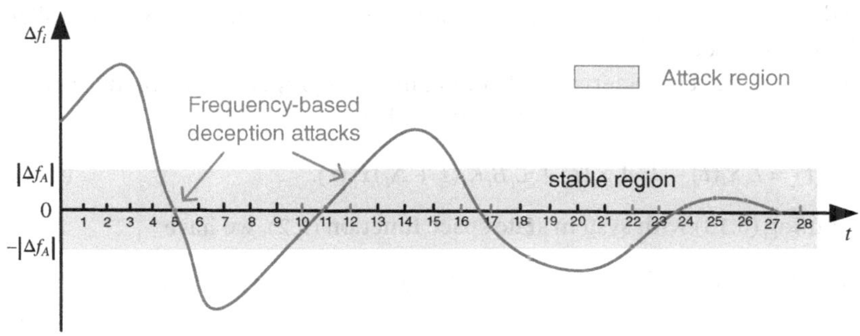

Figure 6.4 Example of frequency-based deception attacks.

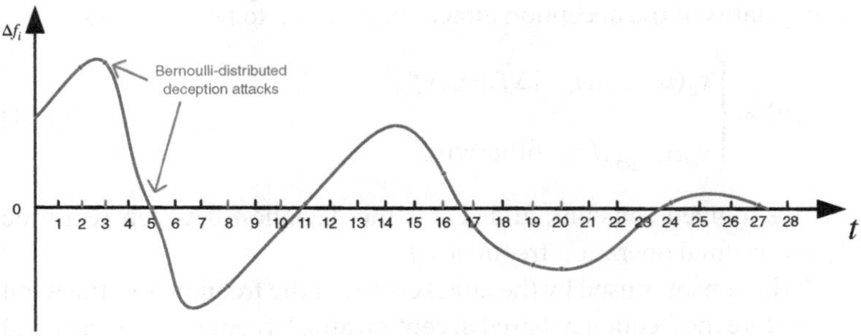

Figure 6.5 Example of Bernoulli distributed deception attacks.

The Bernoulli-distributed deception attacks [24–26] are shown in Figure 6.5. The deception attack signal is injected into the transmitted signal based on the given probability.

Remark 6.5: The attack mode with the Bernoulli distribution is presented in Figure 6.5, which means that the Bernoulli-distributed deception attacks randomly release the aggressive signal. In the previous studies [24–26], it was assumed that Bernoulli distributed attacks may happen at any time. However, frequency-based deception attacks only contaminate the transmitted signal that appears in the stable region. In fact, deception attacks occurring in unstable regions cannot give the attacker the benefit of worsening the system performance in stability. The proposed frequency-based deception attacks can make more significant impacts on practical power systems than Bernoulli-distributed deception attacks.

We then consider the impact of deception attacks. The quantized sliding mode surface function is given as

$$\sigma_i(t) = L_i x_i(t) - L_i \int_0^t \tilde{A}_i x_i(s) ds + L_i \int_0^t \tilde{B}_i K_i (I + \Delta_i) \hat{x}_a(s) ds, \tag{6.12}$$

where the matrix L_i is selected to guarantee the nonsingularity of $L_i \tilde{B}_i$. Controller gain K_i will be discussed later.

According to the characteristics of sliding mode, a proper sliding mode surface should satisfy $\dot{\sigma}_i(t) = 0$ and $\sigma_i(t) = 0$. Thus we have

$$\dot{\sigma}_i(t) = L_i \dot{x}_i(t) - L_i \tilde{A}_i x_i(t) + L_i \tilde{B}_i K_i (I + \Delta_i) \hat{x}_a(t). \tag{6.13}$$

Combining (6.13) with system state space function (6.2), we have

$$\begin{aligned} 0 &= L_i [\tilde{A}_i x_i(t) + \tilde{B}_i u_i(t) + \tilde{w}_i(t)] - L_i \tilde{A}_i x_i(t) + L_i \tilde{B}_i K_i (I + \Delta_i) \hat{x}_a(t) \\ &= L_i \tilde{B}_i u_i(t) + L_i \tilde{w}_i(t) + L_i \tilde{B}_i K_i (I + \Delta_i) \hat{x}_a(t). \end{aligned} \tag{6.14}$$

Further, we have

$$L_i\tilde{B}_iu_i(t) = -L_i\tilde{B}_iK_i(I + \Delta_i)\hat{x}_a(t) - L_i\tilde{\omega}_i(t). \tag{6.15}$$

Therefore, the equivalent SMC law is obtained as

$$u_{\text{eqi}}(t) = -K_i(I + \Delta_i)\hat{x}_a(t) - (L_i\tilde{B}_i)^{-1}L_i\tilde{\omega}_i(t). \tag{6.16}$$

Considering the AMD-based event-triggered scheme (6.8), signal quantization, and the equivalent SMC law (6.16), the system state space function under the frequency-based deception attacks can be obtained

$$\begin{cases} \dot{x}_i(t) = \tilde{A}_ix_i(t) - \sum_{p=1}^{m}\tilde{B}_iK_i(I + \Delta_i)\hat{x}_a(t_{k-p-1}h) + \hat{F}_i\tilde{\omega}_i(t), \\ y_i(t) = C_ix_i(t), \end{cases} \tag{6.17}$$

where $\hat{F}_i = (L_i\tilde{B}_i)^{-1}L_i$.

6.3 Main Results

This section presents the reliable performance analysis of the power system (6.17). Significantly, the reliable AMD event-triggered QSMLFC schemes are designed to stabilize the $i-th$ area wind power systems according to (i) reliable region and (ii) reliability condition, respectively. Before starting on addressing the problem, the following preparations are given.

Assumption 3 The error $e_i(t)$ in the AMD event-trigger scheme satisfies $||e_i(t)|| \leq x_{\max}$, where $x_{\max} = \max\{||x_i(t)||, i = 1, \cdots, N\}$.

Assumption 4 The deception attack signal is assumed as $||f_a(x_i(t))||_2 \leq ||\tilde{G}_ix_i(t)||_2$. Considering event-triggered device and signal quantization, we have the following

$$||f_a(x_Q(t_{k-p+1}h))||_2 \leq ||G_ix_i(t)||_2, \tag{6.18}$$

where $\tilde{G}_i$ and G_i are constant matrices.

Lemma 6.1 [33] *For $x(t) \in R^n$ and positive constant $\upsilon \in R$, there exist $P > 0$ and matrix U with appropriate dimensions, such that the following inequality holds*

$$-\upsilon\int_{t-\upsilon}^{t}\dot{x}^{\top}(s)P\dot{x}(s)ds \leq \begin{bmatrix} x(t) \\ x(t-\upsilon) \end{bmatrix}^{\top}\begin{bmatrix} -P & U^{\top} \\ U & -P \end{bmatrix}\begin{bmatrix} x(t) \\ x(t-\upsilon) \end{bmatrix}.$$

Lemma 6.2 [33] *For $x(t) \in R^n$ and positive constant $\upsilon \in R$, there exists $P > 0$, such that the following inequalities hold*

$$-\upsilon \int_{t-\upsilon}^{t} x^{\mathsf{T}}(s)Px(s)ds \leq -\int_{t-\upsilon}^{t} x^{\mathsf{T}}(s)dsP \int_{t-\upsilon}^{t} x(s)ds,$$

$$-\frac{\upsilon^2}{2} \int_{-\upsilon}^{0} \int_{t+\beta}^{\beta} x^{\mathsf{T}}(s)Px(s)dsd\beta \leq -\int_{-\upsilon}^{0} \int_{t+\beta}^{\beta} x^{\mathsf{T}}(s)dsd\beta P \int_{-\upsilon}^{0} \int_{t+\beta}^{\beta} x(s)dsd\beta.$$

Definition 6.1 (Reliable region) [10] The $i-th$ area wind power system is stable for $x_i(t)$ located in $\bar{S}(\upsilon)$, and $\bar{S}(\upsilon)$ is the complement of closed ball $S(\upsilon)$, if there exist $P_i > 0$ and $Q_i > 0$, such that the radius for closed ball $S(\upsilon)$ centered at $x_i(t) = 0$ satisfies the following

$$\upsilon = \frac{\alpha_i \|P_i\|}{\lambda_{\min}(Q_i)},$$

where α_i is a positive constant, and the eigenvalue of Q_i is $\lambda_{\min}(Q_i)$. Then we term $\bar{S}(\upsilon)$ as a reliable region.

6.3.1 Reliable Region

The reliable regions of the $i-th$ area wind power system is explained in the following.

Theorem 6.1 *For given constants $\tilde{a}_i > 0$, $\rho(0) > 0$, $\varepsilon > 0$, $\alpha \in (0, 1)$, and matrix K_i, the $i-th$ area wind power system (6.17) is regarded as reliable with AMD event-triggered QSMLFC for $x_i(t) \in \bar{S}(\nu)$, if there exist $P_i > 0$, $Q_i > 0$, and matrix G_i with appropriate dimensions, such that the radius for closed ball $S(\upsilon)$ centered at $x_i(t) = 0$ satisfies the following*

$$\nu = \frac{\Xi}{(1-\alpha)\lambda_{\min}(Q_i)}, \tag{6.19}$$

where $\Xi = 2(1-\alpha)\tilde{a}_i\|P_i\| - 4(1-\alpha)\|B\|\|P_i\|x_{\max} - (\rho(0)-\varepsilon)\|\Phi\|x_{\max} - 2\|G_i\| \|P_i\|x_{\max}$, $B = (I + \Delta_i) \sum_{p=1}^{m} \tilde{B}_i K_i$.

Proof. The Lyapunov candidate function is chosen as

$$V(t) = x_i^{\mathsf{T}}(t)P_i x_i(t). \tag{6.20}$$

According to the derivation on $V(t)$, it follows that

$$\begin{aligned}\dot{V}(t) &= 2\dot{x}_i^{\mathsf{T}}(t)P_i x_i(t)\\ &= x_i^{\mathsf{T}}(t)(\tilde{A}_i^{\mathsf{T}}P_i + P_i\tilde{A}_i)x_i(t) - \hat{x}_a^{\mathsf{T}}(t_{k-p+1}h)(B_iK_i)^{\mathsf{T}}P_i x_i(t)\\ &\quad - x_i^{\mathsf{T}}(t)P_i(B_iK_i)\hat{x}_a(t_{k-p+1}h)\\ &\quad + (\hat{F}_i\tilde{\omega}_i(t))^{\mathsf{T}}P_i x_i(t) + x_i^{\mathsf{T}}(t)P_i(\hat{F}_i\tilde{\omega}_i(t)).\end{aligned} \tag{6.21}$$

Combining the quantification and AMD event-trigger device, the following can be derived

$$\begin{aligned}\dot{V}(t) &< x_i^{\mathsf{T}}(t)(\tilde{A}_i^{\mathsf{T}}P_i + P_i\tilde{A}_i)x_i(t) - x_i^{\mathsf{T}}(t-\tau(t))(\tilde{B}_iK_i)^{\mathsf{T}}(\Delta_i + I)P_i x_i(t)\\ &\quad - x_i^{\mathsf{T}}(t)P_i(\tilde{B}_iK_i)(\Delta_i + I)x_i(t-\tau(t)) - e_{ti}^{\mathsf{T}}(t)(\tilde{B}_iK_i)^{\mathsf{T}}(\Delta_i + I)P_i x_i(t)\\ &\quad - x_i^{\mathsf{T}}(t)P_i(\tilde{B}_iK_i)(\Delta_i + I)e_{ti}(t) - f_a(x_i(t))^{\mathsf{T}}P_i x_i(t) - x_i^{\mathsf{T}}(t)P_i f_a(x_i(t))\\ &\quad + (\hat{F}_i\tilde{\omega}_i(t))^{\mathsf{T}}P_i x_i(t) + x_i^{\mathsf{T}}(t)P_i(\hat{F}_i\tilde{\omega}_i(t))\\ &\quad - 2\alpha x_i^{\mathsf{T}}(t)P_i\dot{x}_i(t) - \varepsilon e_{ti}^{T}(t)\Phi e_{ti}(t) + \rho(t)\hat{x}_i^{T}(t)\Phi\hat{x}_i(t).\end{aligned} \tag{6.22}$$

Focusing on $\rho(t)\hat{x}_i^T(t)\Phi\hat{x}_i(t)$, one has

$$\rho(0)\hat{x}_i^T(t)\Phi\hat{x}_i(t) \le \rho(0)\|\Phi\|\|x_i(t)\|x_{\max}. \tag{6.23}$$

By using **Assumption 3** and (6.10), the following is got

$$\begin{aligned}&\|x_i(t-\tau(t))\| \le x_{\max}, \|e_i(t)\| \le x_{\max},\\ &\hat{F}_i\tilde{\omega}_i(t) \le \tilde{a}_i = \|\hat{F}_i\|d_i, \|f_a(x_i(t))\| \le \|G_i\|x_{\max}.\end{aligned} \tag{6.24}$$

Thus, it yields that

$$\begin{aligned}\dot{V}(t) &\le (1-\alpha)\big[-\lambda_{\min}(Q_i)2\|x_i(t)\|^2 + 2\tilde{a}_i\|P_i\|\|x_i(t)\|\\ &\quad - 4\|B\|\|P_i\|\|x_i(t)\|x_{\max}\big] - 2\|G_i\|\|P_i\|\|x_i(t)\|x_{\max}\\ &\quad + (\rho(0)-\varepsilon)\|\Phi\|\|x_i(t)\|x_{\max}.\end{aligned} \tag{6.25}$$

Further, we have

$$\begin{aligned}
&(1-\alpha)[-\lambda_{\min}(Q_i)2\|x_i(t)\|^2+2\tilde{a}_i\|P_i\|\|x_i(t)\| \\
&-4\|B\|\|P_i\|\|x_i(t)\|x_{\max}]-2\|G_i\|\|P_i\|\|x_i(t)\|x_{\max} \\
&+(\rho(0)-\varepsilon)\|\Phi\|\|x_i(t)\|x_{\max} \\
&=\|x_i(t)\|[-2(1-\alpha)\lambda_{\min}(Q_i)\|x_i(t)\|+2(1-\alpha)\tilde{a}_i\|P_i\| \\
&-4(1-\alpha)\|B\|\|P_i\|x_{\max}-2\|G_i\|\|P_i\|x_{\max} \\
&+(\rho(0)-\varepsilon)\|\Phi\|\|x_i(t)\|x_{\max}]
\end{aligned} \tag{6.26}$$

Defining υ as the radius of closed ball $S(\upsilon)$, (6.25) can be rewritten as

$$\dot{V}(t)\leq\|x_i(t)\|(-\|x_i(t)\|+\upsilon) \tag{6.27}$$

wherein $\upsilon=\dfrac{\Xi}{(1-\alpha)\lambda_{\min}(Q_i)}$, $\Xi=2(1-\alpha)\tilde{a}_i\|P_i\|-4(1-\alpha)\|B\|\|P_i\|x_{\max}-(\rho(0)-\varepsilon)\|\Phi\|x_{\max}-2\|G_i\|\|P_i\|x_{\max}$, $B=(I+\Delta_i)\sum_{p=1}^{m}\tilde{B}_iK_i$.

It can be seen that $\dot{V}(t)\leq 0$ is established if $(-\|x_i(t)\|+\upsilon)\leq 0$. Thus, we can conclude that the wind power system is stable if the state trajectory is located in $\bar{S}(\nu)$, where $\bar{S}(\nu)$ is the complement of $S(\nu)$. Based on **Definition 6.1**, the wind power system is reliable. This completes the proof.

Theorem 6.1 proves that the wind power system is reliable when the system state locates in the reliable region $\bar{S}(\nu)$.

Remark 6.6: **Theorem 6.1** provides a method for identifying a reliable region of the power system. With the pole assignment, the controller gain can be obtained, which makes the eigenvalues of the system matrix of (6.17). The results obtained, however, are highly conservative. We shall provide less conservative results in the next subsection.

6.3.2 Reliability Analysis

The following theorem is to give the reliable condition of the system in Figure 6.1. The control gain is calculated by the proposed linear matrix inequality.

Theorem 6.2 *For given scalars $\tau_M>0$, $m>0$, $\rho(t)>0$, $\varepsilon>0$, $\bar{\gamma}_i>0$, $\alpha\in(0,1)$ and matrix K_i, the wind power system (6.17) is reliable under zero initial condition if there exist $Q_1>0$, $Q_2>0$, $Q_3>0$, $\Phi>0$, $P>0$, $G_i>0$, and matrices U, Δ_i with appropriate dimensions, such that the following inequalities hold*

$$\tilde{J} = \begin{bmatrix} \chi + C_i^{\top} C_i & \hat{F} \\ * & -\bar{\gamma}_i I \end{bmatrix} \prec 0, \tag{6.28}$$

$$\begin{bmatrix} Q_2 & U^{\top} \\ U & Q_2 \end{bmatrix} \succ 0, \tag{6.29}$$

where

$$\begin{aligned}
\chi &= \mathrm{sym}\{\Psi_1^{\top} P \Psi_2\} + \varsigma_1^{\top} Q_1 \varsigma_1 - \varsigma_3^{\top} Q_2 \varsigma_3 + \tau_M^2 o^{\top} Q_2 o \\
&+ \frac{\tau_M^4}{4} o^{\top} Q_3 o - \varsigma_5^{\top} P \varsigma_5 + \varsigma_1 G_i^{\top} P G_i \varsigma_1 - \rho \varsigma_2^{\top} \Phi \varsigma_2 \\
&+ \frac{\rho(t)}{m} \varsigma_2^{\top} \Phi \varsigma_4 + \frac{\rho(t)}{m} \varsigma_4^{\top} \Phi \varsigma_2 - (\varepsilon - \frac{\rho(t)}{m^2}) \varsigma_4 \Phi \varsigma_4 \\
&- \begin{bmatrix} \eta_1 \\ \eta_2 \end{bmatrix}^{\top} \begin{bmatrix} Q_2 & U^{\top} \\ U & Q_2 \end{bmatrix} \begin{bmatrix} \eta_1 \\ \eta_2 \end{bmatrix} - \frac{\tau_M^2}{2} \eta_3 Q_3 \eta_3^{\top},
\end{aligned}$$

$$\Psi_1 = [(1-\alpha)\varsigma_1^{\top}, \varsigma_6^{\top}]^{\top}, \Psi_2 = [o^{\top}, \varsigma_1^{\top} - \varsigma_3^{\top}]^{\top},$$

$$o = \tilde{A}_i \varsigma_1 - \check{B} \varsigma_2 - \check{B} \varsigma_4 - \check{B} \varsigma_5, \check{B} = \tilde{B}_i K_i (I + \Delta_i),$$

$$\eta_1 = x_i(t) - x_i(t - \tau(t)), \eta_2 = x_i(t - \tau(t)) - x_i(t - \tau_M),$$

$$\eta_3 = x_i(t) - \int_{t-\tau_M}^{t} x_i(s) ds.$$

Proof. Construct the Lyapunov–Krasovskii candidate function like

$$V(t) = V_1(t) + V_2(t) + V_3(t) + V_4(t), \tag{6.30}$$

where

$$\begin{aligned}
V_1(t) &= x_i^{\top}(t) P x_i(t) + \int_{t-\tau_M}^{t} x_i^{\top}(s) ds P \int_{t-\tau_M}^{t} x_i(s) ds, \\
V_2(t) &= \int_{t-\tau_M}^{t} x_i^{\top}(s) Q_1 x_i(s) ds, \\
V_3(t) &= \tau_M \int_{-\tau_M}^{t} \int_{t+u}^{t} \dot{x}_i^{\top}(s) Q_2 \dot{x}_i(s) ds du, \\
V_4(t) &= \frac{\tau_M^2}{2} \int_{-\tau_M}^{t} \int_{u}^{0} \int_{t+v}^{t} \dot{x}_i^{\top}(s) Q_3 \dot{x}_i(s) ds du dv.
\end{aligned}$$

Taking the derivation of $V(t)$ with system (6.17) yields

$$\begin{aligned}\dot{V}(t) &= \mathrm{sym}\{x_i^{\mathsf{T}}(t)P\dot{x}_i(t)\}\\&+\mathrm{sym}\{(x_i^{\mathsf{T}}(t)-x_i^{\mathsf{T}}(t-\tau_M))\int_{t-\tau_M}^{t} x_i(s)\}\\&+x_i^{\mathsf{T}}(t)Q_1x_i(t)-x_i^{\mathsf{T}}(t-\tau_M)Q_1x_i(t-\tau_M)\\&+\tau_M^2\dot{x}_i^{\mathsf{T}}(t)Q_2\dot{x}_i(t)-\tau_M\int_{t-\tau_M}^{t}\dot{x}_i^{\mathsf{T}}(s)Q_2\dot{x}_i(s)ds\\&+\frac{\tau_M^4}{4}\dot{x}_i^{\mathsf{T}}(t)Q_3\dot{x}_i(t)-\frac{\tau_M^2}{2}\int_{-\tau_M}^{t}\int_{t+u}^{t}\dot{x}_i^{\mathsf{T}}(s)Q_3\dot{x}_i(s)dsdu.\end{aligned} \tag{6.31}$$

According to $e_{ti}(t) = x_i(t_k + lh) - x_i(t_{k-i+1}h)$, we have $x_i(t_{k-i+1}h) = x_i(t-\tau(t)) - e_{ti}(t)$. Then, for $t \in \mathfrak{T}_l$, we have

$$\alpha e_{di}(t)+\sum_{p=1}^{m}\varepsilon e_{ti}^T(t)\Phi e_{ti}(t) < \rho(t_kh)\hat{x}_i^T(t_kh)\Phi\hat{x}_i(t_kh), \tag{6.32}$$

where matrix Φ is designed later, $\rho(t_kh)$ stands for the adaptive parameter.

Considering the frequency-based deception attacks, a symmetric positive definite matrix P exists, making the following inequality holds

$$f_a^{\mathsf{T}}(x_i(t))Pf_a(x_i(t)) - x^{\mathsf{T}}(t)G_i^{\mathsf{T}}PG_ix(t) \le 0. \tag{6.33}$$

Combining Equations (6.31), (6.32), and (6.33), we obtain

$$\begin{aligned}\dot{V}(t) &< 2x_i^{\mathsf{T}}(t)P\dot{x}_i(t)-2\alpha x_i^{\mathsf{T}}(t)P\dot{x}_i(t)-\varepsilon e_{ti}^T(t)\Phi e_{ti}(t)+\rho(t)\hat{x}_i^T(t)\Phi\hat{x}_i(t)\\&+\mathrm{sym}\{(x_i^{\mathsf{T}}(t)-x_i^{\mathsf{T}}(t-\tau_M))\int_{t-\tau_M}^{t} x_i(s)\}\\&+x_i^{\mathsf{T}}(t)Q_1x_i(t)-x_i^{\mathsf{T}}(t-\tau_M)Q_1x_i(t-\tau_M)\\&+\tau_M^2\dot{x}_i^{\mathsf{T}}(t)Q_2\dot{x}_i(t)-\tau_M\int_{t-\tau_M}^{t}\dot{x}_i^{\mathsf{T}}(s)Q_2\dot{x}(s)ds\\&+\frac{\tau_M^4}{4}\dot{x}_i^{\mathsf{T}}(t)Q_3\dot{x}_i(t)-\frac{\tau_M^2}{2}\int_{t-\tau_M}^{t}\int_{u}^{t}\dot{x}_i^{\mathsf{T}}(s)Q_3\dot{x}(s)dsdu\\&-f_a^{\mathsf{T}}(x_i(t))Pf_a(x_i(t))+x_i^{\mathsf{T}}(t)G_i^{\mathsf{T}}PG_ix_i(t).\end{aligned} \tag{6.34}$$

Noting that

$$
\begin{aligned}
&\rho(t)\hat{x}_i^\top(t)\Phi\hat{x}_i(t) \\
&= \rho(t)[x_i(t-\tau(t)) - \frac{1}{m}e_{ti}(t)]^\top\Phi[x_i(t-\tau(t)) - \frac{1}{m}e_{ti}(t)] \\
&= \rho(t)[x_i(t-\tau(t))^\top\Phi x_i(t-\tau(t)) - \frac{1}{m}x_i(t-\tau(t))^\top\Phi e_{ti}(t) \\
&\quad - \frac{1}{m}e_{ti}(t)^\top\Phi x_i(t-\tau(t)) + \frac{1}{m^2}e_{ti}(t)^\top\Phi e_{ti}(t)].
\end{aligned} \tag{6.35}
$$

The following inequality is obtained

$$
\begin{aligned}
\dot{V}(t) &< \mathrm{sym}[(1-\alpha)[x_i^\top(t)A^\top Px_i(t) - e_{ti}^\top\check{B}^\top Px_i(t) \\
&\quad - x_i^\top(t-\tau(t))\check{B}^\top Px_i(t) - f_a(x_i(t))^\top\check{B}^\top Px_i(t)] \\
&\quad + x_i^\top(t)P\int_{t-\tau_M}^{t} x_i(s)ds - x_i(t-\tau_M)^\top P\int_{t-\tau_M}^{t} x_i(s)ds] \\
&\quad - \varepsilon e_{ti}^T(t)\Phi e_{ti}(t) + \rho(t)[x_i(t-\tau(t))^\top\Phi x_i(t-\tau(t)) \\
&\quad - \frac{1}{m}x_i(t-\tau(t))^\top\Phi e_{ti}(t) - \frac{1}{m}e_{ti}(t)^\top\Phi x_i(t-\tau(t)) \\
&\quad + \frac{1}{m^2}e_{ti}(t)^\top\Phi e_{ti}(t)] + x_i^\top(t)Q_1x_i(t) \\
&\quad - x_i^\top(t-\tau_M)Q_1x_i(t-\tau_M) + \tau_M^2\dot{x}_i^\top(t)Q_2\dot{x}_i(t) \\
&\quad - \tau_M\int_{t-\tau_M}^{t}\dot{x}_i(s)Q_2\dot{x}_i(s)ds + \frac{\tau_M^4}{4}\dot{x}_i^\top(t)Q_3\dot{x}_i(t) \\
&\quad - \frac{\tau_M^2}{2}\int_{t-\tau_M}^{t}\int_{t+u}^{t}\dot{x}_i^\top(s)Q_3\dot{x}_i(s)dsdu \\
&\quad - f_a^\top(x_i(t))Pf_a(x_i(t)) + x_i^\top(t)G_i^\top PG_ix_i(t).
\end{aligned} \tag{6.36}
$$

By defining $\varsigma_{\tilde{i}} = [0_{n\times(\tilde{i}-1)n}, I_n, 0_{n\times(8-\tilde{i})n}] \in \mathbb{R}^{n\times 3n}$, for $\tilde{i} = 1, 2, \cdots, 8$, we have

$$
\dot{x}_i^\top(t)Q_2\dot{x}_i(t) = \Theta^\top(t)o^\top Q_2 o\Theta(t), \tag{6.37}
$$

where $o = \tilde{A}_i\varsigma_1 - \check{B}\varsigma_2 - \check{B}\varsigma_4 - \check{B}\varsigma_5$, $\check{B} = \tilde{B}_iK_i(I+\Delta_i)$,

$$
\begin{aligned}
\Theta = [&x_i^\top(t), x_i^\top(t-\tau(t)), x_i^\top(t-\tau_M), e_{ti}^\top(t), f_a^\top(x_i(t)), \\
&\int_{t-\tau_M}^{t} x_i^\top(s)ds]^\top.
\end{aligned} \tag{6.38}
$$

Applying **Lemma 6.1**, we have

$$-\tau_M \int_{t-\tau_M}^{t} \dot{x}_i^{\mathrm{T}}(s) Q_2 \dot{x}_i(s) ds \leq \begin{bmatrix} \eta_1 \\ \eta_2 \end{bmatrix}^{\mathrm{T}} \begin{bmatrix} Q_2 & U^{\mathrm{T}} \\ U & Q_2 \end{bmatrix} \begin{bmatrix} \eta_1 \\ \eta_2 \end{bmatrix}, \tag{6.39}$$

where $\eta_1 = x_i(t) - x_i(t-\tau(t)), \eta_2 = x_i(t-\tau(t)) - x_i(t-\tau_M)$.

Applying **Lemma 6.2**, we have

$$-\frac{\tau_M^4}{4} \int_{t-\tau_M}^{t} \int_{t+u}^{t} \dot{x}_i^{\mathrm{T}}(s) Q_3 \dot{x}_i(s) ds du \leq -\frac{\tau_M^2}{2} \eta_3 Q_3 \eta_3^{\mathrm{T}}, \tag{6.40}$$

where $\eta_3 = x_i(t) - \int_{t-\tau_M}^{t} x_i(s)ds$.

To analyze the condition with $\tilde{\omega}_i(t) \neq 0$, we define the attenuation level $\bar{\gamma}_i > 0$, then

$$\dot{V}(t) + y_i^{\mathrm{T}}(t) y_i(t) - \bar{\gamma}_i^2 \tilde{\omega}_i(t)^{\mathrm{T}} \tilde{\omega}_i(t) \leq y_i^{\mathrm{T}}(t) y_i(t) - \bar{\gamma}_i^2 \tilde{\omega}_i(t)^{\mathrm{T}} \tilde{\omega}_i(t). \tag{6.41}$$

The following inequality can be obtained under initial condition $V(0) = 0$,

$$\tilde{\Theta}^{\mathrm{T}}(t) \tilde{J} \tilde{\Theta}(t) \leq -y_i^{\mathrm{T}}(t) y_i(t) + \bar{\gamma}_i^2 \tilde{\omega}_i(t) \tilde{\omega}_i^{\mathrm{T}}(t) \tag{6.42}$$

where

$$\begin{aligned} \tilde{\Theta}(t) = [& x_i^{\mathrm{T}}(t), x_i^{\mathrm{T}}(t-\tau(t)), x_i^{\mathrm{T}}(t-\tau_M), e_{ti}^{\mathrm{T}}(t), f_a^{\mathrm{T}}(x_i(t)), \\ & \int_{t-\tau_M}^{t} x_i^{\mathrm{T}}(s) ds, \tilde{\omega}_i^{\mathrm{T}}(t)]^{\mathrm{T}}. \end{aligned} \tag{6.43}$$

Under zero initial conditions, the following inequality can be obtained

$$\int_0^{\infty} y_i^{\mathrm{T}}(s) y_i(s) ds \leq \int_0^{\infty} \bar{\gamma}_i^2 \tilde{\omega}_i(s) \tilde{\omega}_i^{\mathrm{T}}(s) ds \tag{6.44}$$

that is $\|y_i(t)\|_2 \leq \bar{\gamma} \|\tilde{\omega}_i(t)\|_2$.

Combining the results, we have $\dot{V}(t) \leq 0$ for $\tilde{J} < 0$. Thus, the desired reliable AMD event-triggered QSMLFC guarantees the stability of the $i-th$ wind power subsystem. The proof is completed.

Theorem 6.2 describes the reliable performance of the wind power system (6.17), but it is difficult to directly solve the controller gain K_i. Thus, we further convert **Theorem 6.2** into **Corollary 6.1** to obtain the controller gain.

Remark 6.7: In **Theorem 6.1**, we construct multiple Lyapunov–Krasovskii functions to prove the reliable condition for the $i-th$ wind power subsystem under frequency-based deception attacks. Compared with a single Lyapunov–Krasovskii function, the proposed function leads to a less conservative result.

Corollary 6.1 *For given scalars $\tau_M > 0$, $\rho(t) > 0$, $m > 0$, $\bar{\gamma}_i > 0$, $\varepsilon > 0$, $\bar{c} > 0$, and $\alpha \in (0,1)$, the wind power system (6.17) is reliable under zero initial condition, if there exist $X_i > 0$, $\tilde{Q}_1 > 0$, $\tilde{Q}_2 > 0$, $\tilde{Q}_3 > 0$, $\Phi > 0$, $G_i > 0$, and matrices Δ_i, $\tilde{U}$, Y_i with appropriate dimensions, such that the following linear matrix inequalities (LMIs) hold*

$$\hat{\tilde{J}} = [a_{\tilde{m}*\tilde{n}}] \prec 0, \tag{6.45}$$

$$\begin{bmatrix} \tilde{Q}_2 & \tilde{U}^\top \\ \tilde{U} & \tilde{Q}_2 \end{bmatrix} \succ 0, \tag{6.46}$$

where

$a_{1,1} = (1-\alpha)A_iX_i + (1-\alpha)X_iA_i^\top + \tilde{Q}_1 - \tilde{Q}_2 - \tau_M^2\tilde{Q}_3$,
$a_{1,2} = -(1-\alpha) * (I+\Delta_i)BY + \tilde{Q}_2 - \tilde{U}$, $a_{1,3} = \tilde{U}$,
$a_{1,4} = -(1-\alpha)(I+\Delta_i)B_iY_i$, $a_{1,5} = -(1-\alpha)(I+\Delta_i)B_iY_i$
$a_{1,6} = X + \tau_M\tilde{Q}_3$, $a_{1,7} = \hat{F}$, $a_{1,8} = \tau_M A_iX_i$,
$a_{1,9} = \frac{\tau_M^2}{2}A_iX_i$ $a_{1,10} = X_iC_i$ $a_{1,11} = mG_iX_i$,
$a_{2,2} = \rho(t)\tilde{\Phi} - 2\tilde{Q}_2 + \tilde{U} + \tilde{U}^\top$, $a_{2,3} = -\tilde{Q}_2 - \tilde{U}$,
$a_{2,4} = -\frac{\rho(t)}{m}\tilde{\Phi}$, $a_{2,8} = -\tau_M(I+\Delta_i)B_iY_i$,
$a_{2,9} = -\frac{\tau_M^2}{2}(I+\Delta_i)B_iY_i$, $a_{3,3} = -\tilde{Q}_1 - \tilde{Q}_2$, $a_{3,6} = -X_i$,
$a_{4,4} = -\varepsilon\tilde{\Phi} + \frac{\rho(t)}{m^2}$, $a_{4,8} = -\tau_M(I+\Delta_i)B_iY_i$,
$a_{4,9} = -\frac{\tau_M^2}{2}(I+\Delta_i)B_iY_i$, $a_{5,5} = -X_i$,
$a_{5,8} = -\tau_M(I+\Delta_i)B_iY_i$, $a_{5,9} = -\frac{\tau_M^2}{2}(I+\Delta_i)B_iY_i$,
$a_{6,6} = -\tilde{Q}_3$, $a_{7,7} = \bar{\gamma}_i$, $a_{8,8} = \bar{c}^2\tilde{Q}_2 - 2\bar{c}X_i$,
$a_{9,9} = \bar{c}^2\tilde{Q}_3 - 2\bar{c}X_i$, $a_{10,10} = -I$, $a_{11,11} = -X_i$,
$\tilde{m}, \tilde{n} = 1, 2, \ldots, 11$.
The controller gain is $K_i = Y_iX_i^{-1}$

Proof. Define $X_i = P^{-1}$, $\tilde{Q}_1 = X_iQ_1X_i$, $\tilde{\Phi}_i = X_i\Phi_iX_i$, $\tilde{U}_i = X_iU_iX_i$, by multiplying $\mathrm{diag}\{\underbrace{X_i, \ldots, X_i}_{7}, Q_2^{-1}, Q_3^{-1}, I, X_i\}$ on both sides and replace nonlinear term $K_iX_i = Y_i$, we can obtain (6.45). The controller gain can be calculated as $K_i = Y_iX_i^{-1}$. This completes the proof.

Remark 6.8: In $\tilde{J}$, there exist some complex items. To obtain the controller gains, we need to solve them first. By multiplying $\mathrm{diag}\{\underbrace{X_i, \ldots, X_i}_{7}, Q_2^{-1}, Q_3^{-1}, I, X_i\}$

on both sides of $\tilde{J}$, there will appear some cross terms like $X_iQ_1X_i, X_iU_iX_i, X_iQ_2^{-1}X_i$, $X_iQ_3^{-1}X_i$. By defining $\tilde{Q}_1 = X_1Q_1X_1$, $\tilde{U}_i = X_iU_iX_i$, and $\tilde{\Phi}_i = X_i\Phi_iX_i$, these cross terms are solvable. And then, by linear approach method, $X_iQ_2^{-1}X_i$ can be solved as $Q_2 - 2X_i$, for $(X_i - Q_2)Q_2^{-1}(X_i - Q_2) > 0$.

In the multi-area wind power system, the reachable condition of SMC for each area can be established by the following theorem under the provided SMC law.

Theorem 6.3 *To assure that the system trajectory of (6.17) can be attracted toward the sliding mode surface and stay on it thereafter, the SMC law is designed as*

$$\begin{aligned} u_i(t) &= (I + \Delta_i)K_i\hat{x}_a(t_{t-p+1}) \\ &\quad - (L_i\tilde{B})^{-1}\left[(||L_i||\tilde{\omega}_m(t) + \eta)\text{sgn}(\sigma(t)) + \varphi\sigma(t)\right], \end{aligned} \tag{6.47}$$

where $\tilde{\omega}_m(t) \geq ||w_i(t)||$, $\text{sgn}(\sigma(t))$ *is a signum function, and* η *and* φ *are the positive constants.*

Proof. Choose the Lyapunov function as

$$V(t) = ||\sigma(t)||. \tag{6.48}$$

Taking the deviation of $V(t)$ function

$$\begin{aligned} \dot{V}(t) &= \frac{\sigma^T(t)}{||\sigma(t)||}[L_i\tilde{B}_iK_i(I + \Delta_i)\hat{x}_a(t_{t-p+1}) + L_i\tilde{B}_iu_i(t) + L_i\tilde{\omega}_i(t)] \\ &= \frac{\sigma^T(t)}{||\sigma(t)||}[(-||L_i||\tilde{\omega}_m - \eta)\text{sgn}(\sigma(t)) - \varphi\sigma(t) + L_i\tilde{\omega}_i] \\ &= -||L_i||\frac{\sigma^T(t)\text{sgn}(\sigma(t))}{||\sigma(t)||}\tilde{\omega}_m - \eta\frac{\sigma^T(t)\text{sgn}(\sigma(t))}{||\sigma(t)||} \\ &\quad - \varphi\frac{\sigma^T(t)\sigma(t)}{||\sigma(t)||} + L_i\frac{\sigma^T(t)}{||\sigma(t)||}\tilde{\omega}_i. \end{aligned} \tag{6.49}$$

Noticing that $\sigma^T(t)\text{sgn}(\sigma(t))$ can be transformed into 1-norm $||\sigma(t)||_1$, and $||\sigma(t)||^2 = \sigma^T(t)\sigma(t)$. Then we have

$$\begin{aligned} \dot{V}(t) &= -||L_i||\frac{||\sigma(t)||_1}{||\sigma(t)||}\tilde{\omega}_m - \eta\frac{||\sigma(t)||_1}{||\sigma(t)||} - \varphi\frac{||\sigma(t)||^2}{||\sigma(t)||} + L_i\frac{\sigma^T(t)}{||\sigma(t)||}\tilde{\omega}_i \\ &\leq -||L_i||(\tilde{\omega}_m - \tilde{\omega}_i)\frac{||\sigma(t)||_1}{||\sigma(t)||} - \eta\frac{||\sigma(t)||_1}{||\sigma(t)||} - \varphi||\sigma(t)||. \end{aligned}$$

Thus, we have $V > 0$ and $\dot{V} < 0$, which means that the system state is ensured to reach the predetermined sliding surface. This completes the proof.

6.4 Numerical Examples

To verify the usefulness of the presented strategy, numerical simulations is tested for wind power systems under frequency-based deception attacks.

Example 6.1 *The* AMD *event-triggered* QSMLFC *technique is simulated for a three-area wind power system under frequency-based deception attacks.*

In the wind power generator model, the output wind power is determined by the wind speed. The wind power generator parameters are the same as those in [34]. The three-area interconnected power systems parameters utilized in the simulation are mostly the same as those in [13].

The upper bound for the time delay is $\tau_M = 0.15s$. The AMD event-triggered strategy parameters are given as $m = 3, \alpha = 0.1$ and $\varepsilon = 0.1$, respectively. The attack signal is assumed as $f_a(x_i(t)) = [0, 0, 0, 1.2, 0]^{\mathrm{T}}$. The attacks' frequency deviation is $\Delta f_A = 0.2$. According to **Corollary 6.1**, the controller gain is calculated as

$$K_1 = \begin{bmatrix} 0.0197 & 0.6717 & -0.2988 & 5.0691 & 0.3532 \end{bmatrix},$$

$$K_2 = \begin{bmatrix} 0.2011 & 1.3373 & -0.9291 & 4.3980 & 0.3032 \end{bmatrix},$$

$$K_3 = \begin{bmatrix} 0.1022 & 1.3527 & -0.7914 & 5.4608 & 0.3616 \end{bmatrix}.$$

The change in wind speed and fluctuation of wind power output are shown in Figure 6.6. To demonstrate the scalability of the proposed method, simulations are conducted on three-area interconnected wind power systems with different parameters. Different system parameters indicate that the size of each area is different. The state response of the three-area wind power systems is illustrated in Figure 6.7. Suppose that there is no disturbance in the first 10s, the system is self-stable and there is no control effect. Due to the uncertain wind power supply, the frequency of the power system will fluctuate, as shown in Figure 6.7. From Figure 6.7, it can be observed that all the three-area power systems can reach zero at about $t = 50s$. The changes in the tie-line power are presented in Figure 6.7. Results show that the deviation of the exchange power got close to zero at about $t = 60s$. Therefore, with the action of the proposed AMD event-triggered QSMLFC, the wind power system under frequency-based deception attacks remains stable, which evidently illustrates the usefulness and scalability of the proposed AMD event-triggered QSMLFC.

Example 6.2 *This part shows the difference among the proposed method and the other event-triggering strategies. Also, we made a comparison between different quantification methods.*

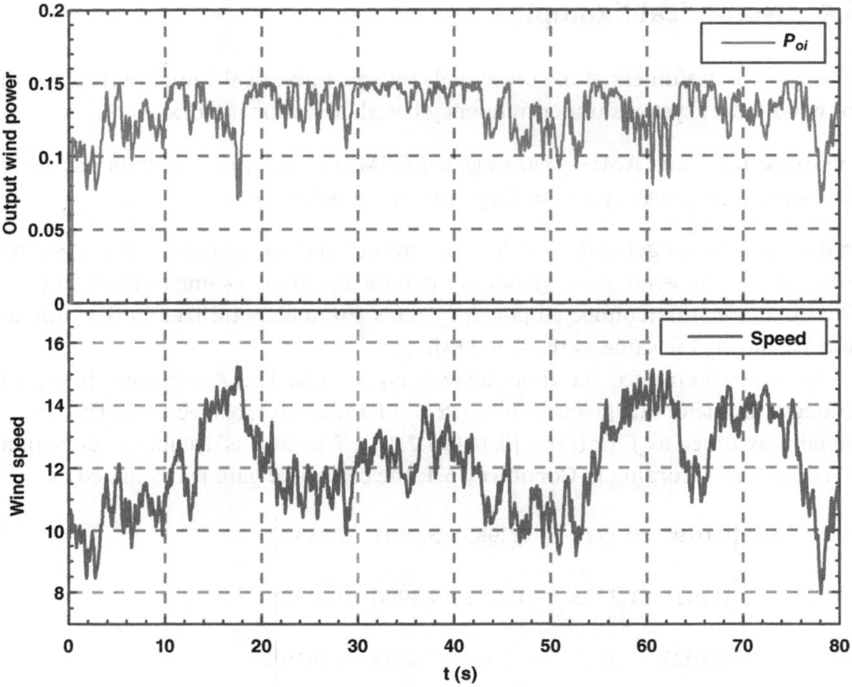

Figure 6.6 Changes of wind speed and output power.

Comparisons are made among the proposed method, AMET method [15], MET method [13], and event-trigger method [37]. All the triggered strategies are implemented in the same environment. The attack frequency deviation is $\Delta f_A = 0.1$. The tested system parameters are the same as area-1.

The simulation results are given in Figure 6.8. From Figure 6.8, we can see that the wind power system under frequency-based deception attacks comes to be stabilized by the different event-triggered schemes. The system comes into the stable state at about $t = 50s$ with the proposed method and AMET scheme [15], and at about $t = 60s$ with the MET method [13]. While under the frequency-based deception attacks, the event-triggered scheme [37] makes small disturbances all the test time. Therefore, the proposed method makes the system stabilization earlier than that in [13] and [37]. This shows the superiority of the AMD event-triggered QSMLFC method. Furthermore, the results evidently show that the added derivative term relaxes the triggering condition compared to that in [15]. Simulation results show that the added derivative term successfully adjusts the triggering rule while maintaining excellent performance.

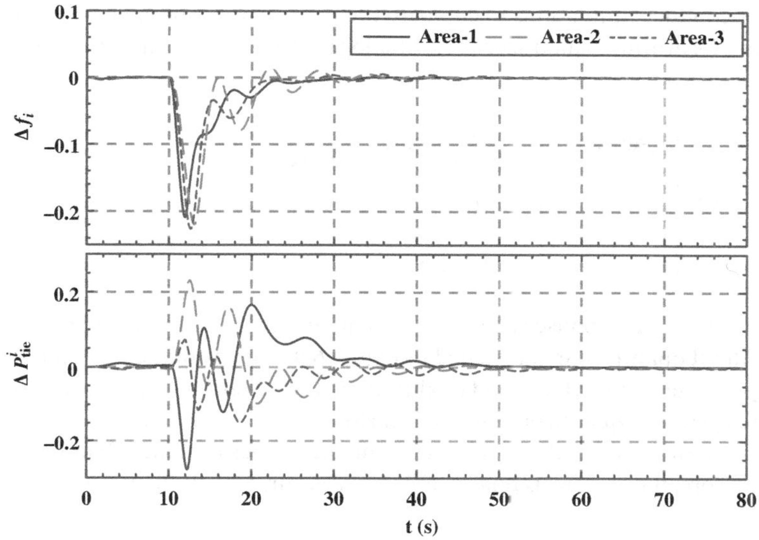

Figure 6.7 State response of power system.

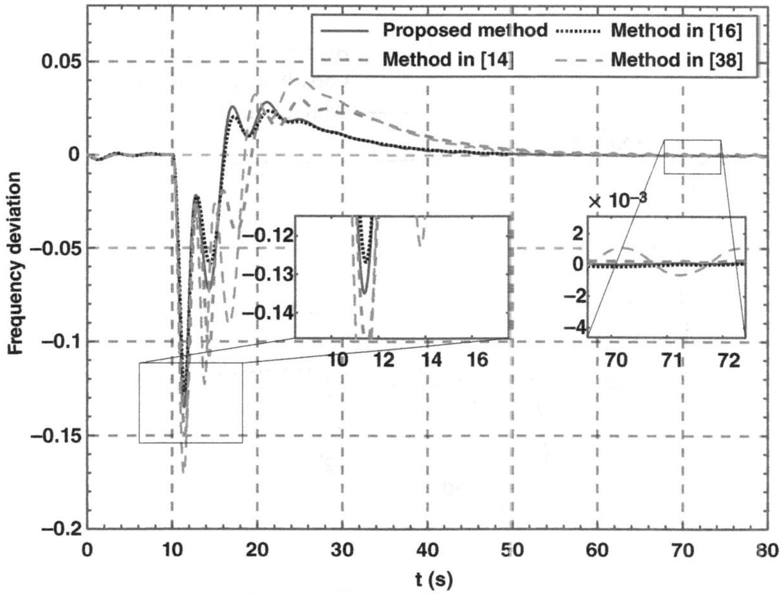

Figure 6.8 State response with different event-triggered scheme.

We further conduct simulations to compare the performances of the hysteresis quantizer and logarithmic quantizer [17]. The logarithmic quantizer is modeled as follows

$$Q_L(x_i(t)) = \begin{cases} \hat{x}_{\hat{i}}\text{sgn}(x_i(t)), & \frac{\hat{x}_{\hat{i}}}{1+\theta} < x_i(t) \leq \frac{\hat{x}_{\hat{i}}}{1-\theta}, x_i(t) > 0 \\ -Q_L(-x_i(t)), & x_i(t) < 0. \\ 0, & x_i(t) = 0 \end{cases}$$

The results of signal response with hysteresis quantizer, logarithmic quantizer [17], and without quantizer are presented in Figure 6.9. The results show that the hysteresis quantizer can work in good quality. And, the system state reaches zero earlier in the hysteresis quantizer than the logarithmic quantizer. The results also show that the hysteresis quantizer offers good quantizer performance than logarithmic. Therefore, the hysteresis quantizer is more accurate than the logarithmic quantizer.

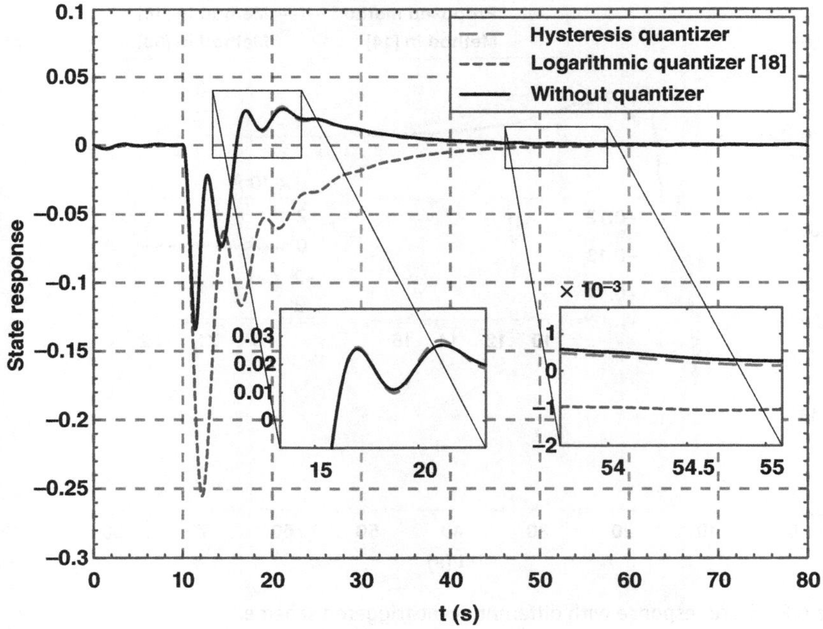

Figure 6.9 System state with different quantizers.

Example 6.3 *The following example simulates the influence and the occurrence of frequency-based deception attacks with different attack thresholds and different attack signals.*

The initial adaptive parameter is designed as $\rho(0) = 0.001$. The changing of the adaptive rule is given in Figure 6.10. The working conditions of AMD event-triggering devices is shown in Figure 6.11.

When the real-time frequency is in the stable region, the transmitted signal will be contaminated by the frequency-based deception attacks. The attack signal is given as $f_a(x_i(t)) = [0, 0, 0, 1.2, 0]^{\mathrm{T}}$. The thresholds Δf_A are set as 0.02, 0.05, and 0.1, respectively. As analyzed above, the frequency-based deception attacks are carried out when $\Delta \bar{f}_i < \Delta f_A$. To show this characteristic of frequency-based deception attacks, the following definition is given

$$\begin{cases} 1 & \text{Attacks} \\ 0 & \text{Without attacks.} \end{cases}$$

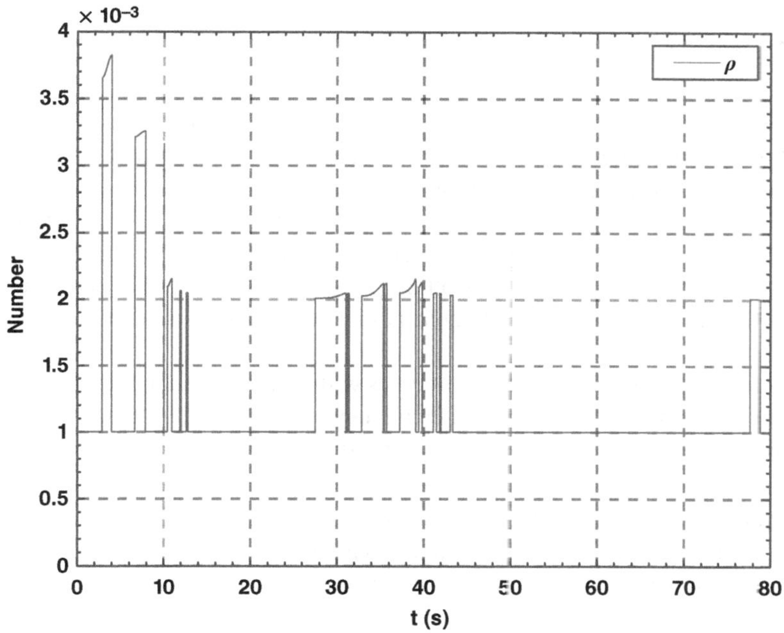

Figure 6.10 Changing of adaptive rule.

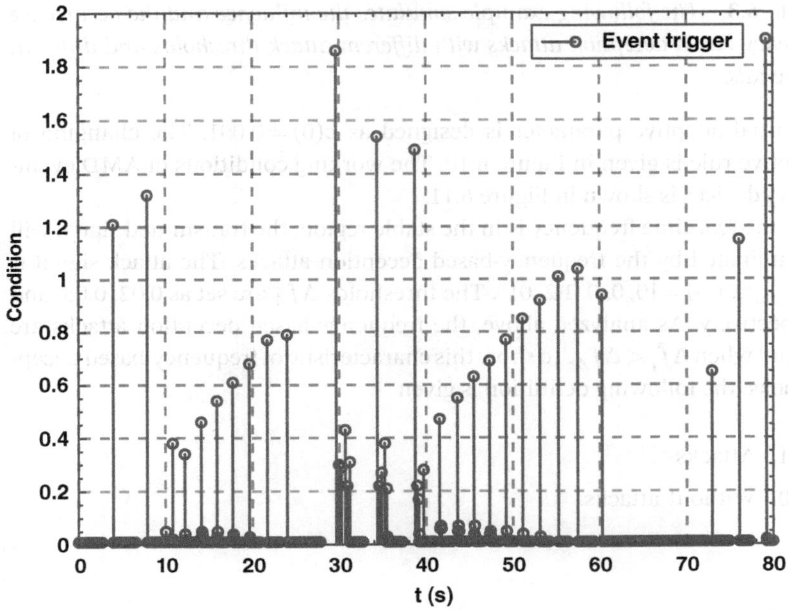

Figure 6.11 Adaptive rule and event-triggering instant.

The results of different attack thresholds are simulated in Figures 6.12–6.14. The results show that the impacts of the deception attacks with different attack thresholds on the system are different. The larger attack frequency threshold is capable of providing more times of attacks. However, a too larger threshold may lead the attacks to occur even if the power system is in huge fluctuations. A smaller threshold can cause difficulty in controlling performance. As can be seen from Figures 6.12–6.14, as the attack threshold is lowered, the system stability time increases.

Furthermore, we provide four different cases with different attack signals. The attack frequency deviation is $\Delta f_A = 0.1$. Frequency-based deception attack signals are set as $f_{a1}(x_i(t)) = [0, 0, 0, 0.2, 0]^{\mathrm{T}}$, $f_{a2}(x_i(t)) = [0, 0, 0, 0.6, 0]^{\mathrm{T}}$, $f_{a3}(x_i(t)) = [0, 0, 0, 1.2, 0]^{\mathrm{T}}$, $f_{a4}(x_i(t)) = [0, 0, 0, 1.8, 0]^{\mathrm{T}}$.

The system state evolution for the four cases is shown in Figure 6.15. The results show that the impacts of the deception attacks with different aggressive signals on the system are different. With the intervention of the controller actions, the system under frequency-based deception attacks gradually becomes stable. Simulated results also show that as the attack signal increases, the time to system stability increases.

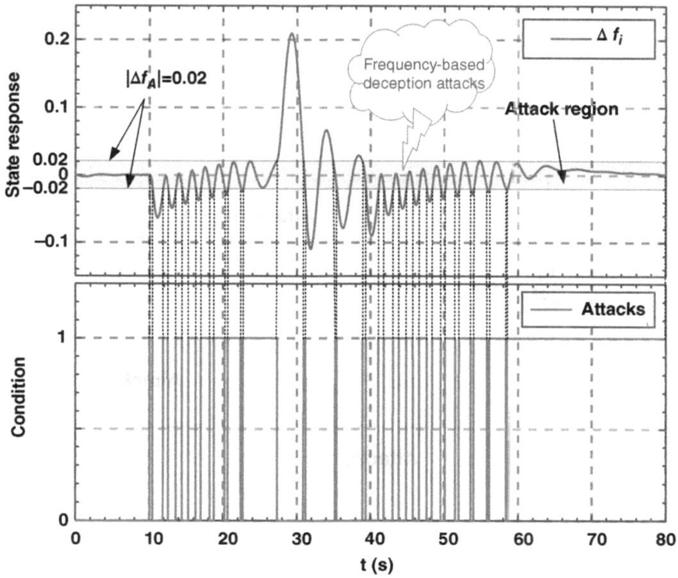

Figure 6.12 State response for $|\Delta f_A| = 0.02$.

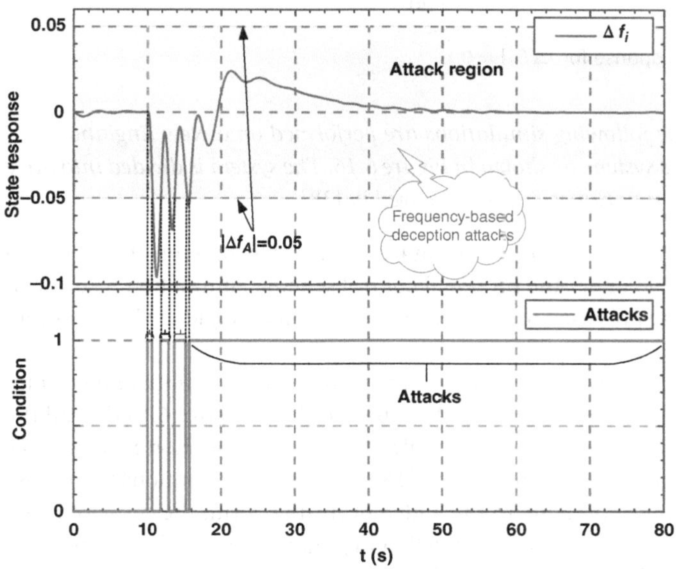

Figure 6.13 State response for $|\Delta f_A| = 0.05$.

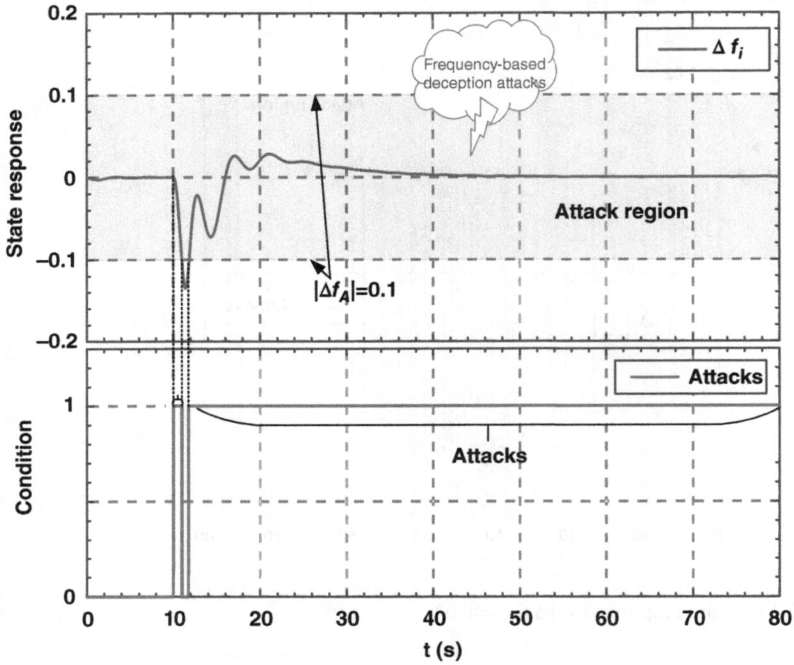

Figure 6.14 State response for $|\Delta f_A| = 0.1$.

Example 6.4 *The following simulations are performed on a New England IEEE 10-generator 39-bus system, as shown in Figure 6.16. The system is divided into three control areas. System parameters can be viewed in* [38].

Simulation results are presented in Figure 6.17. It can be seen that Δf_i and ΔP^i_{tie} tend to be stable under the proposed control scheme. Simulation results evidently show that the proposed method is effective against frequency-based deception attack.

To demonstrate the superiority in control performance, we compare the event-triggering mechanisms in [13, 15, 37]. One observes from Figure 6.18 that all the four triggering schemes make system stable at $t = 50s$. Simulation results show that the proposed method and the method [15] have fewer fluctuations than those [13] and [37]. Furthermore, due to the added derivative term, the proposed triggering scheme has a larger threshold than that in [15]. As presented in Figure 6.18, the proposed method adjusts the triggering rule while maintaining excellent performance. The comparative simulation results show the significant advantages of the proposed method.

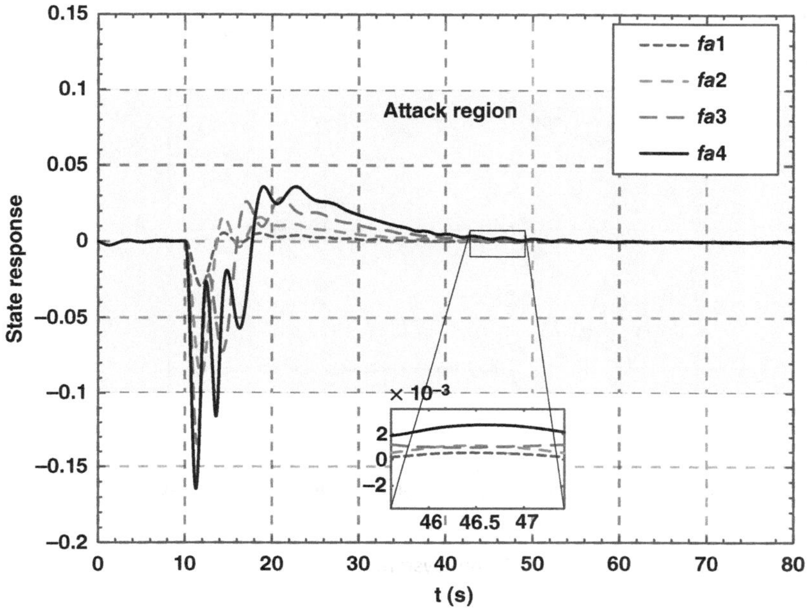

Figure 6.15 State response for different attack signals.

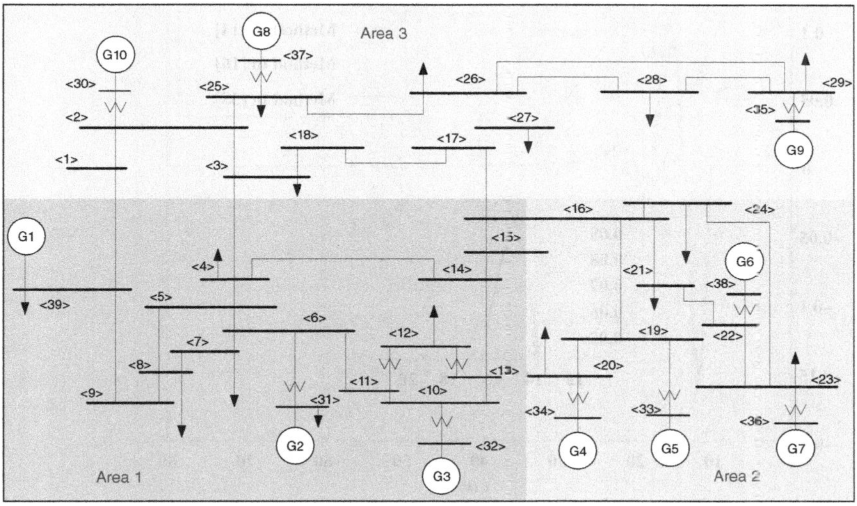

Figure 6.16 New England IEEE 39-bus system.

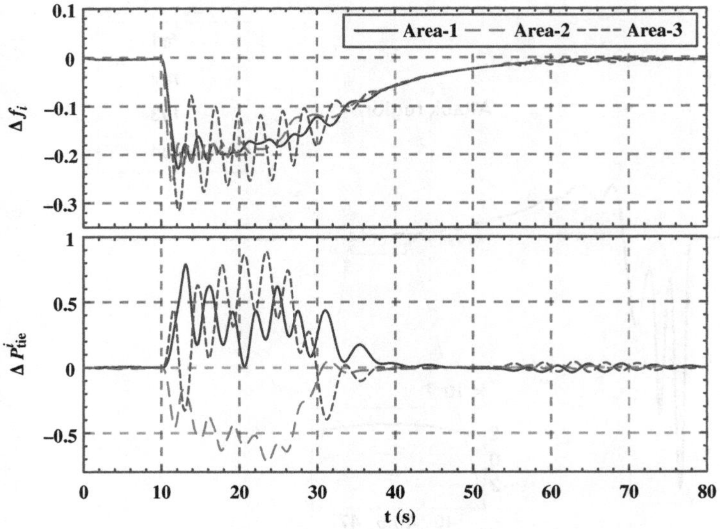

Figure 6.17 State response for IEEE 39-bus system.

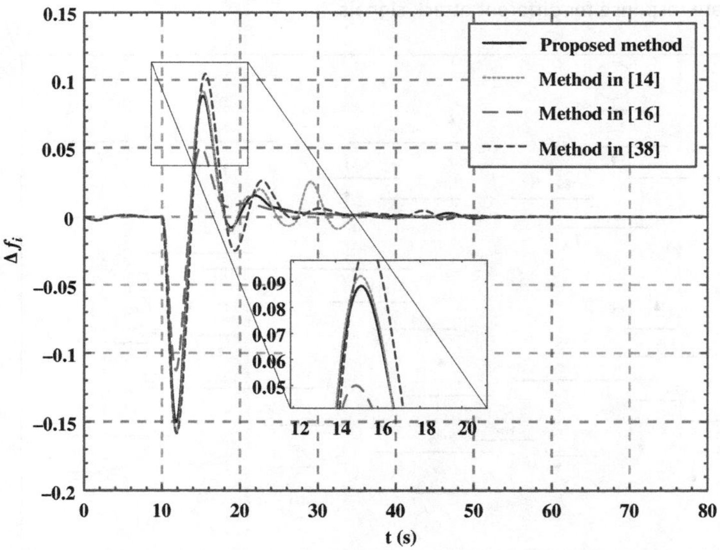

Figure 6.18 State response for different methods.

6.5 Conclusion

We have proposed a reliable AMD event-triggered QSMLFC for multi-area wind power systems under frequency-based deception attacks in this chapter. An AMD event-triggered strategy has been developed to promote the wind power system operation while saving the network resources, and the reliable AMD event-triggered QSMLFC method has been applied to reduce the frequency deviations of the wind power systems. A frequency-based deception attack model has been developed for analyzing the security issues in network communications for wind power systems. The hysteresis quantizer was used to lower the communication rate. To validate the correctness of the control method, a sufficient reliability criterion has been derived to prove the applicability of the AMD event-triggered QSMLFC. Numerical simulations have been applied to justify the correctness of the proposed method.

References

1 G. Luderer, S. Madeddu, L. Merfort, et al. Impact of declining renewable energy costs on electrification in low-emission scenarios. *Nature Energy*, 7(1):32–42, 2022.

2 E. Byon, L. Ntaimo, and Y. Ding. Optimal maintenance strategies for wind turbine systems under stochastic weather conditions. *IEEE Transactions on Reliability*, 59(2):393–404, 2010.

3 J. Teh and I. Cotton. Reliability impact of dynamic thermal rating system in wind power integrated network. *IEEE Transactions on Reliability*, 65(2):1081–1089, 2015.

4 J. Olauson, M. N. Ayob, M. Bergkvist, et al. Net load variability in Nordic countries with a highly or fully renewable power system. *Nature Energy*, 1(12):1–8, 2016.

5 S. Feng, K. Wang, X. Wu, et al. Stability analysis of integrated wind power system based on zeros identification of reduced-order impedance model in sub-bands. *IEEE Transactions on Sustainable Energy*, 13(1):427–439, 2022.

6 H. M. Wang and Z. J. Li. Multi-area load frequency control in power system integrated with wind farms using fuzzy generalized predictive control method. *IEEE Transactions on Reliability*, 72(2):737–747, 2022.

7 Y. J. Zhang, C. Peng, C. L. Cheng, and Y. L. Wang. Attack intensity dependent adaptive load frequency control of interconnected power systems under malicious traffic attacks. *IEEE Transactions on Smart Grid*, 14(2):1223–1235, 2022.

8 P. C. Chen, D. Zhang, L. Yu, and H. C. Yan. Dynamic event-triggered output feedback control for load frequency control in power systems with multiple cyber attacks. *IEEE Transactions on Systems, Man, and Cybernetics: Systems*, 52(10):6246–6258, 2022.

9 P. Mani and Y. H. Joo. Fuzzy logic-based integral sliding mode control of multi-area power systems integrated with wind farms. *Information Sciences*, 545:153–169, 2021.

10 Y. Mi, Y. Fu, C. S. Wang, and P. Wang. Decentralized sliding mode load frequency control for multi-area power systems. *IEEE Transactions on Power Systems*, 28(4):4301–4309, 2013.

11 H. M. Zhou, Q. S. Zhong, S. Y. Hu, et al. Dynamic event-triggered h_∞ quantized load frequency control for interconnected wind power systems under stochastic delay deception attack. *Sustainable Energy & Fuels*, 7(17):4317–4331, 2023.

12 X. Sun, Z. Gu, D. Yue, and X. P. Xie. Event-triggered H_∞ filtering for cyber–physical systems against DoS attacks. *IEEE Transactions on Systems, Man, and Cybernetics: Systems*, 53(5):2705–2715, 2023.

13 S. Yan, Z. Gu, and J. H. Park. Memory-event-triggered H_∞ load frequency control of multi-area power systems with cyber-attacks and communication delays. *IEEE Transactions on Network Science and Engineering*, 8(2):1571–1583, 2021.

14 L. L. Li, Y. Zhang, and T. S. Li. Memory-based event-triggered output regulation for networked switched systems with unstable switching dynamics. *IEEE Transactions on Cybernetics*, 52(11):12429–12439, 2021.

15 Y. Shi, E. G. Tian, S. B. Shen, and X. Zhao. Adaptive memory-event-triggered H_∞ control for network-based T-S fuzzy systems with asynchronous premise constraints. *IET Control Theory & Applications*, 15:534–544, 2021.

16 Y. Wang, B. Shen, and L. Zou. Recursive fault estimation with energy harvesting sensors and uniform quantization effects. *IEEE/CAA Journal of Automatica Sinica*, 9(5):926–929, 2022.

17 X. W. Yu and Y. Lin. Adaptive quantized tracking control for a class of nonlinear systems. *IEEE Transactions on Systems, Man, and Cybernetics: Systems*, 53(4):2554–2559, 2023.

18 H. D. Mo and G. Sansavini. Dynamic defense resource allocation for minimizing unsupplied demand in cyber-physical systems against uncertain attacks. *IEEE Transactions on Reliability*, 66(4):1253–1265, 2017.

19 H. F. Gu, J. N. Zhang, T. Liu, et al. DIAVA: A traffic-based framework for detection of SQL injection attacks and vulnerability analysis of leaked data. *IEEE Transactions on Reliability*, 69(1):188–202, 2019.

20 D. Tang, Y. Feng, S. Q. Zhang, and Z. Qin. FR-RED: Fractal residual based real-time detection of the LDoS attack. *IEEE Transactions on Reliability*, 70(3):1143–1157, 2020.

21 E. Tian and C. Peng. Memory-based event-triggering h_∞ load frequency control for power systems under deception attacks. *IEEE Transactions on Cybernetics*, 50:4610–4618, 2020.

22 G. Rinaldi, P. P. Menon, C. Edwards, et al. Adaptive dual-layer super-twisting sliding mode observers to reconstruct and mitigate disturbances and communication attacks in power networks. *Automatica*, 129:109656, 2021.

23 X. Cai, K. B. Shi, K. She, et al. Reliable sampling mechanism for Takagi–Sugeno fuzzy NCSs under deception cyberattacks for the application of the inverted pendulum system. *IEEE Transactions on Reliability*, 72(3):1220–1231, 2023.

24 H. Y. Zhang, N. Zhao, S. Y. Wang, and R. K. Agarwal. Improved event-triggered dynamic output feedback control for networked T–S fuzzy systems with actuator failure and deception attacks. *IEEE Transactions on Cybernetics*, 53:7989–7999, 2023.

25 Y. Q. Luo, Z. D. Wang, J. Hu, et al. Security-guaranteed fuzzy networked state estimation for 2-D systems with multiple sensor arrays subject to deception attacks. *IEEE Transactions on Fuzzy Systems*, 31(10):3624–3638, 2023.

26 D. Zhao, Z. D. Wang, Q. L. Han, and G. L. Wei. Proportional-integral observer design for uncertain time-delay systems subject to deception attacks: An outlier-resistant approach. *IEEE Transactions on Systems, Man, and Cybernetics: Systems*, 52(8):5152–5164, 2022.

27 S. W. Qiao, X. H. Liu, Y. R. Liang, et al. Event-triggered sliding mode load frequency control of multiarea power systems under periodic denial-of-service attacks. *IEEE Systems Journal*, 17(2):2803–2814, 2023.

28 Y. Z. Zhu and W. X. Zheng. Observer-based control for cyber-physical systems with periodic DoS attacks via a cyclic switching strategy. *IEEE Transactions on Automatic Control*, 65(8):3714–3721, 2019.

29 X. Wang, E. G. Tian, W. X. Zheng, and X. P. Xie. Important-data-based DoS attack mechanism and resilient H_∞ filter design for networked T-S fuzzy systems. *IEEE Transactions on Cybernetics*, 54(5):3352–3362, 2023. https://doi.org/10.1109/TCYB.2023.3285526.

30 J. Liu, Y. Gu, L. Zha, et al. Event-triggered H∞ load frequency control for multi-area power systems under hybrid cyber attacks. *IEEE Transactions on Systems, Man, and Cybernetics: Systems*, 49(8):1665–1678, 2019.

31 B. C. Zheng, X. H. Yu, and Y. M. Xue. Quantized feedback sliding-mode control: An event-triggered approach. *Automatica*, 91:126–135, 2018.

32 Q. Zhong, J. Yang, K. Shi, et al. Event-triggered H∞ load frequency control for multi-area nonlinear power systems based on non-fragile proportional integral control strategy. *IEEE Transactions on Intelligent Transportation Systems*, 23(8):12191–12201, 2021.

33 A. Farnam and R. M. Esfanjani. Improved linear matrix inequality approach to stability analysis of linear systems with interval time-varying delays. *Journal of Computational and Applied Mathematics*, 294:49–56, 2016.

34 T. Senjyu, R. Sakamoto, N. Urasaki, et al. Output power leveling of wind turbine generator for all operating regions by pitch angle control. *IEEE Transactions on Energy Conversion*, 21(2):467–475, 2006.

35 F. Ceragioli, De P. C., and P. Frasca. Discontinuities and hysteresis in quantized average consensus. *Automatica*, 47(9):1916–1928, 2011.

36 P. Cheng, S. P. He, and W. D. Zhang. A genetic algorithm-assisted fault detection observer for networked systems under denial-of-service attacks. *IEEE Transactions on Reliability*, 73(1):304-316, 2023. https://doi.org/10.1109/TR.2023.3311400.

37 X. H. Liu, D. D. Bai, Y. D. Wu, et al. Event-triggered load frequency control of smart grids under deception attacks. *IET Control Theory & Applications*, 15(10):1335–1345, 2021.

38 Y. C. Wang, Z. Y. Wang, J. Gan, et al. Switched observer-based adaptive event-triggered load frequency control for networked power systems under aperiodic dos attacks. *IEEE Transactions on Smart Grid*, 14(6):4816–4826, 2023.

7

Event-triggered Memory Sliding Mode Load Frequency Control of Power System with BESSs Against Frequency-based Deception Attacks

After reading this chapter, you should be able to:

- List the benefits of BESSs in the modern power system.
- Explain the improvement of Lyapunov-based term in the event-triggering strategy.
- Identify the difference between the historical frequency-based deception attacks and frequency-based deception attacks.

7.1 Background

Load frequency control (LFC) is widely applied in multi-area power systems to balance the load demand and frequency in all subareas, eliminating the frequency deviations and the tie-line power deviations between different areas [1–3], such that the frequency and tie-line power in the system can be maintained at a predetermined value [4–6]. Along with the development of power systems, factors such as the fast growth of renewable energy and increasing threats of cyberattacks are imposing challenges on LFC. Battery energy storage systems (BESSs) have been proven to be a useful solution for helping handle such critical challenges [7].

BESSs play an important role in modern power systems by providing fast active power compensation. The BESSs offer a promising capability of a fast frequency control for power systems, especially under severe disturbance [8]. And they can also enhance the LFC performance by charging and discharging actions [9]. Benefiting from the above advantages, BESSs have recently attracted much attention in the operation of the power system. In [10], a rule-based plan was proposed to employ the BESSs and demand response in the LFC task in which different states associated with power system frequency response as well as BESSs state of charge (SOC) are taken into account. To tackle the issue of load frequency fluctuation of the electrical power systems with energy storage when it is disturbed, a load frequency controller which bottom on disturbance observer was devised in [11]. A tube-based distributed model predictive control approach was investigated for power system with energy storage system [12].

Cybersecurity has become a hot topic in multi-area power systems due to frequent cyberattacks. While enjoying efficient data communication in network, one has to face the threat of malicious cyberattacks [13–15]. In the multi-area LFC power systems, deception attacks and DoS attacks are the most common cyberattacks [16]. The deception attacks can tamper with the normally transmitted signal by injecting false information, and further affect the judgment of the control center [17]. Researches on LFC power systems under deception attacks have received considerable attention in the past few years. The PI-based security control issue is addressed by considering randomly occurring joint attacks on both the sensor-to-observer and controller-to-actuator channels [18]. To minimize the negative impacts of cyberattacks and time-varying delays, a H_∞ LFC scheme for multi-area power systems was investigated in [19]. And a dynamic event-triggered decentralized output control scheme was presented for the LFC power system under periodic DoS attacks and deception attacks in [20]. It should be noted that the deception attacks are evaluated based on the stochastic models [18–20]. That is, with a given probability, deception attacks may occur at any sampling time.

More and more different types of cyberattacks have occurred in the real world. To develop effective strategies preemptively, researchers need to consider possible potential cyberattack models in advance. Research on different attack models can help us describe the actual cyberattack behavior and provide timely response measures when an attack occurs. This motivates us to propose the historical frequency-based deception attacks. Although a lot of studies in existing literature have noticed the harm of cyberattacks that appear in the power systems, cybersecurity issues of the power system are mostly considered from the perspective of network control systems, without further connection to the power system itself. To further search for the impact of cyberattacks on the power system, a frequency-based attack model is proposed. The new deception attacks are named historical frequency-based deception attacks. The historical frequency-based deception attacks model is modeled by combining the attribute deception attacks, the triggering mechanism, and the historical frequency information. The historical frequency-based deception attack signal is the same as [18–20]. Differently, an attack shall start only when a historical frequency released condition is satisfied, rather than being carried out following a given probability in the time domain. The attackers use multiple records of historical frequency information to design the attack release mechanism.

Compared to the Bernoulli-distributed deception attacks [18–20], the historical frequency-based deception attacks have the following advances. Due to the utilization of power system frequency, the connection between cyberattacks and power systems is deep, which helps investigate the contamination of deception attacks on power system cybersecurity. Deception attacks only occur when the power system signal meets the historical frequency released condition, which gives cyberattacks specific targets, while Bernoulli-distributed deception attacks happen randomly. With a properly designed historical frequency release condition, the historical frequency-based deception attacks will only happen when the released condition is satisfied. Thus, saving the energy and allowing the attacker to fulfill the purpose of disrupting system stability more efficiently than the Bernoulli-distributed deception attacks [18–20]. Therefore, one of the motivations of this study is to investigate the contamination caused by historical frequency-based deception attacks. Further, provide a proper control strategy to ensure the stability of the multi-area power system with BESSs under such types of attacks.

Event-triggered devices have been applied in networked system to lighten the communication pressure [21, 22]. Based on the designed triggering mechanism, sampled data that does not meet the triggering conditions will not be transmitted. The application of event-triggering devices promotes the transmission capacity of communication networks and helps improve power system performance [23]. Traditional event-triggering strategies with fixed thresholds can already reduce the communication pressure; however, some important but neighbor data may be lost [24]. Before long, an adaptive or dynamic event-triggered mechanism has been developed, such that the triggering threshold changes adaptively or dynamically [25]. In most cases, the adaptive or dynamic parameters are changed by the pre-designed scheme. To obtain an event-triggering strategies that can adapt to more complex situations, we introduce a Lyapunov-based term into the triggering scheme. Compared to the relevant literature [26], the proposed event-triggering scheme has a simpler structure. Besides, different event-triggering conditions can be obtained by adjusting the parameters of the Lyapunov-based term in the triggering threshold. The addition of Lyapunov-based term was provable to ensure the stability of the system while changing the triggering mechanism [27]. The proposed event-triggered mechanism has tunable capability while meeting the general event-triggering requirements.

SMC is seen as one of the most competent approaches in controlling uncertain systems owing to its robustness and insensitivity to perturbations [28]. Thus, sliding mode load frequency control (SMLFC) has been introduced to ensure the stability of multi-area power systems. In [29], a fractional-order sliding mode controller was proposed to ensure the stability of a multi-area power system with a time-varying delay. In [30], a second-order SMC via double integrated sliding surface was applied to the multi-area power system. A robust H_∞ SMLFC was constructed to guarantee a fast response and robust performance in [31]. Furthermore, SMLFC can also be applied to solving cyberattack problems. A distributed resilient multi-event cooperative triggered mechanism-based SMLFC approach was investigated for wind power systems under DoS attacks in [32]. The event-triggered adaptive SMC for a discrete nonlinear systems subject to false data injection attack was investigated in [33]. In this study, an event-triggered memory SMLFC scheme is proposed to realize efficient utilization of BESSs and stabilization of multi-area power systems under historical frequency-based deception attacks.

The main contributions are summarized as follows.

1. A historical frequency-based deception attack model is constructed by combining memory-event-triggered mechanism, system frequency, and the attributes of deception attacks. Different from the deception attacks in [18–20], the proposed attack will contaminate the sampled date, which is close to the stable state.
2. An event-triggered mechanism is applied to lighten the communication pressure. Without destroying the stability of the system, a Lyapunov-based term is added to the triggering strategy to extend the triggering conditions [27].
3. A memory SMC approach is investigated for the multi-area power system with BESSs to address the stability problems caused by deception attacks. Unlike the sliding surface in [29–31], a memory-based sliding surface is designed to further improve the control performance.

7.2 Problem Formulation

We focus on a simplified subsystem i consisting of the governor, turbine, power generator, and so on. The power system model is given in Figure 7.1.

7.2.1 Battery Energy Storage System

In related literature [9], the input or output power of a BESS can be described by a first-order model. Then, the dynamics of the exchanged power between a BESS and the power system can be represented by the following function

$$\Delta \dot{P}_{bi} = -\frac{1}{T_{bi}}\Delta P_{bi} + \frac{K_{bi}}{T_{bi}}u_i(t), \tag{7.1}$$

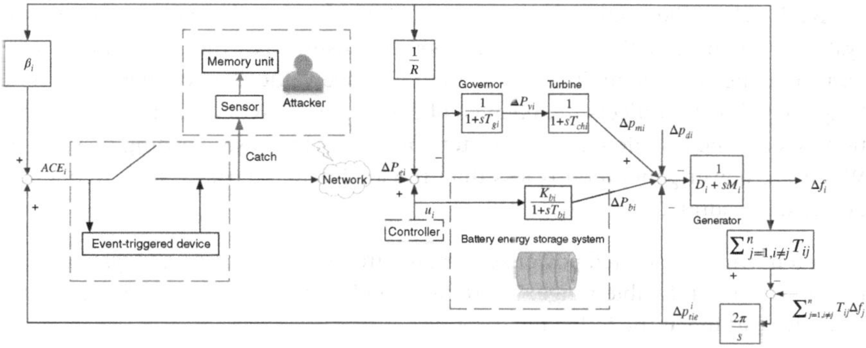

Figure 7.1 Model of i-th area power system under historical frequency-based deception attacks.

where T_{bi} is the time constant, $u_i(t)$ is the control signal received from the control center. K_{bi} is the BESS controller gain, which is determined by

$$K_{bi} = \begin{cases} K_{di} = \dfrac{p_{di}}{\Delta f_{\max}^{+} - \Delta f_{\min}^{+}}, & \text{Discharging} \\ K_{ci} = \dfrac{p_{ci}}{\Delta f_{\max}^{-} - \Delta f_{\min}^{-}}, & \text{Charging} \end{cases} \tag{7.2}$$

where K_{di} and K_{ci} are the discharging/charging controller gains, p_{di} and p_{ci} are the desired power rates in discharging mode and charging mode, respectively. $\Delta f_{\max}^{+}$ and $\Delta f_{\max}^{-}$ are the positive and negative thresholds of Δf_i, respectively, and $\Delta f_{\min}^{+}$ and $\Delta f_{\min}^{-}$ are the positive and negative dead zone thresholds of Δf_i, respectively.

For a multi-area power system with BESSs, the frequency of the power system can be adjusted by changing the charging and discharging state of the battery. To realize this, the main issue is to confirm the situation of charging or discharging. The i-th area power system frequency deviation $\Delta f_i = f_{\text{ref}} - f_i$ is applied to determine the situation of charging or discharging, where f_{ref} is the reference frequency and f_i is the system frequency. Hence

$$\begin{cases} \Delta f_i > 0, & \text{Discharging} \\ \Delta f_i = 0, & \text{Disconnected} \\ \Delta f_i < 0, & \text{Charging} \end{cases} \tag{7.3}$$

Remark 7.1: The total active power generated by the generator in the power system is balanced with the total active power consumed by the load, and the system frequency can be maintained at the reference value. The change in system frequency directly reflects the balance of active power. When the power generation exceeds the load demand, the system frequency increases, leading to $\Delta f_i < 0$. When the power generation is less than the load demand, the system frequency decreases, leading to $\Delta f_i > 0$.

For a normally operated power system, the frequency of the power system fluctuates within a tolerable range. Therefore, combining with (7.2), BESSs are predictably working at the following cases

1. For $\Delta f_i > \Delta f^+_{\min}$, the BESSs are discharging and $P_{bi} > 0$;
2. For $\Delta f^-_{\min} \le \Delta f_i \le \Delta f^+_{\min}$, the BESSs are disconnected and $P_{bi} = 0$. Note that $[\Delta f^-_{\min}, \Delta f^+_{\min}]$ is the dead zone with tolerable frequency deviation;
3. For $\Delta f_i < \Delta f^-_{\min}$, the BESSs are charging and $P_{bi} < 0$.

Figure 7.2 shows the characteristics of BESSs power exchange with frequency deviation Δf_i.

Traditionally, SOC is applied to analyze the working performance of BESS. By using the Coulomb counting method, one has

$$SOC_i(t) = SOC(0) - \frac{1}{C_{soc}} \int I_{bi} dt, \tag{7.4}$$

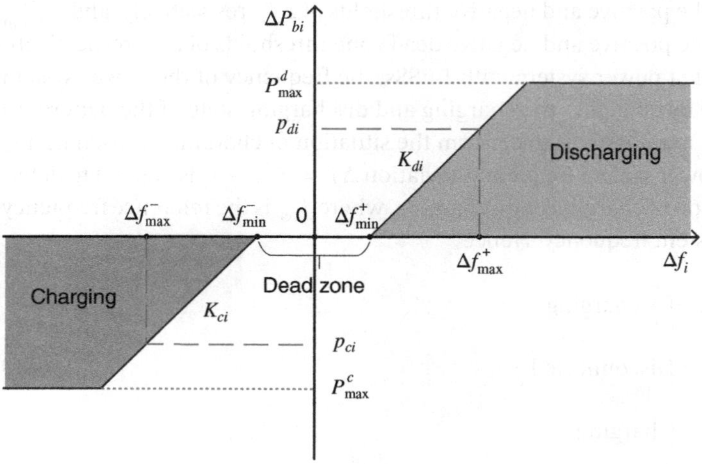

Figure 7.2 Power exchange of BESS with Δf_i.

where $SOC(0)$ is the initial SOC value, C_{soc} is the capacity of BESS, and I_{bi} is the output current. Notice that the power output of BESS is

$$P_{bi} = V_{bi} I_{bi}. \tag{7.5}$$

For convenience of discussion, the output voltage V_{bi} of each battery can be assumed to remain constant for a large range of SOC. Then, (7.4) can be rewritten as

$$SOC_i(t) = SOC(0) - \frac{1}{C_{\text{soc}}} \int \frac{P_{bi}}{V_{bi}} dt. \tag{7.6}$$

Equation (7.1) describes the exchanged power between BESSs and the power system. Equations (7.1–7.6) demonstrate the connection between the operating state of the BESS and the power system frequency. The charge/discharge behavior of BESSs has been discussed in detail. We have provided instructions on how to obtain the parameters in Equation (7.1). By combining (7.1) with the dynamic changes of the power system, the state space function of the power system integrated with BESSs can be obtained.

7.2.2 Power System Model

The dynamic equations of the system based on the system model and information transfer can be described as

$$\begin{cases} \Delta \dot{P}_{mi}(t) = -\dfrac{\Delta P_{mi}(t)}{T_{\text{chi}}} + \dfrac{\Delta P_{vi}(t)}{T_{\text{chi}}} \\ \Delta \dot{P}_{\text{tiei}}(t) = 2\pi \displaystyle\sum_{j=1, i \neq j}^{n} T_{ij}(\Delta f_i(t) - \Delta f_j(t)) \\ \Delta \dot{P}_{vi}(t) = -\dfrac{\Delta f_i(t)}{R_i T_{gi}} - \dfrac{\Delta P_{vi}(t)}{T_{gi}} - \dfrac{\Delta P_{ei}(t)}{T_{gi}} \\ \Delta \dot{f}_i(t) = \dfrac{\Delta P_{mi}(t)}{M_i} - \dfrac{D_i \Delta f_i(t)}{M_i} - \dfrac{\Delta P^i_{\text{tie}}(t)}{M_i} - \dfrac{\Delta P_{di}(t)}{M_i} \end{cases} \tag{7.7}$$

Hence, by combining (7.1) with (7.7), the state vector space model of the multi-area power systems with BESSs can be represented as

$$\dot{x}_i(t) = A_i x_i(t) + B_i u_i(t) + F_i w_i(t), \tag{7.8}$$

where $w(t)$ is the bounded disturbance of the system, which is generally defined as $w(t) = \Delta P_{di} < \epsilon$, where ϵ is a positive constant.

And, we have definitions as follows

$$x_i(t) = [\Delta P_{mi} \Delta P_{vi} \Delta P_{\text{tiei}} \Delta f_i \Delta P_{bi}]^{\text{T}},$$

$$A_i = \begin{bmatrix} -\frac{1}{T_{\text{chi}}} & \frac{1}{T_{\text{chi}}} & 0 & 0 & 0 \\ 0 & -\frac{1}{T_{gi}} & 0 & -\frac{1}{R_i T_{gi}} & 0 \\ 0 & 0 & 0 & 2\pi \sum_{j=1, j\neq i}^{n} T_{ij} & 0 \\ \frac{1}{M_i} & 0 & -\frac{1}{M_i} & -\frac{D_i}{M_i} & 0 \\ 0 & 0 & 0 & 0 & -\frac{1}{T_{bi}} \end{bmatrix},$$

$$B_i = \begin{bmatrix} 0 & \frac{1}{T_{gi}} & 0_{1\times 2} & \frac{K_{bi}}{T_{bi}} \end{bmatrix}, F_i = \begin{bmatrix} 0_{1\times 3} & -\frac{1}{M_i} & 0 \end{bmatrix}.$$

7.2.3 Preliminaries

Lemma 7.1 [34] *For $x(t) \in R^n$ and positive constant $\upsilon \in R$, there exist $P > 0$ and matrix U with appropriate dimensions, such that the following inequality holds*

$$-\upsilon \int_{t-\upsilon}^{t} \dot{x}^{\top}(s) P \dot{x}(s) ds \leq \begin{bmatrix} x(t) \\ x(t-\upsilon) \end{bmatrix}^{\top} \begin{bmatrix} -P & U^{\top} \\ U & -P \end{bmatrix} \begin{bmatrix} x(t) \\ x(t-\upsilon) \end{bmatrix}.$$

Lemma 7.2 [34] *For $x(t) \in R^n$ and positive constant $\upsilon \in R$, there exists $P > 0$, such that the following inequality holds*

$$-\upsilon \int_{t-\upsilon}^{t} x^{\top}(s) P x(s) ds \leq -\int_{t-\upsilon}^{t} x^{\top}(s) ds P \int_{t-\upsilon}^{t} x(s) ds.$$

Assumption 5 [16] The aggressive signals of deception attacks are assumed as $||f_a(x_i(t))||_2 \leq ||\tilde{G}_i x_i(t)||_2$. Considering the memory signal $\bar{x}_i(t)$ in the controller design, the following inequality is established

$$||f_a(x_i(t))||_2 \leq ||G_i \bar{x}_i(t)||_2,$$

where $\tilde{G}_i$ and G_i are constant matrices.

Proof. In related literature [16], the aggressive signals of deception attacks $f_a(x_i(t))$ are assumed as

$$||f_a(x_i(t))||_2 \leq ||\tilde{G}_i x_i(t)||_2$$

where $\tilde{G}_i$ is a constant matrix.

In this chapter, a memory SMC approach is designed for multi-area power system under deception attacks. The memory signal is designed as

$$\bar{x}_i(t) = \frac{1}{n}\sum_{q=1}^{n} x_i(t).$$

where n is the number of multiple signal, satisfy $n \geq 1$.

Obviously, we have

$$x_i(t) = n\bar{x}_i(t)$$

Combining with the aggressive signals of deception attacks $f_a(x_i(t))$, we have

$$||f_a(x_i(t))||_2 \leq ||n\tilde{G}_i\bar{x}_i(t)||_2$$

Defining $n\tilde{G}_i = G_i$, one can be obtained

$$||f_a(x_i(t))||_2 \leq ||G_i\bar{x}_i(t)||_2$$

Therefore, **Assumption 5** is proved to be established. This completes the proof.

7.3 Main Results

The following parts are to show our main results. The frequency-based deception attack model, event-triggered MSMLFC design and stability analysis are described in this section.

7.3.1 Frequency-based Deception Attacks

A frequency-based deception attack is considered as the contamination of a communication network. The attackers are assumed to use the recorded historical frequency of the power systems with BESSs by a frequency sensor and memory storage. By combining the event-triggered mechanism with the historical frequency information, deception attacks are reconfigured. The released conditions of frequency-based deception attacks are given as

$$\begin{cases} |\Delta\bar{f}_i| < |\Delta\bar{f}_A|, & \text{Released} \\ |\Delta\bar{f}_i| \geq |\Delta\bar{f}_A|, & \text{not attack} \end{cases} \tag{7.9}$$

where $\Delta\bar{f}_A$ is defined as the released threshold and used to judge whether the deception attacks should happen. The frequency-based parameters are described as

$$\Delta\bar{f}_A = \frac{1}{m}\sum_{p=1}^{m} f_p, \tag{7.10}$$

where, f_p denotes the historical frequency, m denotes the number of recorded historical frequencies. Obviously, $\Delta\bar{f}_A$ changes with the frequency of the power systems with BESSs.

Then, we need to determine whether the current system frequency meets the attack released condition. The frequency deviation is given as

$$\Delta\bar{f}_i = f_N - f_s, \tag{7.11}$$

where f_s is defined as the real-time frequency measured by frequency sensor, f_N is the frequency under normal operating of power system.

For an ideal operating power system, the stable state is confirmed when $\Delta\bar{f}_i = 0$. However, from the attacker's perspective, the deception attack signal is to affect the normal operation of the power system or destroys the normal work of the power system. In the proposed deception attack strategy, deception attacks occur when the system approaches or reaches a stable state. Taking into consideration of the power system stability criterion, the purpose of the deception attacks, and (7.9), the system state can be divided into two cases

$$\begin{cases} |\Delta\bar{f}_i| \leq |\Delta\bar{f}_A|, & \text{system under attack} \\ |\Delta\bar{f}_i| > |\Delta\bar{f}_A|, & \text{without attack} \end{cases} \tag{7.12}$$

Figure 7.3 illustrates the changes of (7.9) and frequency-based deception attacks contamination region.

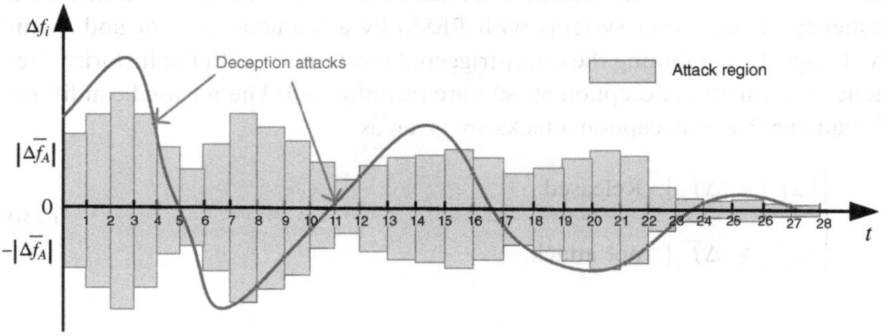

Figure 7.3 Example of historical frequency-based deception attacks.

Remark 7.2: Based on the design of (7.9), the proposed frequency-based deception attacks happened when the system frequency satisfies $|\Delta\bar{f}_i| < |\Delta\bar{f}_A|$. This means the frequency-based deception attacks only happen in a certain condition. From Figure 7.3, all the historical frequency-based deception attacks happen in the attack region, which is obtained based on (7.12). When the power system frequency gets close to the stable state, there will be frequency-based deception attacks at the sampling instant.

Remark 7.3: The attack releasing instant is decided by the designed mechanisms (7.12). In [3], the attack releasing mechanisms is designed with constant threshold. Differently, a dynamic attack threshold $\bar{f}_A$ is designed in (7.12). As can be seen in [3], the choice of attack threshold is difficult. In this chapter, the dynamic attack threshold $\bar{f}_A$ is designed with historical frequency.

The system state $x_i(t)$ is assumed to be caught by the attacker and replaced with the attack signal $x_a(t)$. The attack signal consists of system state $x_i(t)$ and interference signal $f_a(x_i(t))$. Based on **Assumption 5**, the deception signal is supposed to meet the following mathematical characters

$$\begin{aligned} &x_a(t) = x_i(t) + f_a(x_i(t)), \\ &||f_a(x_i(t))||_2 \leq ||G_i\bar{x}_i(t)||_2, \end{aligned} \tag{7.13}$$

where G_i is a constant matrix.

Combining (7.9), (7.12), and (7.13), all the frequency-based deception attacks are guaranteed to contaminate the sampled signal, which is located in the attack region. Thus, we have the following equations

$$\hat{x}_a(t) = \begin{cases} x_a(t), & |\Delta\bar{f}_i| < |\Delta\bar{f}_A|, \\ x_i(t), & \text{otherwise}. \end{cases} \tag{7.14}$$

7.3.2 Event-triggered SMC Scheme

An event-triggered mechanism is proposed to reduce the communication pressure and promote the power system performance. The event-triggered scheme is given as

$$t_{k+1}h = t_k h + \min_{l\in\mathbb{N}} \{lh | f_d(t) + e_i^{\mathrm{T}}(t)\Pi e_i(t) - \vartheta x_i^{\mathrm{T}}(t_k h)\Pi x_i(t_k h) < 0\}, \tag{7.15}$$

where h is the sampling interval, and $\vartheta \in (0, 1)$ is the triggering parameter. $f_d(t) = 2\xi x_i^{\mathrm{T}}(t)P_1\dot{x}_i(t)$, and ξ is an adjustable parameter and will be designed later, P_1 represents the Lyapunov matrix. $e_i(t) = x_i(t) - x_i(t_k)$, and Π will be designed in the following.

Remark 7.4: The event-triggered device is applied to reduce communication pressure by the designed triggering condition. However, the triggering condition may need to be adjusted to make it suitable for different environments. Random changes in the triggering condition may affect the power system performance. A Lyapunov-based term $f_d(t)$ is introduced into the event-triggered mechanism. The unique design ensures that $f_d(t)$ does not affect the stability performance of the system under the scaling of triggering conditions.

Remark 7.5: To exclude the Zeno phenomenon (infinite triggering events in finite period) in continuous-time event-triggered mechanism, a positive term is added in the event-triggered scheme. With the help of this term, a positive lower bound of the interevent time can be obtained even though the added term equals to 0, which guarantees no Zeno behavior.

The sliding mode surface in [35–37] is designed based on the currently transmitted signal. In this chapter, a memory SMC is developed by using the historical signals for the power systems with BESSs.

Firstly, the memory signal is defined as

$$\bar{x}_i(t) = \frac{1}{n}\sum_{q=1}^{n} x_i(t). \tag{7.16}$$

Then, a memory sliding mode surface can be developed as

$$\sigma = L_i x_i(t) - \int_0^t L_i A_i x_i(s)ds - \int_0^t L_i B_i K_i \bar{x}_i(s)ds, \tag{7.17}$$

where L_i is selected to ensure that $L_i B_i$ is nonsingular.

The memory signal under frequency-based deception attacks (7.13) can be expressed as

$$\bar{\hat{x}}_a(t) = \frac{1}{n}\sum_{q=1}^{n} \hat{x}_a(t). \tag{7.18}$$

Remark 7.6: An event-triggered memory SMLFC approach is proposed to address the cybersecurity issue of multi-area power systems with BESSs. To improve the control performance, the sliding surface is designed using multiple signals. The number of multiple signals is determined by n. When $n = 1$, the sliding surface degenerates into a general form. With the increase of n, it increases the burden of memory and computation while improving the controller performance.

Remark 7.7: A historical-frequency-based deception attack model is proposed from the attacker's perspective. The number of historical frequencies m can affect the working condition of historical-frequency-triggered deception attacks.

The trajectory of $\Delta\bar{f}_A$ comes close to the power system frequency along with the increases of m. A larger m will trigger more attack behavior and cause more contamination. As the size of A increases, the required stabilization time for the power system becomes longer [38].

Taking the deviation on (7.17), we have

$$\dot{\sigma} = L_i\dot{x}_i(t) - L_iA_ix_i(t) - L_iB_iK_i\bar{\hat{x}}_a(t). \tag{7.19}$$

Based on the SMC theorem, a proper sliding mode surface should satisfy $\sigma = \dot{\sigma} = 0$. Therefore, we have the following equation

$$L_i\dot{x}_i(t) - L_iA_ix_i(t) - L_iB_iK_i\bar{\hat{x}}_a(t) = 0. \tag{7.20}$$

Combining with power system state function (7.8), we have

$$L_i(A_ix_i(t) + B_iu_i(t) + F_iw_i(t)) - L_iA_ix_i(t) - L_iB_iK_i\bar{\hat{x}}_a(t) = 0. \tag{7.21}$$

Thus, the equivalent memory sliding mode controller can be derived as

$$u_{eq}(t) = -K_i\bar{\hat{x}}_a(t) - (L_iB_i)^{-1}L_iF_iw_i(t). \tag{7.22}$$

Then, we can derive the state space function of the event-triggered MSMLFC power systems with BESSs under frequency-based deception attacks as

$$\begin{cases} \dot{x}_i(t) = A_ix_i(t) - B_iK_i\bar{\hat{x}}_a(t_k) + \hat{F}_iw_i(t), \\ y_i(t) = C_ix_i(t), \end{cases} \tag{7.23}$$

where $\hat{F}_i = F_i - B_i(L_iB_i)^{-1}L_iF_i$.

7.3.3 Stability Analysis

Theorem 7.1 *For given scalars $h > 0$, $\gamma > 0$, $\xi < 1$, and $\vartheta \in (0, 1)$, power system (7.23) is stable if there exist $P_1 > 0, P_2 > 0, Q_1 > 0, Q_2 > 0, R_1 > 0, R_2 > 0, \Pi > 0$, and matrices U_i, G_i with appropriate dimensions, such that*

$$\Phi = \begin{bmatrix} \Psi_{11} & \Psi_{21}^{\mathrm{T}} \\ \Psi_{21} & \Psi_{22} \end{bmatrix} \prec 0, \tag{7.24}$$

$$\begin{bmatrix} -Q_2 & U_i^{\mathrm{T}} \\ U_i & -Q_2 \end{bmatrix} \succ 0, \tag{7.25}$$

where $\Psi_{11} = [\psi_{\check{i}\check{j}}]_{7\times 7}$, $\psi_{11} = (1-\xi)A_i^{\mathrm{T}}P_1 + (1-\xi)P_1^{\mathrm{T}}A_i + Q_1 + h^2R_1 - R_2 + C_i^{\mathrm{T}}C_i + \vartheta\Pi$, $\psi_{12} = -(1-\xi)K_i^{\mathrm{T}}B_i^{\mathrm{T}}P_1$, $\psi_{22} = G_i^{\mathrm{T}}P_1G_i$, $\psi_{13} = U_i$, $\psi_{33} = -R_2$, $\psi_{14} = P_2 + Q_2$, $\psi_{34} = -P_2 - Q_2$, $\psi_{44} = -R_1$, $\psi_{15} = -(1-\xi)K_i^{\mathrm{T}}B_i^{\mathrm{T}}P_1$, $\psi_{55} = -P_1$, $\psi_{16} = -\vartheta\Pi$, $\psi_{66} = (\vartheta - 1)\Pi$, $\psi_{17} = \hat{F}_iP_1$, $\psi_{77} = -\gamma^2 I$, $\Psi_{21} = h\Omega$, $\Omega = [A_i^{\mathrm{T}}P_1, -K_i^{\mathrm{T}}B_i^{\mathrm{T}}P_1, 0, 0, 0, 0, F_i^{\mathrm{T}}P_1]$, $\Psi_{22} = -R_2^{-1}$.

Proof. The Lyapunov–Krasovskii candidate function is given as

$$V(t) = V_1(t) + V_2(t) + V_3(t), \tag{7.26}$$

where

$$V_1(t) = x_i^{\mathrm{T}}(t)P_1x_i(t) + \int_{t-h}^{t} x_i^{\mathrm{T}}(s)dsP_2\int_{t-h}^{t} x_i(s)ds,$$

$$V_2(t) = \int_{t-h}^{t} x_i^{\mathrm{T}}(s)Q_1x_i(s)ds + \int_{t-h}^{t} x_i^{\mathrm{T}}(s)dsQ_2\int_{t-h}^{t} x_i(s)ds,$$

$$V_3(t) = h\int_{-h}^{t}\int_{t+\alpha}^{t} x_i^{\mathrm{T}}(s)R_1x_i(s)dsd\alpha + h\int_{-h}^{t}\int_{t+\alpha}^{t} \dot{x}_i^{\mathrm{T}}(t)R_2\dot{x}_i(t)dsd\alpha.$$

Then, the deviation of $V(t)$ is expressed as

$$\begin{aligned}\dot{V}_1(t) &= \dot{x}_i^{\mathrm{T}}(t)P_1x_i(t) + x_i^{\mathrm{T}}(t)P_1\dot{x}_i(t)\\ &\quad + x_i^{\mathrm{T}}(t)P_2\int_{t-h}^{t} x_i(s)ds - x_i^{\mathrm{T}}(t-h)P_2\int_{t-h}^{t} x_i(s)ds\\ &\quad + \int_{t-h}^{t} x_i^{\mathrm{T}}(s)dsP_2x_i(t) - \int_{t-h}^{t} x_i^{\mathrm{T}}(s)dsP_2x_i(t-h),\end{aligned}$$

$$\begin{aligned}\dot{V}_2(t) &= x_i^{\mathrm{T}}(t)Q_1x_i(t) - x_i^{\mathrm{T}}(t-h)Q_1x_i(t-h)\\ &\quad + x_i^{\mathrm{T}}(t)Q_2\int_{t-h}^{t} x_i(s)ds - x_i^{\mathrm{T}}(t-h)Q_2\int_{t-h}^{t} x_i(s)ds\\ &\quad + \int_{t-h}^{t} x_i^{\mathrm{T}}(s)dsQ_2x_i(t) - \int_{t-h}^{t} x_i^{\mathrm{T}}(s)dsQ_2x_i(t-h),\end{aligned}$$

$$\begin{aligned}\dot{V}_3(t) &= h^2x_i^{\mathrm{T}}(s)R_1x_i(s) - h\int_{t-h}^{t} x_i(s)R_1x_i(s)ds\\ &\quad + h^2\dot{x}_i^{\mathrm{T}}(s)R_2\dot{x}_i(s) - h\int_{t-h}^{t} \dot{x}_i^{\mathrm{T}}(s)R_2\dot{x}_i(s)ds.\end{aligned}$$

By applying **Lemma 7.1**, we have

$$-h\int_{t-h}^{t}\dot{x}_i^{\top}(s)R_2\dot{x}_i(s)ds \le -x_i^{\top}(t)R_2x_i(t)+x_i^{\top}(t-h)U_ix_i(t) + x_i^{\top}(t)U_ix_i(t-h)-x_i^{\top}(t-h)R_2x_i(t-h). \tag{7.27}$$

By applying **Lemma 7.2**, we have

$$-h\int_{t-h}^{t}x_i(s)R_1x_i(s)ds \le -\int_{t-h}^{t}x_i(s)dsR_1\int_{t-h}^{t}x_i(s)ds. \tag{7.28}$$

Then, $\dot{V}_3(t)$ can be rewritten as

$$\begin{aligned}\dot{V}_3(t) \le{}& h^2x_i^{\top}(s)R_1x_i(s)-\int_{t-h}^{t}x_i(s)dsR_1\int_{t-h}^{t}x_i(s)ds+h^2\dot{x}_i^{\top}(s)R_2\dot{x}_i(s)\\ &-x_i^{\top}(t)R_2x_i(t)+x_i^{\top}(t-h)U_ix_i(t)+x_i^{\top}(t)U_ix_i(t-h)\\ &-x_i^{\top}(t-h)R_2x_i(t-h).\end{aligned} \tag{7.29}$$

Furthermore, notice the frequency-based deception attacks in (7.13), we have

$$\bar{x}_i^{\top}(t)G_i^{\top}P_1G_i\bar{x}_i(t)-f_a^{\top}(x_i(t))P_1f_a(x_i(t)) \ge 0. \tag{7.30}$$

For $e_i(t)=x_i(t)-x_i(t_kh)$, we can obtain $x_i(t_k)=x_i(t)-e_i(t)$. Then, there exist

$$\vartheta[x_i(t)-e_i(t)]^{\top}\Pi[x_i(t)-e_i(t)]-2\xi x_i^{\top}(t)P_1\dot{x}_i(t)-e_i^{\top}(t)\Pi e_i(t)>0. \tag{7.31}$$

Combining with the power system model (7.23), we have

$$\begin{aligned}\dot{V}(t) \le{}& 2(1-\xi)[A_ix_i(t)-B_iK_i\overline{x}_i(t)-B_iK_if_a(x_i(t))+\hat{F}w_i(t)]^{\top}P_1x_i(t)\\ &+x_i^{\top}(t)P_2\int_{t-h}^{t}x_i(s)ds-x_i^{\top}(t-h)P_2\int_{t-h}^{t}x_i(s)ds\\ &+\int_{t-h}^{t}x_i^{\top}(s)dsP_2x_i(t)-\int_{t-h}^{t}x_i^{\top}(s)dsP_2x_i(t-h)\\ &+x_i^{\top}(t)Q_1x_i(t)-x_i^{\top}(t-h)Q_1x_i(t-h)\\ &+x_i^{\top}(t)Q_2\int_{t-h}^{t}x_i(s)ds-x_i^{\top}(t-h)Q_2\int_{t-h}^{t}x_i(s)ds\end{aligned}$$

$$
\begin{aligned}
&+\int_{t-h}^{t} x_i^{\mathrm{T}}(s)dsQ_2x_i(t)-\int_{t-h}^{t} x_i^{\mathrm{T}}(s)dsQ_2x_i(t-h)\\
&+h^2x_i^{\mathrm{T}}(s)R_1x_i(s)-\int_{t-h}^{t} x_i(s)dsR_1\int_{t-h}^{t} x_i(s)ds\\
&+h^2\dot{x}_i^{\mathrm{T}}(s)R_2\dot{x}_i(s)-x_i^{\mathrm{T}}(t)R_2x_i(t)+x_i^{\mathrm{T}}(t-h)U_ix_i(t)\\
&+x_i^{\mathrm{T}}(t)U_ix_i(t-h)-x_i^{\mathrm{T}}(t-h)R_2x_i(t-h)\\
&+\overline{x}_i^{\mathrm{T}}(t)G_i^{\mathrm{T}}P_1G_i\overline{x}_i(t)-f_a^{\mathrm{T}}(x_i(t))P_1f_a(x_i(t))\\
&+\vartheta[x_i(t)-e_i(t)]^{\mathrm{T}}\Pi[x_i(t)-e_i(t)]-e_i^{\mathrm{T}}(t)\Pi e_i(t).
\end{aligned}
$$

Obviously, there appears new aggregation item

$$
2(1-\xi)[A_ix_i(t)-B_iK_i\bar{x}_i(t)-B_iK_if_a(x_i(t))+\hat{F}w_i(t)]^{\mathrm{T}}P_1x_i(t)
$$

To ensure that the introduced derivative term does not affect the sign of the original aggregate term, $2(1-\xi)>0$ needs to be guaranteed. Thus, the triggering parameters should satisfy $\xi<1$.

Besides, for nonzero $w_i(t)\in L_2(0,+\infty)$, there exists disturbance attenuation level $\Upsilon=\gamma^2$ to ensure that the following inequality holds

$$
\dot{V}(t)-y_i^{\mathrm{T}}(t)y_i(t)+\gamma^2w_i^{\mathrm{T}}(t)w_i(t)\le -y_i^{\mathrm{T}}(t)y_i(t)+\gamma^2w_i^{\mathrm{T}}(t)w_i(t). \tag{7.32}
$$

Define the following vector

$$
\zeta^{\mathrm{T}}(t)=[x_i(t),\bar{x}_i(t),x_i(t-h),\int_{t-h}^{t} x_i(s)ds,f_a(x_i(t)),e_i(t),w_i(t)]^{\mathrm{T}}.
$$

To handle complex term $h^2\dot{x}_i^{\mathrm{T}}(s)R_2\dot{x}_i(s)$, we define $\Omega=[A_i^{\mathrm{T}}P_1,-K_i^{\mathrm{T}}B_i^{\mathrm{T}}P_1,0,0,0,0,F_i^{\mathrm{T}}P_1]$, such that

$$
h^2\dot{x}_i^{\mathrm{T}}(s)R_2\dot{x}_i(s)=h^2\zeta^{\mathrm{T}}(t)\Omega^{\mathrm{T}}R_2\Omega\zeta(t). \tag{7.33}
$$

Then, (7.32) can be rewritten as

$$
\zeta(t)^{\mathrm{T}}\Phi\zeta(t)<0. \tag{7.34}
$$

Under zero initial condition, we have

$$
\int_0^{+\infty} y_i^{\mathrm{T}}(t)y_i(t)\le\int_0^{+\infty}\gamma^2w_i^{\mathrm{T}}(t)w_i(t), \tag{7.35}
$$

which means $\|y_i(t)\| \le \gamma\|w_i(t)\|$ for nonzero $w_i(t) \in L_2[0, +\infty)$. Therefore, the event-triggered MSMLFC power systems with BESSs under frequency-based deception attacks are asymptotically stable with an H_∞ performance. The proof is completed.

Remark 7.8: The stability condition for the multi-area power system with BESSs under frequency-based deception attacks is provided in **Theorem 7.1**. Due to the presence of nonlinear terms $K_i^\top B_i^\top P_1$, the controller gain cannot be obtained straightly. The following theorem is proposed to obtain the controller gain K_i.

Corollary 7.1 *For given scalars $h > 0$, $\gamma > 0$, $\xi < 1$ and $\vartheta \in (0, 1)$, power system with BESSs (7.23) is stable if there exist $X_i > 0$, $\tilde{P}_2 > 0$, $\tilde{Q}_1 > 0$, $\tilde{Q}_2 > 0$, $\tilde{R}_1 > 0$, $\tilde{R}_2 > 0$, $\tilde{\Pi} > 0$ and matrices $\tilde{U}_i$ G_i, Y_i with appropriate dimensions, such that the following linear matrix inequalities (LMIs) hold*

$$\tilde{\Phi} = [\tilde{\psi}_{\bar{i}\bar{j}}]_{9\times 9} \prec 0, \tag{7.36}$$

$$\begin{bmatrix} -\tilde{Q}_2 & \tilde{U}_i^\top \\ \tilde{U}_i & -\tilde{Q}_2 \end{bmatrix} \succ 0, \tag{7.37}$$

where $\tilde{\psi}_{11} = (1-\xi)X_iA_i^\top + (1-\xi)A_iX_i + \tilde{Q}_1 + h^2\tilde{R}_1 - \tilde{R}_2 + \vartheta\Pi$, $\tilde{\psi}_{12} = -(1-\xi)Y_i^\top B_i^\top + X_iG_i$, $\tilde{\psi}_{22} = -X_i$, $\tilde{\psi}_{13} = \tilde{U}_i$, $\tilde{\psi}_{33} = -\tilde{R}_2$, $\tilde{\psi}_{14} = \tilde{P}_2 + \tilde{Q}_2$, $\tilde{\psi}_{34} = -\tilde{P}_2 - \tilde{Q}_2$, $\tilde{\psi}_{44} = -\tilde{R}_1$, $\tilde{\psi}_{15} = -(1-\xi)Y_iB_i$, $\tilde{\psi}_{55} = -X_i$, $\psi_{16} = -\vartheta\Pi$, $\psi_{66} = (\vartheta - 1)\Pi$, $\tilde{\psi}_{17} = \hat{F}$, $\tilde{\psi}_{77} = -\gamma^2 I$, $\tilde{\psi}_{18} = X_iA_i^\top$, $\tilde{\psi}_{28} = -Y_i^\top B_i^\top$, $\tilde{\psi}_{78} = \hat{F}_i^\top$, $\tilde{\psi}_{88} = -\tilde{R}_2^{-1}$, $\tilde{\psi}_{19} = X_iC_i^\top$, $\tilde{\psi}_{99} = -I$.

The controller gain can be calculated by $K_i = Y_iX_i^{-1}$.

Proof. In **Theorem 7.1**, there exist some complex items in inequality matrix (7.24). To make the inequality solvable, complex items need to be transformed as linear terms. We define $X_i = P_1^{-1}$ to achieve the linearization.

Focus on the dimension of Φ, pre- and post-multiplying both sides of Φ with $\text{diag}[X_i, X_i, X_i, X_i, X_i, X_i, X_i, X_i]$. Complex item $K_i^\top B_i^\top P_1$ can be transformed as

$$K_i^\top B_i^\top P_1 = X_iK_i^\top B_i^\top. \tag{7.38}$$

We have $Y_iB_i^\top$ after replacing nonlinear term $K_iX_i = Y_i$. The controller gain can be calculated as $K_i = Y_iX_i^{-1}$.

There exists complex term $X_iC_i^{\mathrm{T}}C_iX_i$ after linearization. Based on Schur complement theorem, we have the following

$$\begin{bmatrix} \tilde{\psi}_{11} + X_iC_i^{\mathrm{T}}C_iX_i & \cdots & \tilde{\psi}_{17} \\ \vdots & \ddots & \vdots \\ \tilde{\psi}_{71} & \cdots & \tilde{\psi}_{77} \end{bmatrix} \prec 0 \Leftrightarrow \begin{bmatrix} \tilde{\psi}_{11} & \cdots & \tilde{\psi}_{17} & X_iC_i^{\mathrm{T}} \\ \vdots & \ddots & \vdots & \vdots \\ \tilde{\psi}_{71} & \cdots & \tilde{\psi}_{77} & 0 \\ C_iX_i & \cdots & 0 & -I \end{bmatrix} \prec 0.$$

Besides, there will be some cross terms like $X_iP_2X_i$, $X_iQ_1X_i$, $X_iU_iX_i$, $X_iQ_2^{-1}X_i$. By defining $\tilde{P}_2 = X_iP_2X_i$, $\tilde{Q}_1 = X_iQ_1X_i$ $\tilde{U}_i = X_iU_iX_i$, and $\tilde{Q}_2 = X_iQ_2^{-1}X_i$, these cross terms are solvable. Therefore, the power system is stabilized such that the LMIs (7.36) and (7.37) hold. This completes the proof.

Remark 7.9: In **Corollary 7.1**, the proposed MSMLFC gain K_i can be calculated by using Matlab LMI toolbox. Compared with complex terms in **Theorem 7.1**, new complex terms in **Corollary 7.1** are easy to solve.

7.3.4 Reachability Condition

The reachable condition of SMC for each area can be established by the following theorem under the provided SMC law.

Theorem 7.2 *To assure that the system trajectory of (7.23) can be attracted toward the sliding mode surface and stay on it thereafter, the SMC law is designed as*

$$u_i(t) = -K_i\bar{\hat{x}}_a(t) - \epsilon(L_iB_i)^{-1}|L_i\hat{F}_i|\mathrm{sgn}(\sigma_i(t)), \tag{7.39}$$

where $||w_i(t)|| < \epsilon$, $\mathrm{sgn}(\sigma(t))$ *is a signum function.*

Proof. The Lyapunov function is chosen as

$$V_l = \sigma_i^{\mathrm{T}}(t)\sigma_i(t). \tag{7.40}$$

Take deviation on V_l, we have

$$\begin{aligned} \dot{V}_l &= 2\sigma_i^{\mathrm{T}}(t)\dot{\sigma}_i(t) \\ &= 2\sigma_i^{\mathrm{T}}(t)[L_i\dot{x}_i(t) - L_iA_ix_i(t) - B_iK_i\bar{x}_i(t)] \\ &= 2\sigma_i^{\mathrm{T}}(t)\{L_i[A_ix_i(t) + B_iu_i(t) + F_iw_i(t)] - L_ix_i(t)A_i - B_iK_i\bar{\hat{x}}_a(t)\} \end{aligned}$$

$$= 2\sigma_i^{\mathrm{T}}(t)[-\epsilon|L_iF_i|\mathrm{sgn}(\sigma_i(t)) + L_iF_iw_i(t)]$$

$$\leq 2\{-\epsilon|L_iF_i||\sigma_i(t)| + |\sigma_i(t)||L_iF_i||w_i(t)|\}$$

$$\leq 2|L_iF_i||\sigma_i(t)|[-\epsilon + |w_i(t)|] < 0$$

Thus, the designed control law can ensure the reaching condition. This completes the proof.

7.4 Numerical Examples

The usefulness of the proposed approach is illustrated by a three-area interconnected power system with BESSs. Power system parameters are similar to those in [19], and BESSs parameters are similar to those in [9]. For convenience, they are presented in Table 7.1. The BESSs are allowed to work in 10 ~ 90%. The initial SOCs are set as 30, 50, and 70%, respectively. The number of memory state in controller is given as $n = 5$. The number of historical frequency is given as $m = 3$.

Example 7.1 *The event-triggered MSMLFC technique is simulated for a three-area power system with BESSs under historical frequency-based deception attacks. Also, the state response of BESSs is simulated in this example.*

The triggering instant of event-triggered is shown in Figure 7.4. The historical frequency-based deception attack signal is assumed to be $f_a(x_i(t)) = 1.2 * \bar{x}_i(t)$. According to **Corollary 7.1**, the controller gains are calculated as

$$K_1 = \begin{bmatrix} -0.0476 & 0.0895 & -0.0039 & -0.3910 & 0.0266 \end{bmatrix},$$

$$K_2 = \begin{bmatrix} -0.0179 & 0.0706 & -0.0065 & -0.6019 & 0.0327 \end{bmatrix},$$

$$K_3 = \begin{bmatrix} -0.0090 & 0.0795 & -0.0034 & -0.3098 & 0.0416 \end{bmatrix}.$$

Table 7.1 Parameters of each area and BESSs.

Area-1	Parameters	D_1	T_{ch1}	T_{g1}	R_1	T_{12}	M_1
	Value	1.7	0.1	0.1	0.06	0.2π	12
Area-2	Parameters	D_2	T_{ch2}	T_{g2}	R_2	T_{13}	M_2
	Value	1.5	0.2	0.1	0.06	1.2π	12
Area-3	Parameters	D_3	T_{ch3}	T_{g3}	R_3	T_{23}	M_3
	Value	1.2	0.3	0.1	0.06	0.3π	12
BESSs	Parameters	p_{di}	p_{ci}	V_{bi}	T_{bi}	C_{SOC}	$\Delta f^+_{\min}$
	Value	0.3	0.3	2.4	0.3	0.96	0.02

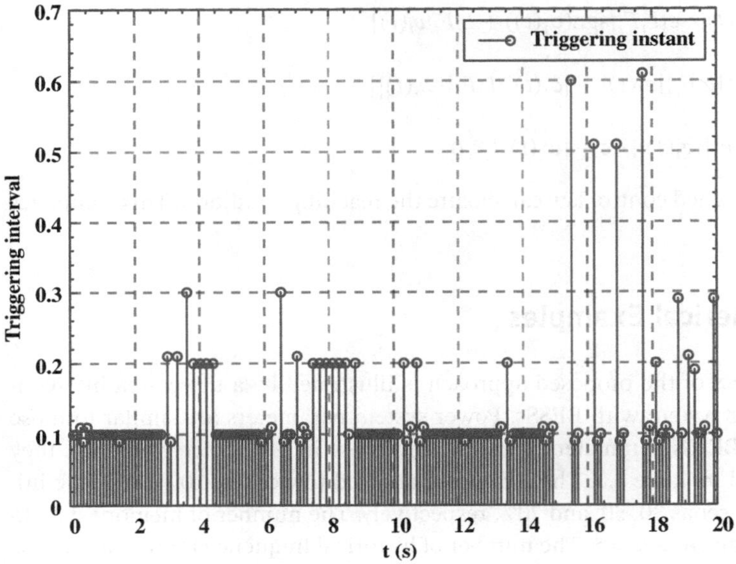

Figure 7.4 Event-triggered interval.

Three-area power system state responses are shown in Figure 7.5. From Figure 7.5, it can be observed that all the three-area power systems have small fluctuations at about $t = 10s$. Figure 7.5 also presents the changes in the tie-line power exchange. Results show that the deviation of the exchange power becomes stable at about $t = 12s$. Thus, with the action of the proposed approach, the power system with BESSs under historical-frequency-based deception attacks remains to be stable. Simulated results illustrate the usefulness of the proposed method. Besides, different frequency deviations lead to different actions of BESSs, which are shown as changes in Figure 7.6.

Example 7.2 *Comparisons among different control strategies are presented in this part. Also, the usage of BESSs enhances the LFC performance is simulated.*

Choosing area-1 as the test power system. The comparison is made among different ξ in the Lyapunov term. When $\xi = 0$, the event-triggered mechanism reduces to traditional form [33]. When $0 < \xi < 1$, the proposed event-triggered mechanism is the same as [27]. The difference among the proposed method, method in [27], method in [33], and different choices of ξ are shown in Figure 7.7. The results show that with the changes of Lyapunov-based term, the state response also changes.

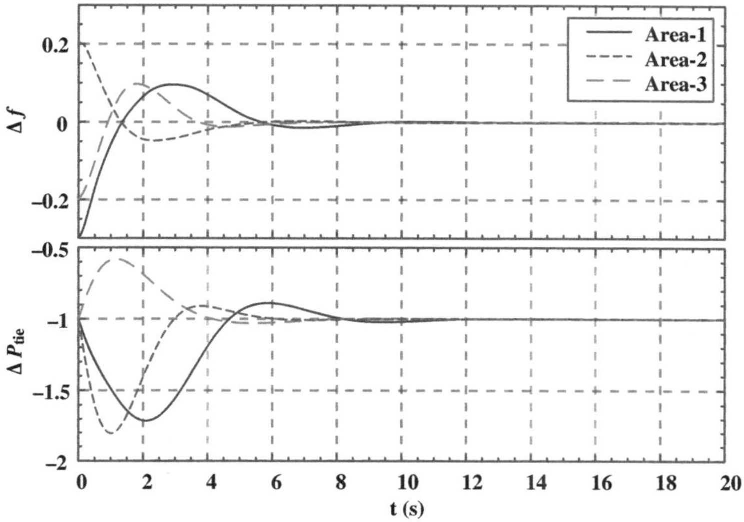

Figure 7.5 State response of three-area power system with BESSs.

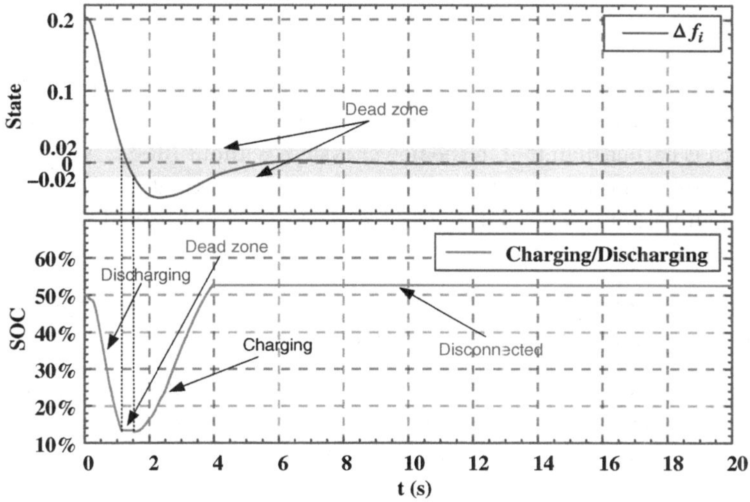

Figure 7.6 BESSs of area-2 with 50% SOC.

According to the result in Figure 7.7, when $\xi < 0$, the range of event-triggered state rises. When $0 < \xi < 1$, the proposed event-triggered mechanism reduces the triggering range. Simulated result proves the feasibility of Lyapunov-based term in adjusting the triggering mechanism.

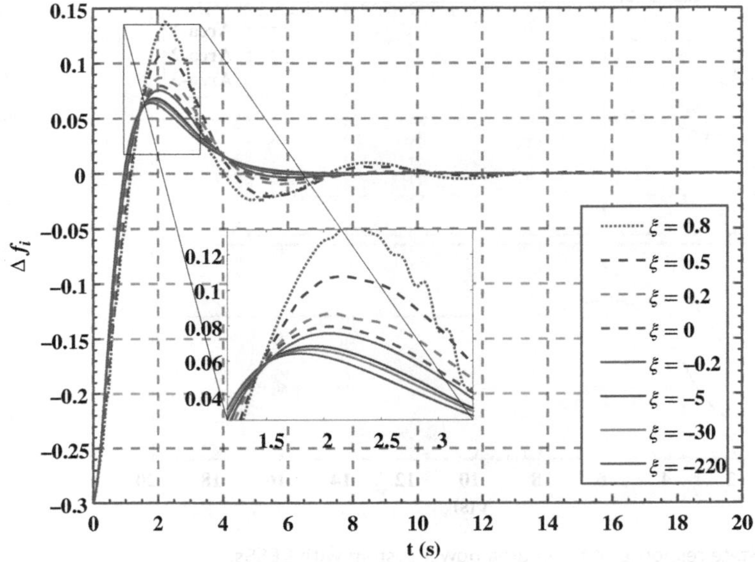

Figure 7.7 Frequency deviation for different size of ξ.

The effect of BESSs on the stability of power system frequency is shown in Figure 7.8. At about $t = 8s$, the power system frequency with BESSs almost reaches stability, which is earlier than that in the power system without BESSs. Also, the power system tie-line power deviation with BESSs reaches zero at about $t = 6s$, while the system without BESSs still fluctuated. The results confirmed the supportive effect of BESSs on the stability of the power system.

Example 7.3 *By testing different sizes of attack signals, the harm of the frequency-based deception attacks is simulated in this example.*

Choosing area-1 as the test power system. The frequency-based deception attack signals are set as $f_{a1}(x_i(t)) = 0.2 * \bar{x}_i(t)$, $f_{a2}(x_i(t)) = 0.5 * \bar{x}_i(t)$, $f_{a3}(x_i(t)) = 1.2 * \bar{x}_i(t)$, $f_{a4}(x_i(t)) = 3.7 * \bar{x}_i(t)$, $f_{a5}(x_i(t)) = 5.3 * \bar{x}_i(t)$, respectively.

State response under different sizes of attack signals is shown in Figure 7.9. Simulated results show that with the increase of attack signals, the stable time rises. The result proves that an attacker can increase the destructive power of cyberattacks by increasing the strength of attack signals.

The result of triggered state and attack threshold is presented in Figure 7.10. The attack releasing threshold is designed by using three historical frequencies. With the changes in system frequency, the dynamic attack threshold can be obtained.

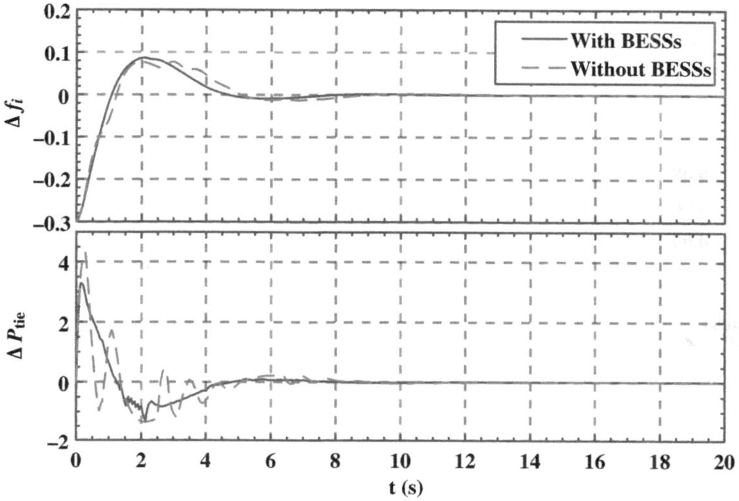

Figure 7.8 Frequency deviation with BESS and without BESS.

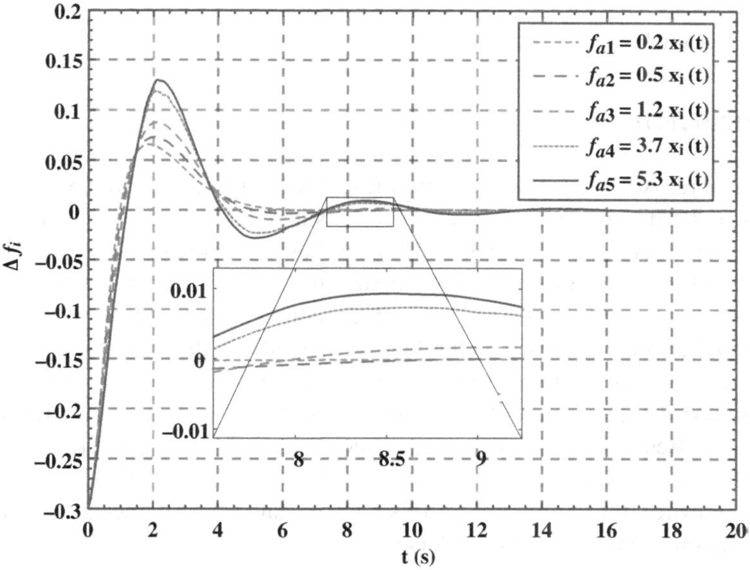

Figure 7.9 Δf_i with different attack signals.

Simulated results show that when the Δf_i satisfies (5.6), historical frequency-based deception attacks shall happen. This illustrates the characteristics of the historical frequency-based deception attacks with dynamic attack threshold.

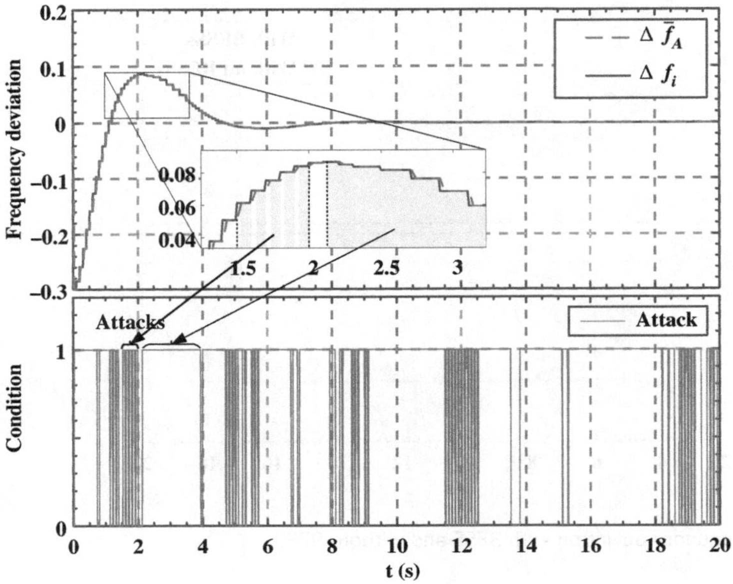

Figure 7.10 Triggered state and attacks threshold.

7.5 Conclusion

An event-triggered MSMLFC has been investigated for multi-area power systems with BESSs under deception attacks. We have considered a historical frequency-based deception attack model which is designed based on historical frequency information and triggering mechanism. Compared to Bernoulli-distributed deception attacks, the historical frequency-based deception attacks have a closer connection to the power system, which is helpful for studying the impact of cyber-attacks in the context of the power system. An event-triggered MSMLFC strategy has been proposed to guarantee the stability of the power systems with BESSs under historical frequency-based deception attacks. The effectiveness of the protection method and the promotion of BESSs have been demonstrated by simulation results. Furthermore, simulated results have revealed the characteristics of historical frequency-based deception attacks.

References

1 S. Zou, X. Shi, X. Zhao, and Z. Ma. Distributed optimal load frequency control for multi-area power systems with controllable loads. *Journal of the Franklin Institute*, 361(13):107007, 2024.

2 Z. Wu, L. Wu, X. Guan, et al. Model-free load frequency control of nonlinear power systems based on deep reinforcement learning. *IEEE Transactions on Industrial Informatics*, 20(4):6825–6833, 2024.

3 Y. Wang, G. Xiao, P. Wang, et al. H_∞ load frequency control of power system integrated with EVs under DoS attacks: Non-fragile output sliding mode control approach. *IEEE Transactions on Intelligent Transportation Systems*, 25(5):4565–4577, 2024.

4 B. Li, H. Li, B. Liu, et al. Decentralized secure load frequency control for multi-area power systems under complex cyber attacks. *IEEE Transactions on Smart Grid*, 15(5):5043–5054, 2024. https://doi.org/10.1109/TSG.2024.3379411.

5 J. Lu, J. Cao, J. Yu, et al. False data injection attacks on LFC systems: An AI-based detection and countermeasure strategy. *IEEE Transactions on Circuits and Systems I: Regular Papers*, 71(5):1969–1977, 2024.

6 Y. Wang, G. Liu, J. Li, and H. Feng. A novel failure-distribution-dependent non-fragile H_∞ fault-tolerant load frequency control for faulty multi-area power systems. *IEEE Transactions on Power Systems*, 39(2):2936–2946, 2024.

7 B. Chen, L. Yu, P. Chen, and S. Liu. Multi-agent reinforcement learning for decentralized resilient secondary control of energy storage systems against dos attacks. *IEEE Transactions on Smart Grid*, 13(3):1739–1750, 2022.

8 K. Shi, Y. Yu, S. Zhong, et al. Stability and stabilization for T–S fuzzy load frequency control power system with energy storage system. *IEEE Transactions on Fuzzy Systems*, 32(3):893–905, 2024. https://doi.org/10.1109/TFUZZ.2023.3311925.

9 Y. Tian, G. Ledwich, H. Su, et al. Dual-consensus-based distributed frequency control for multiple energy storage systems. *IEEE Transactions on Smart Grid*, 10(6):6396–6403, 2019.

10 A. Ashouri-Zadeh, A. M. Ranjbar, S. A. Hosseini, and M. Toulabi. Battery energy storage systems and demand response applied to power system frequency control. *International Journal of Electrical Power & Energy Systems*, 136:107680, 2022.

11 H. Jiang, M. Wei, S. Yu, et al. Load frequency control of power system with energy storage based on disturbance observer. *Energy Reports*, 8:615–622, 2022.

12 X. Kong, Y. Zhang, W. Wang, et al. Tube-based distributed MPC for load frequency control of power system with high wind power penetration. *IEEE Transactions on Power Systems*, 39(2):3118–3129, 2024.

13 H. Zhang, J. H. Park, X. Liu, et al. Security load frequency control of networked power systems via round-robin protocol under denial of service attacks. *Journal of the Franklin Institute*, 361(16):107155, 2024.

14 C. Zhang, W. Yao, Y. Zhao, et al. Resilient load frequency control of power systems to compensate random time delays and time-delay attacks. *IEEE Transactions on Industrial Electronics*, 70(5):5115–5128, 2023.

15 D. Ding, Y. Zhang, and Y. Ren. Resilient control co-design for cyber-physical systems with DoS attacks via a successive convex optimization approach. *Journal of the Franklin Institute*, 360(9):6253–6274, 2023.

16 J. Yang, Q. Zhong, K. Shi, and S. Zhong. Dynamic-memory event-triggered $H\infty$ load frequency control for reconstructed switched model of power systems under hybrid attacks. *IEEE Transactions on Cybernetics*, 53(6):3913–3925, 2022.

17 L. Zha, X. Xie, E. Tian, et al. Reinforcement learning-based tracking control for networked control systems with DoS attacks. *IEEE Transactions on Information Forensics and Security*, 19:4188–4197, 2024. https://doi.org/10.1109/TIFS.2024.3376250.

18 X. L. Wang, D. R. Ding, H. L. Dong, and X. J. Yi. Pi-based security control against joint sensor and controller attacks and applications in load frequency control. *IEEE Transactions on Systems, Man, and Cybernetics: Systems*, 53(2):970–980, 2022.

19 O. Bamisile, Y. Xing, D. Cai, et al. An H_∞ load frequency control scheme for multi-area power system under cyber-attacks and time-varying delays. *IEEE Transactions on Power Systems*, 38(2):1336–1349, 2023.

20 P. C. Chen, D. Zhang, L. Yu, and H. C. Yan. Dynamic event-triggered output feedback control for load frequency control in power systems with multiple cyber attacks. *IEEE Transactions on Systems, Man, and Cybernetics: Systems*, 52(10):6246–6258, 2022.

21 Z. Zuo and G. Wang. Consensus control of multi-agent systems with different state constraints and event-triggered communication. *IEEE Transactions on Circuits and Systems II: Express Briefs*, 71(2):817–821, 2023.

22 X. Lv, W. Li, and Y. Niu. Dynamic event-triggering sliding mode resilient control for multi-agent systems. *Journal of the Franklin Institute*, 360(3): e2182–e2199, 2023.

23 P. Li, G. Wang, and Z. Zuo. Distributed control in uncertain nonlinear multiagent systems under event-triggered communication and general directed graphs. *IEEE Transactions on Signal and Information Processing over Networks*, 10:599–609, 2024.

24 G. Wang, X. Tian, and Z. Ai. Switching-like dynamic event-triggered load frequency control for multi-area power systems with electric vehicles under dos attacks. *Journal of the Franklin Institute*, 361(1):513–525, 2024.

25 Y. Song, C. Peng, D. Du, et al. Secure adaptive event-triggered control for cyber-physical power systems under denial-of-service attacks. *IEEE Transactions on Cybernetics*, 54(3):1722–1733, 2024.

26 W. Li, J. Zhang, D. Yang, et al. Resilient output control of multiagent systems with DoS attacks and actuator faults: Fully distributed event-triggered approach. *IEEE Transactions on Cybernetics*, 54(12):7681–7690, 2024.

27 Z. Gu, S. Yan, and F. Yang. Derivative-based event-triggered control for networked systems with quantization. *Applied Mathematics and Computation*, 383:125359, 2020.

28 M. Sarcinelli-Filho, D. KD. Villa, and A. S. Brandão. Adaptive sliding mode control applied to quadrotors-a practical comparative study. *Journal of the Franklin Institute*, 360(16):11578–11599, 2023.

29 D. Li, S. Lin, S. M. Muyeen, et al. Fractional-order sliding mode load frequency control and stability analysis for interconnected power systems with time-varying delay. *IEEE Transactions on Power Systems*, 39(1):1006–1018, 2024.

30 C. S. T. Dong, B. D. Hoang, O. Kaynak, et al. Sliding surface design for sliding mode load frequency control of multiarea multisource power system. *IEEE Transactions on Industrial Informatics*, 20:7797–7809, 2024. https://doi.org/10.1109/TII.2024.3359445.

31 Z. Wei, G. Sun, X. Wu, et al. Robust H_∞ load frequency control of multi-area power system with time delay: A sliding mode control approach. *IEEE/CAA Journal of Automatica Sinica*, 5(2):610–617, 2017.

32 B. Wang and Z. Huo. Distributed resilient multi-event cooperative triggered mechanism based discrete sliding-mode control for wind-integrated power systems under denial of service attacks. *Applied Energy*, 333:120636, 2023.

33 G. Liu, X. Yu, H. Zhang, and J. Hu. Event-triggered secure control of discrete systems under cyber-attacks using an observer-based sliding mode strategy. *Information Sciences*, 587:587–606, 2022.

34 A. Farnam and R. M. Esfanjani. Improved linear matrix inequality approach to stability analysis of linear systems with interval time-varying delays. *Journal of Computational and Applied Mathematics*, 294:49–56, 2016.

35 R. J. Ma, P. Shi, and L. G. Wu. Dissipativity-based sliding-mode control of cyber-physical systems under denial-of-service attacks. *IEEE Transactions on Cybernetics*, 51(5):2306–2318, 2020.

36 W. Qi, G. Zong, and H. R. Karimi. Finite-time observer-based sliding mode control for quantized semi-Markov switching systems with application. *IEEE Transactions on Industrial Informatics*, 16(2):1259–1271, 2019.

37 Y. Mi, Y. Fu, C. S. Wang, and P. Wang. Decentralized sliding mode load frequency control for multi-area power systems. *IEEE Transactions on Power Systems*, 28(4):4301–4309, 2013.

38 G. Xiao, S. Qiao, X. Liu, et al. Security performance of MOSMLFC power system under historical-frequency-triggered dos attacks. *IEEE Transactions on Automation Science and Engineering*, 22:6518–6529, 2025. https://doi.org/10.1109/TASE.2024.3446883.

8

Security Performance of MOSMLFC Power System Under Historical-frequency-triggered DoS Attacks

After reading this chapter, you should be able to:

- Understand the role of Lyapunov method in system stability.
- Explain the impact of added parameters in inequality.
- Identify the difference between the historical frequency-based DoS attacks and frequency-based DoS attacks.

8.1 Background

Power systems have long been in an inseparable relationship with human society [1]. To ensure that the power system is able to provide a high-quality power supply, load frequency control (LFC) has been widely applied in modern power systems to achieve a good balance between load demand and generation [2–4]. Extensive studies have been conducted on the LFC problem. Among them, the non-fragile H_∞ fault-tolerant LFC problem for power systems with a stochastic distributed actuator failure model was studied in [5], and a discrete-time Laguerre functions-based intelligent model predictive control approach was administered in [6]. A model-free LFC method was investigated for nonlinear power systems based on deep deterministic policy gradient framework in [7]. With the increasing importance of communication networks in the power system, it is recognized that cyberattacks may cause unexpected serious damage to the LFC power systems.

In the LFC power system, a communication network is used to transmit decision or control signals. The open nature of the network makes it vulnerable to cyber-attacks [8, 9]. Particularly, the most representative cyberattack is denial-of-service (DoS) attacks. When a DoS attack occurs, the transmission of signals in the network may be blocked, which may further destroy the stability of the system [10]. Different types of attack models have been built to describe the actual behavior or maximize the contamination of the attacks. In [11], the DoS attacks interval was supposed to be known. In [12], the DoS attacks model was designed based on the Bernoulli process. In [13–15], the DoS attack model was designed based on a given attack frequency and duration. In [16], a DoS model was proposed within a known bounds of the sleeping and active period of the attack signal. In [17], DoS attacks are conducted by a type of periodic jamming signal. It is significant to note that different models of DoS attacks have completely different attack characteristics. To improve the contamination of the attacks, an important-data mechanism was designed from the perspective of attackers in [18]. The study of different cyber-attack models can help defenders formulate countermeasures and against the contamination of cyberattacks, thereby reducing the destructive impact of cyberattacks. In this chapter, a new DoS attack model was proposed to describe the actual behavior of cyberattacks that may happen.

In the related studies [11–17], there is no direct relation between the target system and the considered DoS attack models. Smart attacker may notice certain characteristics of the target system, such as the power system frequency. In [18], the attacker's action was judged by comparing the current measured output and the most recent unattacked signal. In this chapter, the attacker's action was judged by comparing the historical frequency in this chapter. The DoS attack model was proposed in the context of the LFC power system. Frequency is a direct measure of the generation-load imbalances, stability, and power quality of the power system [19]. Therefore, we allow the attacker to take into account the power system frequency in the design of the DoS attack to determine when a DoS attack should occur. Besides, an event-triggered strategy is introduced to the proposed cyber-attack model. In most cases, the event-triggered mechanism is applied to decide whether the sampled signal needs to be transmitted [20, 21]. In this chapter, the event-triggered mechanism is applied in the DoS attack model. The occurrence of DoS attack is determined by a historical-frequency-triggered mechanism, which is designed by combining memory-event-triggered strategy and multiple historical frequency information. Therefore, a historical-frequency-triggered DoS attack model is proposed in this chapter of which the main features can be summarized as follows.

a. The consideration of power system frequency in the DoS attack triggering condition enhances the connection between the power system and the cyberattack. The proposed cyberattack model effectively reflects the impact of cyberattacks on the frequency of the power system, which is helpful for researching and analyzing power system security issues.

b. The unknown nature of cyberattacks that occur in the real world makes it difficult for defenders to quickly resolve them. Research on new cyberattacks can help defenders to develop plans, so as to deal with this type of cyberattack in time.

c. Due to the use of an event-triggering mechanism in the DoS attack model, an attack only occurs if the frequency-triggered scheme is satisfied. This is more in line with the aim of cyberattacks to destabilize power systems.

Increasing efforts have been made to develop countermeasures to protect LFC power systems against DoS attacks [22]. In [23], a distributed observer-based event-triggered secure LFC approach was investigated to mitigate the impact of cyberattacks. A security-based event-triggered fuzzy LFC scheme was developed for power systems under cross-layer DoS attacks in [24]. A secure adaptive event-triggered mechanism was proposed for cyber-physical power systems under energy-limited DoS attacks in [25]. Such researches demonstrate that it is a feasible way to deal with the contamination of cyberattacks and ensure the security performance of the power system by devising proper control schemes.

Due to its simple structure and strong robustness, sliding mode control (SMC) has been widely used in systems with unexpected disturbances or external influences [26], it is not a surprise that SMC has been introduced into the LFC power system to improve stable performance. A fractional order parameter was incorporated into the sliding mode surface in [27], which can effectively improve the dynamic performance of time-varying delay LFC power systems. An event-triggered sliding-mode LFC approach was investigated for multi-area power systems under periodic DoS attacks in [28]. A second-order SMC via double integrated sliding surface was meant in [29] to improve the stable performance of multi-area steam-hydropower system. In [27–29], system state information was applied in the sliding mode controller design. In addition, researchers have found that the output signal can also be used for controller design. In [30] a static output feedback SMC was proposed for a class of uncertain systems with mismatched uncertainty in the state matrix. In [31], a dynamic output feedback SMC algorithm was investigated for linear multi-input multi-output systems with mismatched norm-bounded uncertainties. In [32], a memory output SMC law was designed to smooth the vibrations of consensus errors subject to estimation biases, singular perturbations, and input noises. Once the DoS attacks happened, the status of the state and the output may be different. Therefore, this chapter proposes a controller that can simultaneously guarantee the stability of the system state and the output.

In this chapter, a memory output sliding mode load frequency control (MOSMLFC) strategy is developed to ensure the security performance of power systems under malicious historical-frequency-triggered DoS attacks. Both the system state and the memory output are taken into account in the design of the sliding mode surface. If the system trajectory is in the required sliding surface, the entire state and output can be ensured, such that the security performance of the power system under historical-frequency-triggered DoS attacks can be obtained.

The main contributions can be summarized as follows.

1. A historical-frequency-triggered DoS attack model is developed by combining power system historical information, event-triggered scheme, and the attributes of DoS attacks. Due to the application of frequency information, the proposed DoS attack model is helpful for investigating the security problem in the context of power system than that in [11–17].
2. A MOSMLFC strategy is proposed to ensure the stabilization and security performance of multi-area power systems under cyberattacks. Compared with those design methods using only the current transmitted signal [27–29], the proposed approach achieves stronger robustness by properly utilizing system state and the memory-output information.
3. Considering the contamination of DoS attacks, a sufficient condition for the power system security performance is obtained. By utilizing the treatment method of inequality [33], a less conservative condition for stability is obtained compared to that in [28].

8.2 Problem Formulation

In this part, a simplified multi-area power system model is considered in Figure 8.1, which consists of the turbine, power generator, governor, and so on. To investigate the impact of cyberattacks on power system security, a historical-frequency-triggered DoS attack model is proposed.

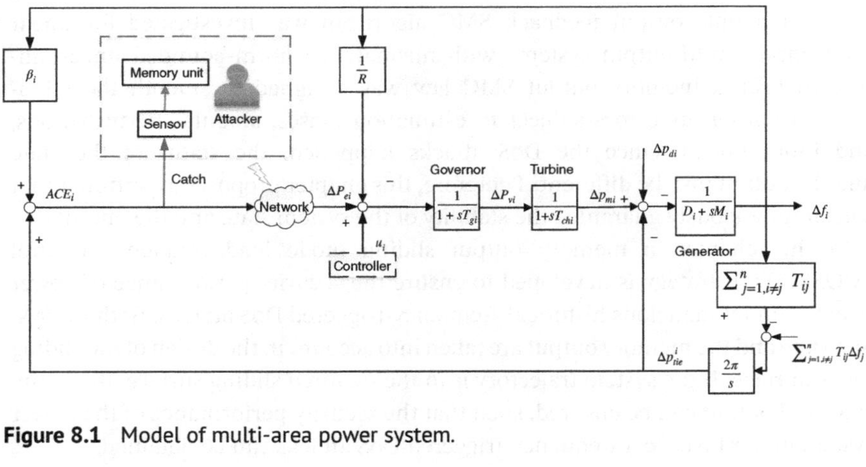

Figure 8.1 Model of multi-area power system.

8.2.1 Power System Model

The dynamic equations of the system based on the system model and information transfer can be described as

$$\begin{cases} \Delta\dot{P}_{mi}(t) = -\dfrac{\Delta P_{mi}(t)}{T_{\text{chi}}} + \dfrac{\Delta P_{vi}(t)}{T_{\text{chi}}} \\ \Delta\dot{P}^{i}_{\text{tie}}(t) = 2\pi \sum\limits_{j=1,i\neq j}^{n} T_{ij}(\Delta f_i(t) - \Delta f_j(t)) \\ \Delta\dot{P}_{vi}(t) = -\dfrac{\Delta f_i(t)}{R_i T_{gi}} - \dfrac{\Delta P_{vi}(t)}{T_{gi}} - \dfrac{\Delta P_{ei}(t)}{T_{gi}} \\ \Delta\dot{f}_i(t) = \dfrac{\Delta P_{mi}(t)}{M_i} - \dfrac{D_i \Delta f_i(t)}{M_i} - \dfrac{\Delta P^{i}_{tie}(t)}{M_i} - \dfrac{\Delta P_{di}(t)}{M_i} \\ ACE_i(t) = \Delta\dot{P}_{\text{tie}i}(t) + \beta_i \Delta f_i(t) \end{cases} \tag{8.1}$$

Hence, the state vector space model of the multi-area power systems can be represented as

$$\begin{cases} \dot{x}_i(t) = A_i x_i(t) + B_i u_i(t) + F_i w_i(t) \\ y_i(t) = C_i x_i(t) \end{cases} \tag{8.2}$$

where $w(t)$ is the bounded disturbance of the system, which is generally defined as $w(t) = \Delta P_{di} < \epsilon$, where ϵ is a positive constant.

And, we have definitions as follows

$$x_i(t) = [\Delta P_{mi}\ \Delta P_{vi}\ \Delta P_{\text{tie}i}\ \Delta f_i]^{\mathsf{T}},\ y_i(t) = ACE_i(t)$$

$$A_i = \begin{bmatrix} -\dfrac{1}{T_{\text{chi}}} & \dfrac{1}{T_{\text{chi}}} & 0 & 0 \\ 0 & -\dfrac{1}{T_{gi}} & 0 & -\dfrac{1}{R_i T_{gi}} \\ 0 & 0 & 0 & 2\pi \sum\limits_{j=1,j\neq i}^{n} T_{ij} \\ \dfrac{1}{M_i} & 0 & -\dfrac{1}{M_i} & -\dfrac{D_i}{M_i} \end{bmatrix},$$

$$B_i = \begin{bmatrix} 0 & \dfrac{1}{T_{gi}} & 0_{1\times 2} \end{bmatrix}, F_i = \begin{bmatrix} 0_{1\times 3} & -\dfrac{1}{M_i} \end{bmatrix}.$$

$$C_i = \begin{bmatrix} 0 & 0 & 1 & \beta_i \end{bmatrix}.$$

8.2.2 DoS Attacks Model

Contamination from cyberattacks is considered to be a type of historical-frequency-triggered DoS attack. The attackers are assumed to use the historical frequency of the power systems by a frequency sensor and memory storage. The action of DoS attack is reconfigured based on the designed historical-frequency-triggered mechanism. The historical-frequency-triggered mechanism is given as

$$\begin{cases} |\Delta \bar{f}_i| < |\Delta \bar{f}_A|, \text{ triggered} \\ |\Delta \bar{f}_i| \geq |\Delta \bar{f}_A|, \text{ not triggered} \end{cases} \tag{8.3}$$

where $\Delta \bar{f}_i$ is the attack frequency deviation and will be designed later, and $\Delta \bar{f}_A$ is defined as the triggering threshold. The occurrence of DoS attacks depends on the triggering scheme. The historical-frequency-triggered parameters are described as

$$\Delta \bar{f}_A = \frac{1}{m} \sum_{p=1}^{m} \Delta f_p \tag{8.4}$$

where, Δf_p denotes the historical frequency deviation, and m denotes the number of recorded historical frequencies. Obviously, $\Delta \bar{f}_A$ changes with the frequency of the power systems.

Remark 8.1: For frequency-triggered cyberattacks, the design of the triggered threshold will directly affect the attacking node of cyberattacks. The selection of a fixed threshold makes the cyberattack only happen within an unchangeable attack condition [34]. However, finding a proper selection of the fixed threshold could be difficult. Too large a threshold will cause cyberattacks to occur at each sampling. Too small a threshold will make network attacks unable to affect system stability. For example, the frequency variation range of the power system under normal operation is 50 ± 0.2. When the threshold is set to 50.1, the system can be under stable operation at 50.15 while not contaminated by frequency-triggered cyberattacks with fixed threshold. To carry out effective attacks, a historical-frequency-triggered scheme applies the historical information of the power systems. The historical-frequency triggering threshold changes along with the variability of power system frequency in operation, which helps attackers obtain a dynamic threshold.

Remark 8.2: In the Bernoulli-distributed cyberattack model, the occurrence of the attack follows the given probability [35], which means that the action of the Bernoulli-distributed DoS attacks is uncontrollable at all times. Even if the power system is already unstable, a DoS attack can occur. To our knowledge, the aim of cyberattack is to destroy the stable operation of the target system. If the system is already unstable, the happening of cyberattacks at this moment is meaningless or the destructive power at this time is insufficient. Differently, the proposed historical-frequency-triggered DoS attack happens when the power system is approaching to stable state. This enhances the destructive power of the attack and prolongs the duration of its existence.

Then, we need to determine whether the current system frequency meets the historical-frequency-triggering condition. The attack frequency deviation is given as

$$\Delta \bar{f}_i = f_N - f_s, \tag{8.5}$$

where f_s is defined as the real-time frequency measured by frequency sensor, and f_N is the frequency under normal operating of power system.

Remark 8.3: Frequency deviation is simultaneously applied to the attack model and control strategy. Attack frequency deviation $\Delta \bar{f}_i$ is used to design the historical-frequency-triggered DoS attack model. LFC power system frequency deviation Δf_i is applied to the system stability analysis.

For an ideal operating power system, the stable state is confirmed when $\Delta \bar{f}_i = 0$. However, from the attacker's perspective, the aim of attack action is to destroy the stable operation of the power system. Taking into consideration of the power system stability criterion, the purpose of the DoS attacks, and (8.3), the system state can be divided into two cases, which are given as follows

$$\begin{cases} |\Delta \bar{f}_i| < |\Delta \bar{f}_A|, & \text{attack region} \\ |\Delta \bar{f}_i| \geq |\Delta \bar{f}_A|, & \text{without attack} \end{cases} \tag{8.6}$$

Figure 8.2 illustrates the changes of (8.3) and historical-frequency-triggered DoS attacks contamination region.

Obviously, only the sampled signals that are close to a stable state are affected by historical-frequency-triggered DoS attacks. Once a DoS attack happens, signal

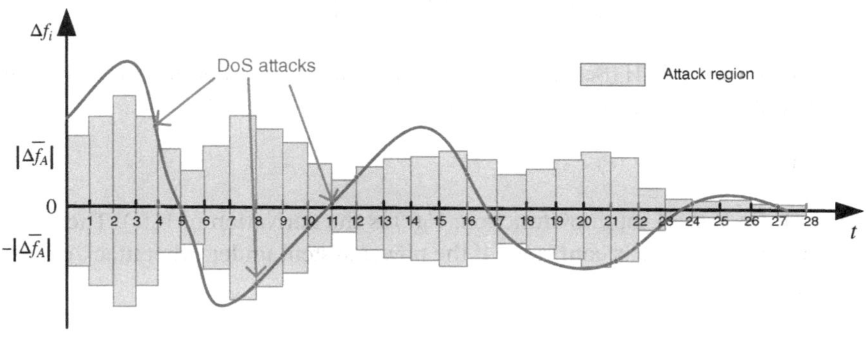

Figure 8.2 Example of historical-frequency-triggered DoS attacks.

transmission will be interrupted. The contamination of DoS attacks is considered as a time-varying delay [36], and given as

$$\hat{x}_a(t) = \begin{cases} x_i(t) & \text{without DoS attacks} \\ x_i(t - \tau_i(t)) & \text{DoS attacks} \end{cases} \tag{8.7}$$

where $\tau_m \leq \tau_i(t) < \tau_M$, and $0 \leq \tau_m < \tau_M$ are the lower and higher bounds of the time delay.

8.3 Main Results

Lemma 8.1 [33] *For $x_i(t) \in R^n$ and $\upsilon \in R$, there exist symmetric matrix $P > 0$, such that the following inequalities hold*

$$-\upsilon \int_{t-\upsilon}^{t} x_i^\top(s) P x_i(s) ds \leq -\int_{t-\upsilon}^{t} x_i^\top(s) ds P \int_{t-\upsilon}^{t} x_i(s) ds,$$

$$-\frac{\upsilon^2}{2} \int_{-\upsilon}^{0} \int_{t+\beta}^{\beta} x_i^\top(s) P x_i(s) ds d\beta \leq$$
$$- \int_{-\upsilon}^{0} \int_{t+\beta}^{t} x_i^\top(s) ds d\beta P \int_{-\upsilon}^{0} \int_{t+\beta}^{t} x_i(s) ds d\beta$$

Lemma 8.2 [33] *For $x_i(t) \in R^n$ and $\upsilon \in R$, there exist symmetric matrix $P > 0$, and matrix U with proper dimensions, such that the following inequality holds*

$$-\upsilon \int_{t-\upsilon}^{t} \dot{x}_i^\top(s) P \dot{x}_i(s) ds \leq$$
$$\begin{bmatrix} x_i(t) \\ x_i(t-\upsilon) \end{bmatrix}^\top \begin{bmatrix} -P & U^\top \\ U & -P \end{bmatrix} \begin{bmatrix} x_i(t) \\ x_i(t-\upsilon) \end{bmatrix}.$$

Definition 8.1 [24] (Security performance) Based on Lyapunov theorem, the power system under cyberattack is stable if

$$V_i(t) > 0 \text{ and } \dot{V}_i(t) < 0$$

where $V_i(t)$ is the Lyapunov function, $\dot{V}_i(t)$ is the derivation of $V_i(t)$. Then, the security performance is confirmed if the power system under cyberattack can be stable.

8.3.1 Memory Output SMC Scheme

The sliding mode surface in [37–39] is designed based on the currently transmitted signal. In this chapter, a memory output SMC is developed by using the historical output information for the power systems. It is worth noticing that $\Delta P_{ei}(t) = \int ACE_i(t)$, and we have $\Delta \dot{P}_{ei}(t) = ACE_i(t) = y_i(t)$.

Then, a memory sliding mode surface can be developed as

$$\begin{aligned}\sigma = L_i x_i(t) - \int_0^t L_i A_i x_i(s) ds \\ - \frac{1}{n}\sum \int_0^t L_i B_i K_i \Delta P_{ei}(s) ds\end{aligned} \tag{8.8}$$

where L_i is selected to ensure that $L_i B_i$ is nonsingular.

Taking the derivation on (8.8), we have

$$\dot{\sigma} = L_i \dot{x}_i(t) - L_i A_i x_i(t) - L_i B_i K_i \bar{y}_A(t) \tag{8.9}$$

where $\bar{y}_A(t)$ is designed later.

Remark 8.4: The sliding mode surface is designed by only using the most recently transmitted signal [37–39] or the current output signal [40]. However, there exists a situation that the system trajectory crosses the stable state. The stability of the power system may be misjudged by using the signal at the moment of contact. Differently, the proposed memory output sliding mode surface is designed based on the historical output signals. The use of multiple information can more accurately ensure the stable state of the system and improve the robustness of the system to a certain degree.

Remark 8.5: System output information is introduced in the sliding mode surface design. In [40], system output $y_i(t)$ is introduced in the sliding surface design and its derivative is $\dot{y}_i(t) = C_i \dot{x}_i(t)$. This increases the complexity of the controller design. In this chapter, the system output is $\Delta P_{ei}(t) = \int ACE_i(t)$ and its derivative is $\int A\dot{C}E_i(t) = C_i x_i(t)$. Thus, the system output is used in the sliding surface design while reducing the complexity.

Firstly, the memory output signal is defined as

$$\bar{y}_i(t) = \frac{1}{n}\sum_{q=1}^{n} C_i x_i(t) \tag{8.10}$$

The memory signal (8.10) under historical-frequency-triggered DoS attacks can be expressed as

$$\bar{y}_A(t) = C_i\bar{\hat{x}}_A(t) = C_i\frac{1}{n}\sum_{q=1}^{n}\hat{x}_a(t) \tag{8.11}$$

Based on the SMC theorem, a proper sliding mode surface should satisfy $\sigma = \dot{\sigma} = 0$. Therefore, we have the following equation

$$L_i\dot{x}_i(t) - L_iA_ix_i(t) - L_iB_iK_i\bar{y}_A(t) = 0 \tag{8.12}$$

Combining with power system state function (8.2), we have

$$\begin{aligned} &L_i(A_ix_i(t) + B_iu_i(t) + F_iw_i(t)) \\ &\qquad - L_iA_ix_i(t) - L_iB_iK_iC_i\bar{\hat{x}}_A(t) = 0 \end{aligned} \tag{8.13}$$

Thus, the equivalent memory sliding mode controller can be derived as

$$u_{eq}(t) = -K_iC_i\bar{\hat{x}}_A(t) - (L_iB_i)^{-1}L_iF_iw_i(t) \tag{8.14}$$

Once the DoS attacks have occurred, the sampled signal $x_i(t_k)$ cannot be transmitted. Control center must use last received signal $x_i(t_{k-1})$. Then, we can derive the state space function of the MOSMLFC power systems under historical-frequency-triggered DoS attacks as

$$\begin{cases} \dot{x}_i(t) = A_ix_i(t) - B_iK_iC_i\bar{\hat{x}}_A(t) + \hat{F}_iw_i(t) \\ y_i(t) = C_ix_i(t) \end{cases} \tag{8.15}$$

where $\hat{F}_i = F_i - B_i(L_iB_i)^{-1}L_iF_i$.

8.3.2 Security Performance Analysis

Theorem 8.1 *For given constants $\tau_M > \tau_m \geq 0$, θ and $\eta \in [0, 1]$, system (8.15) is secure, if there exist symmetric matrices $P_i > 0$, $Q_1 > 0$, $Q_2 > 0$, $Q_3 > 0$, $Q_4 > 0$, $R_1 > 0$, $R_2 > 0$, $S_1 > 0$, $S_2 > 0$, $S_3 > 0$, $S_4 > 0$, $G_1 > 0$, $G_2 > 0$, $G_3 > 0$, $G_4 > 0$, and U_i with appropriate dimensions, such that*

$$\begin{bmatrix} \Phi_{11} & \Phi_{21}^{\mathrm{T}} \\ \Phi_{21} & \Phi_{22} \end{bmatrix} \prec 0 \tag{8.16}$$

$$\begin{bmatrix} -S_2 & U_i^\top \\ U_i & -S_2 \end{bmatrix} \succ 0 \tag{8.17}$$

where $\Phi_{11} = [\phi_{\bar{i},\bar{j}}]_{11\times 11}$,

$$\phi_{1,1} = A_i^\top P_i + P_i^\top A_i + R_1 + \tau_m^2 S_1 - S_2 + (\tau_M - \tau_m)^2 S_3 + C_i^\top C_i + \frac{\tau_m^4}{4} G_1 - \tau_m^2 G_2$$
$$+ \frac{(\tau_2^2 - \tau_1^2)^2}{4} G_3 - 3(\tau_M - \tau_m)^2 G_4,$$
$$\phi_{2,2} = -3S4,\ \phi_{3,3} = -R_1 + R_2 - S_2 - (2-\theta)S_4.$$
$$\phi_{4,4} = -R_2 - (1+\theta)S_4,\ \phi_{5,5} = -S_1 - G_2,$$
$$\phi_{6,6} = -(2-\theta)S_3 - (2-\eta)G_4,$$
$$\phi_{7,7} = -(1+\theta)S_3 - (1+\eta)G_4,\ \phi_{8,8} = -G_1,$$
$$\phi_{9,9} = -(2-\eta)G_3,\ \phi_{10,10} = -(1+\eta)G_3,\ \phi_{11,11} = -\gamma^2 I,$$
$$\phi_{2,1} = \frac{1}{n} C_i^\top K_i^\top B_i^\top P_i,\ \phi_{3,1} = U,\ \phi_{3,2} = (2-\theta)S_4,$$
$$\phi_{4,2} = (1+\theta)S_4,\ \phi_{5,1} = P_i + Q_1 + \tau_m G_2,\ \phi_{5,3} = -P_i - Q_1,$$
$$\phi_{6,1} = (2-\eta)(\tau_M - \tau_m)G_4,\ \phi_{6,3} = P_i + Q_2,$$
$$\phi_{6,4} = -P_i - Q_2,\ \phi_{7,1} = (1+\eta)(\tau_M - \tau_m)G_4,$$
$$\phi_{7,3} = P_i + Q_2,\ \phi_{7,4} = -P_i - Q_2,\ \phi_{8,1} = P_i + Q_3,$$
$$\phi_{8,5} = -P_i - Q_3,\ \phi_{9,1} = (\tau_M - \tau_m)(P_i + Q_4),$$
$$\phi_{9,6} = -P_i - Q_4,\ \phi_{9,7} = -P_i - Q_4,$$
$$\phi_{10,1} = (\tau_M - \tau_m)(P_i + Q_4),\ \phi_{10,6} = -P_i - Q_4,$$
$$\phi_{10,7} = -P_i - Q_4,\ \phi_{11,1} = \hat{F}_i^\top P_i,$$
$$\Phi_{21} = \mathrm{col}[\tau_1 \Lambda, (\tau_2 - \tau_1)\Lambda, \frac{\tau_1^2}{2}\Lambda, \frac{\tau_2^2 - \tau_1^2}{2}\Lambda],$$
$$\Phi_{22} = -\mathrm{diag}\{S_2, S_4, G_2, G_4\},$$
$$\Lambda = [A_i, \frac{1}{n} B_i K_i C_i, 0, 0, 0, 0, 0, 0, 0, 0, \hat{F}_i].$$

Proof. Choose the Lyapunov–Krasovskii functional as

$$V_i(t) = V_1 + V_2 + V_3 + V_4 + V_5 \tag{8.18}$$

where

$$\begin{aligned}
V_1 &= x_i^{\top}(t)P_i x_i(t) + \int_{t-\tau_m}^{t} x_i^{\top}(s)dsP_i \int_{t-\tau_m}^{t} x_i(s)ds \\
&+ \int_{t-\tau_m}^{t-\tau_M} x_i^{\top}(s)dsP_i \int_{t-\tau_m}^{t-\tau_M} x_i(s)ds \\
&+ \int_{-\tau_m}^{0}\int_{t+\beta}^{t} x_i^{\top}(s)dsd\beta P_i \int_{-\tau_m}^{0}\int_{t+\beta}^{t} x_i(s)dsd\beta \\
&+ \int_{-\tau_M}^{-\tau_m}\int_{t+\beta}^{t} x_i^{\top}(s)dsd\beta P_i \int_{-\tau_M}^{-\tau_m}\int_{t+\beta}^{t} x_i(s)dsd\beta,
\end{aligned}$$

$$\begin{aligned}
V_2 &= \int_{t-\tau_m}^{t} x_i^{\top}(s)dsQ_1 \int_{t-\tau_m}^{t} x_i(s)ds \\
&+ \int_{t-\tau_m}^{t-\tau_M} x_i^{\top}(s)dsQ_2 \int_{t-\tau_m}^{t-\tau_M} x_i(s)ds \\
&+ \int_{-\tau_m}^{0}\int_{t+\beta}^{t} x_i^{\top}(s)dsd\beta Q_3 \int_{-\tau_m}^{0}\int_{t+\beta}^{t} x_i(s)dsd\beta \\
&+ \int_{-\tau_M}^{-\tau_m}\int_{t+\beta}^{t} x_i^{\top}(s)dsd\beta Q_4 \int_{-\tau_M}^{-\tau_m}\int_{t+\beta}^{t} x_i(s)dsd\beta
\end{aligned}$$

$$V_3 = \int_{t-\tau_m}^{t} x_i^{\top}(s)R_1 x_i(s)ds + \int_{t-\tau_M}^{t-\tau_m} x_i^{\top}(s)R_2 x_i(s)ds$$

$$\begin{aligned}
V_4 &= \tau_m \int_{-\tau_m}^{0}\int_{t+\beta}^{t} x_i^{\top} S_1 x_i(s)dsd\beta \\
&+ \tau_m \int_{-\tau_m}^{0}\int_{t+\beta}^{t} \dot{x}_i^{\top}(s)S_2 \dot{x}_i(s)dsd\beta \\
&+ (\tau_M - \tau_m) \int_{-\tau_M}^{-\tau_m}\int_{t+\beta}^{t} x_i^{\top}(s)S_3 x_i(s)dsd\beta \\
&+ (\tau_M - \tau_m) \int_{-\tau_M}^{-\tau_m}\int_{t+\beta}^{t} \dot{x}_i^{\top}(s)S_4 \dot{x}_i(s)dsd\beta
\end{aligned}$$

$$V_5 = \frac{\tau_m^2}{2}\int_{-\tau_m}^{0}\int_{\beta}^{0}\int_{t+\alpha}^{t} x_i^\top(s)G_1x_i(s)dsd\alpha d\beta$$
$$+\frac{\tau_m^2}{2}\int_{-\tau_m}^{0}\int_{\beta}^{0}\int_{t+\alpha}^{t} \dot{x}_i^\top(s)G_2\dot{x}_i(s)dsd\alpha d\beta$$
$$+\frac{\tau_M^2-\tau_m^2}{2}\int_{-\tau_M}^{-\tau_m}\int_{\beta}^{0}\int_{t+\alpha}^{t} x_i^\top(s)G_3x_i(s)dsd\alpha d\beta$$
$$+\frac{\tau_M^2-\tau_m^2}{2}\int_{-\tau_M}^{-\tau_m}\int_{\beta}^{0}\int_{t+\alpha}^{t} \dot{x}_i^\top(s)G_4\dot{x}_i(s)dsd\alpha d\beta$$

Take deviation on $V_i(t)$, we have

$$\dot{V}_1 = \text{sym}\left\{\dot{x}_i^\top P_i x_i(t) + \left(x_i^\top(t) - x_i^\top(t-\tau_m)\right) P_i \int_{t-\tau_m}^{t} x_i(s)ds\right.$$
$$+\left(x_i^\top(t-\tau_m) - x_i^\top(t-\tau_M)\right) P_i \int_{t-\tau_M}^{t-\tau_m} x_i(s)ds$$
$$+\left(\tau_m x_i^\top(t) - \int_{t-\tau_m}^{t} x_i(s)ds\right) P_i \int_{-\tau_m}^{0}\int_{t+\beta}^{t} x_i(s)dsd\beta$$
$$+(\tau_M-\tau_m)x_i^\top(t)dsP_i\int_{-\tau_M}^{-\tau_m}\int_{t+\beta}^{t} x_i(s)dsd\beta$$
$$\left.-\int_{t-\tau_M}^{t-\tau_m} x_i(s)dsP_i\int_{-\tau_M}^{-\tau_m}\int_{t+\beta}^{t} x_i(s)dsd\beta\right\},$$

$$\dot{V}_2 = \text{sym}\left\{\left(x_i^\top(t) - x_i^\top(t-\tau_m)\right) Q_1 \int_{t-\tau_m}^{t} x_i(s)ds\right.$$
$$+\left(x_i^\top(t-\tau_m) - x_i^\top(t-\tau_M)\right) Q_2 \int_{t-\tau_M}^{t-\tau_m} x_i(s)ds$$
$$+\left(\tau_m x_i^\top(t) - \int_{t-\tau_m}^{t} x_i^\top(s)ds\right) Q_3 \int_{-\tau_m}^{0}\int_{t+\beta}^{t} x_i(s)dsd\beta$$
$$+(\tau_M-\tau_m)x_i^\top(t)Q_4\int_{-\tau_M}^{-\tau_m}\int_{t+\beta}^{t} x_i(s)dsd\beta$$
$$\left.-\int_{t-\tau_M}^{t-\tau_m} x_i(s)dsQ_4\int_{-\tau_M}^{-\tau_m}\int_{t+\beta}^{t} x_i(s)dsd\beta\right\},$$

$$\dot{V}_3 = x_i^{\mathrm{T}}(t)R_1x_i(t) - x_i^{\mathrm{T}}(t-\tau_m)R_1x_i(t-\tau_m)$$
$$+ x_i^{\mathrm{T}}(t-\tau_m)R_2x_i(t-\tau_m) - x_i^{\mathrm{T}}(t-\tau_M)R_2x_i(t-\tau_M),$$

$$\dot{V}_4 = \tau_m^2 x_i^{\mathrm{T}}(t)S_1x_i(t) - \tau_m \int_{t-\tau_m}^{t} x_i(s)S_1x_i(s)ds$$
$$+ \tau_m^2 \dot{x}_i^{\mathrm{T}}(t)S_2\dot{x}_i(t) - \tau_m \int_{t-\tau_m}^{t} \dot{x}_i^{\mathrm{T}}(s)S_2\dot{x}_i(s)ds$$
$$+ (\tau_M-\tau_m)^2 x_i^{\mathrm{T}}(t)S_3x_i(t) - (\tau_M-\tau_m)\int_{t-\tau_M}^{t-\tau_m} x_i^{\mathrm{T}}(s)S_3x_i(s)ds$$
$$+ (\tau_M-\tau_m)^2 \dot{x}_i^{\mathrm{T}}(t)S_4\dot{x}_i(t) - (\tau_M-\tau_m)\int_{t-\tau_M}^{t-\tau_m} \dot{x}_i^{\mathrm{T}}(s)S_4\dot{x}_i(s)ds,$$

$$\dot{V}_5 = \frac{\tau_m^4}{4}x_i^{\mathrm{T}}(t)G_1x_i(t) - \frac{\tau_m^2}{2}\int_{-\tau_m}^{0}\int_{t+\beta}^{t} x_i^{\mathrm{T}}(s)G_1x_i(s)dsd\beta$$
$$+ \frac{\tau_m^4}{4}\dot{x}_i^{\mathrm{T}}(t)G_2\dot{x}_i(t) - \frac{\tau_m^2}{2}\int_{-\tau_m}^{0}\int_{t+\beta}^{t} \dot{x}_i^{\mathrm{T}}(s)G_2\dot{x}_i(s)dsd\beta$$
$$+ \frac{(\tau_M^2-\tau_m^2)^2}{4}x_i^{\mathrm{T}}(t)G_3x_i(t) + \frac{(\tau_M^2-\tau_m^2)^2}{4}\dot{x}_i^{\mathrm{T}}(t)G_4\dot{x}_i(t)$$
$$- \frac{\tau_M^2-\tau_m^2}{2}\int_{-\tau_M}^{-\tau_m}\int_{t+\beta}^{t} x_i^{\mathrm{T}}(s)G_3x_i(s)dsd\beta$$
$$- \frac{\tau_M^2-\tau_m^2}{2}\int_{-\tau_M}^{-\tau_m}\int_{t+\beta}^{t} \dot{x}_i^{\mathrm{T}}(s)G_4\dot{x}_i(s)dsd\beta,$$

Focusing on $\dot{V}_1$, we have the following

$$(\tau_M-\tau_m)x_i^{\mathrm{T}}(t)dsP_i\int_{-\tau_M}^{-\tau_m}\int_{t+\beta}^{t} x_i(s)dsd\beta$$
$$= (\tau_M-\tau_m)x_i^{\mathrm{T}}(t)dsP_i\left(\int_{-\tau_i(t)}^{-\tau_m}\int_{t+\beta}^{t} x_i(s)dsd\beta \right. \tag{8.19}$$
$$\left. + \int_{-\tau_M}^{-\tau_i(t)}\int_{t+\beta}^{t} x_i(s)dsd\beta\right).$$

Similarly

$$\begin{aligned}
&-\int_{t-\tau_M}^{t-\tau_m} x_i(s)dsP_i \int_{-\tau_M}^{-\tau_m}\int_{t+\beta}^{t} x_i(s)dsd\beta \\
&= -\left(\int_{t-\tau_i(t)}^{t-\tau_m} x_i^{\mathrm{T}}(t)ds + \int_{t-\tau_M}^{t-\tau_i(t)} x_i^{\mathrm{T}}(t)ds\right) P_i \\
&\left(\int_{-\tau_i(t)}^{-\tau_m}\int_{t+\beta}^{t} x_i(s)dsd\beta + \int_{-\tau_M}^{-\tau_i(t)}\int_{t+\beta}^{t} x_i(s)dsd\beta\right).
\end{aligned} \tag{8.20}$$

Focusing on $\dot{V}_4$, we have the following by applying **Lemma 8.1**,

$$-\tau_m \int_{t-\tau_m}^{t} x_i(s)S_1x_i(s)ds \leq -\int_{t-\tau_m}^{t} x_i^{\mathrm{T}}(s)dsS_1 \int_{t-\tau_m}^{t} x_i(s)ds. \tag{8.21}$$

By applying **Lemma 8.2**, we have

$$\begin{aligned}
&-\tau_m \int_{t-\tau_m}^{t} \dot{x}_i(s)S_2\dot{x}_i(s)ds \leq -x_i^{\mathrm{T}}(t)S_2x_i(t) + x_i^{\mathrm{T}}(t-\tau_m)U_ix_i(t) \\
&+ x_i^{\mathrm{T}}(t)U_ix_i(t-\tau_m) - x_i^{\mathrm{T}}(t-\tau_m)S_2x_i(t-\tau_m).
\end{aligned} \tag{8.22}$$

The time-varying delay is defined as $\tau_m \leq \tau_i(t) \leq \tau_M$. By defining $\theta = \dfrac{\tau_i(t)-\tau_m}{\tau_M-\tau_m}$ and applying **Lemma 8.1**, we have

$$\begin{aligned}
&-(\tau_M-\tau_m)\int_{t-\tau_M}^{t-\tau_m} x_i^{\mathrm{T}}(s)S_3x_i(s)ds \\
&= -(\tau_M-\tau_i(t))\int_{t-\tau_M}^{t-\tau_i(t)} x_i^{\mathrm{T}}(s)S_3x_i(s)ds - (\tau_i(t)-\tau_m)\int_{t-\tau_M}^{t-\tau_i(t)} x_i^{\mathrm{T}}(s)S_3x_i(s)ds \\
&-(\tau_M-\tau_i(t))\int_{t-\tau_i(t)}^{t-\tau_m} x_i^{\mathrm{T}}(s)S_3x_i(s)ds - (\tau_i(t)-\tau_m)\int_{t-\tau_i(t)}^{t-\tau_m} x_i^{\mathrm{T}}(s)S_3x_i(s)ds \\
&\leq -\int_{t-\tau_M}^{t-\tau_i(t)} x_i^{\mathrm{T}}(s)dsS_3\int_{t-\tau_M}^{t-\tau_i(t)} x_i(s)ds - \theta\int_{t-\tau_M}^{t-\tau_i(t)} x_i^{\mathrm{T}}(s)dsS_3\int_{t-\tau_M}^{t-\tau_i(t)} x_i(s)ds \\
&-\int_{t-\tau_i(t)}^{t-\tau_m} x_i^{\mathrm{T}}(s)dsS_3\int_{t-\tau_i(t)}^{t-\tau_m} x_i(s)ds - (1-\theta)\int_{t-\tau_i(t)}^{t-\tau_m} x_i^{\mathrm{T}}(s)dsS_3\int_{t-\tau_i(t)}^{t-\tau_m} x_i(s)ds.
\end{aligned} \tag{8.23}$$

Furthermore, we have

$$
\begin{aligned}
&-(\tau_M-\tau_m)\int_{t-\tau_M}^{t-\tau_m}\dot{x}_i^{\mathrm{T}}(s)S_4\dot{x}_i(s)ds\\
\leq&-(1+\theta)\big[x_i^{\mathrm{T}}(t-\tau_i(t))S_4x_i(t-\tau_i(t))\\
&-x_i^{\mathrm{T}}(t-\tau_M)S_4x_i(t-\tau_i(t))-x_i^{\mathrm{T}}(t-\tau_i(t))S_4x_i(t-\tau_M)\\
&+x_i^{\mathrm{T}}(t-\tau_M)S_4x_i(t-\tau_M)\big]\\
&-(2-\theta)\big[x_i^{\mathrm{T}}(t-\tau_m)S_4x_i(t-\tau_m)\\
&-x_i^{\mathrm{T}}(t-\tau_i(t))S_4x_i(t-\tau_m)-x_i^{\mathrm{T}}(t-\tau_m)S_4x_i(t-\tau_i(t))\\
&+x_i^{\mathrm{T}}(t-\tau_i(t))S_4x_i(t-\tau_i(t))\big].
\end{aligned}
\tag{8.24}
$$

Focusing on $\dot{V}_5$, we have

$$
\begin{aligned}
&-\frac{\tau_m^2}{2}\int_{-\tau_m}^{0}\int_{t+\beta}^{t}x_i^{\mathrm{T}}(s)G_1x_i(s)dsd\beta\\
\leq&-\int_{-\tau_m}^{0}\int_{t+\beta}^{t}x_i^{\mathrm{T}}(s)dsd\beta G_1\int_{-\tau_m}^{0}\int_{t+\beta}^{t}x_i^{\mathrm{T}}(s)dsd\beta.
\end{aligned}
\tag{8.25}
$$

For further

$$
\begin{aligned}
&-\frac{\tau_m^2}{2}\int_{-\tau_m}^{0}\int_{t+\beta}^{t}\dot{x}_i^{\mathrm{T}}(s)G_2\dot{x}_i(s)dsd\beta\\
\leq&-\tau_m^2x_i^{\mathrm{T}}(t)G_2x_i(t)\\
&+\tau_m\int_{t-\tau_m}^{t}x_i^{\mathrm{T}}(s)dsG_2x_i(t)+\tau_mx_i^{\mathrm{T}}(t)G_2\int_{t-\tau_m}^{t}x_i(s)ds\\
&-\int_{t-\tau_m}^{t}x_i^{\mathrm{T}}(s)dsG_2\int_{t-\tau_m}^{t}x_i(s)ds.
\end{aligned}
\tag{8.26}
$$

Similarly

$$
\begin{aligned}
&-\frac{\tau_M^2-\tau_m^2}{2}\int_{-\tau_m}^{-\tau_M}\int_{t+\beta}^{t}x_i^{\mathrm{T}}(s)G_3x_i(s)dsd\beta\\
&=-\frac{\tau_M^2-\tau_i^2(t)}{2}\Big\{\int_{-\tau_M}^{-\tau_i(t)}x_i^{\mathrm{T}}(s)G_3x_i(s)dsd\beta+\int_{-\tau_i(t)}^{-\tau_m}x_i^{\mathrm{T}}(s)G_3x_i(s)dsd\beta\Big\}\\
&\quad-\frac{\tau_i^2(t)-\tau_m^2}{2}\Big\{\int_{-\tau_M}^{-\tau_i(t)}x_i^{\mathrm{T}}(s)G_3x_i(s)dsd\beta+\int_{-\tau_i(t)}^{-\tau_m}x_i^{\mathrm{T}}(s)G_3x_i(s)dsd\beta\Big\}\\
&\leq-\int_{-\tau_M}^{-\tau_i(t)}\int_{t+\beta}^{t}x_i^{\mathrm{T}}(s)dsG_3\int_{-\tau_M}^{-\tau_i(t)}\int_{t+\beta}^{t}x_i(s)dsd\beta\\
&\quad-\eta\int_{-\tau_M}^{-\tau_i(t)}\int_{t+\beta}^{t}x_i^{\mathrm{T}}(s)dsG_3\int_{-\tau_M}^{-\tau_i(t)}\int_{t+\beta}^{t}x_i(s)dsd\beta\\
&\quad-\int_{-\tau_i(t)}^{-\tau_m}\int_{t+\beta}^{t}x_i^{\mathrm{T}}(s)dsG_3\int_{-\tau_i(t)}^{-\tau_m}\int_{t+\beta}^{t}x_i(s)dsd\beta\\
&\quad-(1-\eta)\int_{-\tau_i(t)}^{-\tau_m}\int_{t+\beta}^{t}x_i^{\mathrm{T}}(s)dsG_3\int_{-\tau_i(t)}^{-\tau_m}\int_{t+\beta}^{t}x_i(s)dsd\beta,
\end{aligned}
\tag{8.27}
$$

where $\eta=\dfrac{\tau_i^2(t)-\tau_m^2}{\tau_M^2-\tau_m^2}$.

In the same way, we have

$$
\begin{aligned}
&-\frac{\tau_M^2-\tau_m^2}{2}\int_{-\tau_M}^{-\tau_m}\int_{t+\beta}^{t}\dot{x}_i^{\mathrm{T}}(s)G_4\dot{x}_i(s)dsd\beta\\
&\leq-(1+\eta)\Big[(\tau_M-\tau_m)^2x_i^{\mathrm{T}}(t)G_4x_i(t)-(\tau_M-\tau_m)\int_{t-\tau_M}^{t-\tau_i(t)}x_i^{\mathrm{T}}(s)dsG_4x_i(t)\\
&\quad-(\tau_M-\tau_m)x_i^{\mathrm{T}}(t)G_4\int_{t-\tau_M}^{t-\tau_i(t)}x_i(s)ds+\int_{t-\tau_M}^{t-\tau_i(t)}x_i^{\mathrm{T}}(s)dsG_4\int_{t-\tau_M}^{t-\tau_i(t)}x_i(s)ds\Big]\\
&\quad-(2-\eta)\Big[(\tau_M-\tau_m)^2x_i^{\mathrm{T}}(t)G_4x_i(t)\\
&\quad-(\tau_M-\tau_m)\int_{t-\tau_i(t)}^{t-\tau_m}x_i^{\mathrm{T}}(s)dsG_4x_i(t)\\
&\quad-(\tau_M-\tau_m)x_i^{\mathrm{T}}(t)G_4\int_{t-\tau_i(t)}^{t-\tau_m}x_i(s)ds\\
&\quad+\int_{t-\tau_i(t)}^{t-\tau_m}x_i^{\mathrm{T}}(s)dsG_4\int_{t-\tau_i(t)}^{t-\tau_m}x_i(s)ds\Big].
\end{aligned}
\tag{8.28}
$$

Introducing the power system state space function (8.15) for $\dot{x}_i(t)$ in $\dot{V}_1(t)$. For $w_i(t) = 0$, we have

$$\dot{x}_i^{\mathsf{T}} P_i x_i(t) = [A_i x_i(t) - B_i K_i C_i \bar{\hat{x}}_A(t)] P_i x_i(t). \tag{8.29}$$

Combining with (8.19)–(8.29) and introducing the attenuation level γ. For $w_i(t) \neq 0$, we have the following

$$\begin{aligned} &\Upsilon^{\mathsf{T}}(t)(\Phi_{11} - \Phi_{21}^{\mathsf{T}} \Phi_{22} \Phi_{21}) \Upsilon(t) \\ &-y_i^{\mathsf{T}}(t) y_i(t) + \gamma^2 w_i^{\mathsf{T}}(t) w_i(t) \leq 0. \end{aligned} \tag{8.30}$$

where

$$\begin{aligned} \Upsilon^{\mathsf{T}}(t) = \Big[&x_i(t), x_i(t - \tau_i(t)), x_i(t - \tau_m), x_i(t - \tau_M), \\ &\int_{t-\tau_m}^{t} x_i(s)ds, \int_{t-\tau_i(t)}^{t-\tau_m} x_i(s)ds, \int_{t-\tau_M}^{t-\tau_i(t)} x_i(s)ds, \\ &\int_{-\tau_m}^{0} \int_{t+\beta}^{t} x_i(s)dsd\beta, \int_{-\tau_i(t)}^{-\tau_m} \int_{t+\beta}^{t} x_i(s)dsd\beta, \\ &\int_{-\tau_M}^{-\tau_i(t)} \int_{t+\beta}^{t} x_i(s)dsd\beta, w_i(t)\Big]^{\mathsf{T}}. \end{aligned}$$

Then, we have

$$\Upsilon^{\mathsf{T}}(t)(\Phi_{11} - \Phi_{21}^{\mathsf{T}} \Phi_{22} \Phi_{21}) \Upsilon(t) \leq 0. \tag{8.31}$$

Under condition of $V(0) = 0$, on can obtain

$$\int_0^{+\infty} y_i^{\mathsf{T}}(t) y_i(t) \leq \int_0^{+\infty} \gamma^2 w_i^{\mathsf{T}}(t) w_i(t), \tag{8.32}$$

which means $\|y_i(t)\| \leq \gamma \|w_i(t)\|$ for nonzero $w_i(t) \in L_2[0, +\infty)$. Based on **Definition 8.1**, the MOSMLFC power system is security. The proof is complete.

Theorem 8.1 describes the security performance of the power system (8.15), but it is difficult to directly solve the controller gain K_i. Thus, we further convert **Theorem 8.1** into **Theorem 8.2** to obtain the controller gain.

Remark 8.6: **Theorem 8.1** provides the security performance condition for the power system under historical-frequency-triggered DoS attacks. Under the proposed strategy, the power system under historical-frequency-triggered DoS attacks is proven to be stable with conditions (8.16) and (8.17). Based on **Definition 8.1**, security performance is confirmed if the power system can remain stable under cyberattack.

Theorem 8.2 *For given constants $\tau_M > \tau_m \geq 0$, $\gamma > 0$, $\lambda > 0$, θ and $\eta \in [0, 1]$, system (8.15) is security, if there exist symmetric matrices $X_i = P_i^{-1} > 0$, $\tilde{Q}_1 > 0$, $\tilde{Q}_2 > 0$, $\tilde{Q}_3 > 0$, $\tilde{Q}_4 > 0$, $\tilde{R}_1 > 0$, $\tilde{R}_2 > 0$, $\tilde{S}_1 > 0$, $\tilde{S}_2 > 0$, $\tilde{S}_3 > 0$, $\tilde{S}_4 > 0$, $\tilde{G}_1 > 0$, $\tilde{G}_2 > 0$, $\tilde{G}_3 > 0$, $\tilde{G}_4 > 0$ and $\tilde{U}_i$, N_i, M_i with appropriate dimensions, such that the following linear matrix inequalities (LMIs) hold*

$$\begin{bmatrix} \tilde{\Phi}_{11} & * & * \\ \tilde{\Phi}_{21} & \tilde{\Phi}_{22} & * \\ \tilde{\Phi}_{31} & 0 & -I \end{bmatrix} \prec 0 \tag{8.33}$$

$$\begin{bmatrix} -\tilde{S}_2 & \tilde{U}_i^\top \\ \tilde{U}_i & -\tilde{S}_2 \end{bmatrix} \succ 0 \tag{8.34}$$

$$\begin{bmatrix} -\lambda I & * \\ N_i C_i - C_i X_i & -I \end{bmatrix} \prec 0, \lambda \to 0 \tag{8.35}$$

where $\tilde{\Phi}_{11} = [\tilde{\phi}_{i,j}]_{11\times 11}$,

$$\tilde{\phi}_{1,1} = X_i A_i^\top + A_i X_i + \tilde{R}_1 + \tau^2 \tilde{S}_1 - \tilde{S}_2 + (\tau_M - \tau_m)^2 \tilde{S}_3 + \frac{\tau_m^4}{4}\tilde{G}_1 - \tau_m^2 \tilde{G}_2 + \frac{(\tau_2^2 - \tau_1^2)^2}{4}\tilde{G}_3 - 3(\tau_M - \tau_m)^2 \tilde{G}_4,$$

$\tilde{\phi}_{2,2} = -3\tilde{S}_4$, $\tilde{\phi}_{3,3} = -\tilde{R}_1 + \tilde{R}_2 - \tilde{S}_2 - (2-\theta)\tilde{S}_4$

$\tilde{\phi}_{4,4} = -\tilde{R}_2 - (1+\theta)\tilde{S}_4$, $\tilde{\phi}_{5,5} = -\tilde{S}_1 - \tilde{G}_2$,

$\tilde{\phi}_{6,6} = -(2-\theta)\tilde{S}_3 - 3\tilde{G}_4$, $\tilde{\phi}_{7,7} = -(1+\theta)\tilde{S}_3 - (1+\eta)\tilde{G}_4$,

$\tilde{\phi}_{8,8} = -\tilde{G}_1$, $\tilde{\phi}_{9,9} = -(2-\eta)\tilde{G}_3$, $\tilde{\phi}_{10,10} = -(1+\eta)\tilde{G}_3$, $\tilde{\phi}_{11,11} = -\gamma^2 I$,

$\tilde{\phi}_{2,1} = \frac{1}{n} C_i^\top M_i^\top B_i^\top$, $\tilde{\phi}_{3,1} = \tilde{U}_i$, $\tilde{\phi}_{3,2} = (2-\theta)\tilde{S}_4$,

$\tilde{\phi}_{4,2} = (1+\theta)\tilde{S}_4$, $\tilde{\phi}_{5,1} = X_i + \tilde{Q}_1 + \tau_m \tilde{G}_2$,

$\tilde{\phi}_{5,3} = -X_i - \tilde{Q}_1$, $\tilde{\phi}_{6,1} = (2-\eta)(\tau_M - \tau_m)\tilde{G}_4$,

$\tilde{\phi}_{6,3} = X_i + \tilde{Q}_2$, $\tilde{\phi}_{6,4} = -X_i - \tilde{Q}_2$, $\tilde{\phi}_{7,1} = (1+\eta)(\tau_M - \tau_m)\tilde{G}_4$, $\tilde{\phi}_{7,3} = X_i + \tilde{Q}_2$,

$\tilde{\phi}_{7,4} = -X_i - \tilde{Q}_2$, $\tilde{\phi}_{8,1} = X_i + \tilde{Q}_3$,

$\tilde{\phi}_{8,5} = -X_i - \tilde{Q}_3$, $\tilde{\phi}_{9,1} = (\tau_M - \tau_m)(X_i + \tilde{Q}_4)$,

$\tilde{\phi}_{9,6} = -X_i - \tilde{Q}_4$, $\tilde{\phi}_{9,7} = -X_i - \tilde{Q}_4$,

$\tilde{\phi}_{10,1} = (\tau_M - \tau_m)(X_i + \tilde{Q}_4)$, $\tilde{\phi}_{10,6} = -X_i - \tilde{Q}_4$,

$\tilde{\phi}_{10,7} = -X_i - \tilde{Q}_4$, $\tilde{\phi}_{11,1} = \hat{F}_i^{\mathrm{T}}$,

$\tilde{\Phi}_{21} = \mathrm{col}[\tau_1\tilde{\Lambda}, (\tau_2 - \tau_1)\tilde{\Lambda}, \frac{\tau_1^2}{2}\tilde{\Lambda}, \frac{\tau_2^2 - \tau_1^2}{2}\tilde{\Lambda}]$, $\tilde{\Phi}_{22} = -\mathrm{diag}\{\tilde{S}_2, \tilde{S}_4, \tilde{G}_2, \tilde{G}_4\}$,

$\tilde{\Phi}_{31} = [C_iX_i, 0, 0, 0, 0, 0, 0, 0, 0, 0, 0]$,

$\tilde{\Lambda} = [A_iX_i, \frac{1}{n}B_iM_iC_i, 0, 0, 0, 0, 0, 0, 0, 0, \hat{F}_iX_i]$.

The MOSMLFC gain can be calculated by $K_i = M_iN_i^{-1}$.

Proof. Pre- and post-multiplying both sides of (8.16) with $\mathrm{diag}[X_i, X_i, X_i, X_i, X_i, X_i, X_i, X_i, X_i, X_i, I, S_2^{-1}, S_4^{-1}, G_2^{-1}, G_4^{-1}]$.

Let $\tilde{Q}_1 = X_iQ_1X_i$, $\tilde{Q}_2 = X_iQ_2X_i$, $\tilde{Q}_3 = X_iQ_3X_i$, $\tilde{Q}_4 = X_iQ_4X_i$, $\tilde{R}_1 = X_iR_1X_i$, $\tilde{R}_2 = X_iR_2X_i$, $\tilde{S}_1 = X_iS_1X_i$, $\tilde{S}_2 = X_iS_2X_i$, $\tilde{S}_3 = X_iS_3X_i$, $\tilde{S}_4 = X_iS_4X_i$, $\tilde{G}_1 = X_iG_1X_i$, $\tilde{G}_2 = X_iG_2X_i$, $\tilde{G}_3 = X_iG_3X_i$, $\tilde{G}_4 = X_iG_4X_i$. Noticing that there exists cross term $B_iK_iC_iX_i$. Defining

$$\begin{cases} K_iC_iX_i = M_iC_i \\ C_iX_i = N_iC_i \end{cases} \tag{8.36}$$

Then we have

$$K_i = M_iN_i^{-1} \tag{8.37}$$

Besides, $C_iX_i = N_iC_i$ equals $(N_iC_i - C_iX_i)^{\mathrm{T}}(N_iC_i - C_iX_i) = 0$. Based on the treatment in [25] and Schur complement theory, we have

$$\begin{bmatrix} -\lambda I & * \\ N_iC_i - C_iX_i & -I \end{bmatrix} \prec 0, \tag{8.38}$$

where λ is closer to 0. This completes the proof.

Remark 8.7: A new parameter λ is introduced to solve the cross-terms that appear in **Theorem 8.1**. To ensure the accuracy of the results, the parameter should satisfy $\lambda \to 0$. The effect of λ at different scales on the stable performance is analyzed in the simulation part.

8.3.3 Reachability Analysis

Theorem 8.3 *The trajectories of the power system (8.15) under historical-frequency-triggered DoS attacks can be ensured onto the designed sliding surface in a finite time with the following SMC law*

$$u_i(t) = K_i C_i \bar{\hat{x}}_A(t) - d_i (L_i B_i)^{-1} \|L_i F_i\| \mathrm{sgn}(\sigma_i(t)) \tag{8.39}$$

where $d_i > 0$, $\mathrm{sgn}(\cdot)$ *is the switch function.*

Proof. The Lyapunov function is chosen as

$$V_s = \frac{1}{2}\sigma_i^{\mathrm{T}}(t)\sigma_i(t) \tag{8.40}$$

Take deviation of V_s, we have

$$\begin{aligned}
\dot{V}_s &= \sigma_i^{\mathrm{T}}(t)\dot{\sigma}_i(t) \\
&= \sigma_i^{\mathrm{T}}(t)[L_i\dot{x}_i(t) - L_iA_ix_i(t) - L_iB_iK_iC_i\bar{\hat{x}}_A(t)] \\
&= \sigma_i^{\mathrm{T}}(t)\{L_iA_ix_i(t) + L_iB_iu_i(t) + L_i\hat{F}_iw_i(t) \\
&\quad - L_iA_ix_i(t) - L_iB_iK_iC_i\bar{\hat{x}}_A(t)\} \\
&= \sigma_i^{\mathrm{T}}(t)\{-L_iB_iK_iC_i\bar{\hat{x}}_A(t) - d_i|L_i\hat{F}_i|\mathrm{sgn}(\sigma_i(t)) \\
&\quad + \hat{F}_iw_i(t) - L_iB_iK_iC_i\bar{\hat{x}}_A(t)\} \\
&= \sigma_i^{\mathrm{T}}(t)[-d_i|L_iF_i|\mathrm{sgn}(\sigma_i(t)) + L_iF_iw_i(t)] \\
&\le \{-d_i|L_iF_i||\sigma_i(t)| + |\sigma_i(t)||L_iF_i||w_i(t)|\} \\
&\le |L_iF_i||\sigma_i(t)|[-d_i + |w_i(t)|] < 0.
\end{aligned}$$

Thus, the designed control law can ensure the reaching condition. This completes the proof.

Remark 8.8: The trajectories of the power system under historical-frequency-triggered DoS attacks can be ensured to reach onto the predesigned sliding mode surface in a finite time in **Theorem 8.3**. The results prove the correctness and effectiveness of the proposed control strategies.

8.4 Numerical Examples

Numerical examples are provided in this part to verify the usefulness of the proposed method for power systems under historical-frequency-triggered DoS attacks. It also outlines the characteristics of the proposed historical-frequency-triggered DoS attacks. The simulated results are tested on the IEEE 10-generator 39-bus system as shown in Figure 8.3. The system parameters are mostly the same as [17].

Example 8.1 *The mode and characteristics of the historical-frequency-triggered DoS attacks are discussed in this part.*

There are two main factors that affect the triggering threshold of the attack. One is the number of memory signals m, the other one is the attack's sampling interval h_a. The number of historical frequencies influences the size of Δf_A. The number of historical frequencies is set to $m = 3$ and $m = 5$, respectively. The sampling interval of the attack is set to $h_a = 0.5$.

Figures 8.4 and 8.5 show the characteristics of the historical-frequency-triggered DoS attacks. Obviously, a dynamic attack triggering threshold is obtained. Different attack regions can be built by applying different numbers of historical frequency information. Unlike the Bernoulli-distributed DoS attacks, the proposed DoS attack only occurs when the power system frequency meets the attack trigger conditions (8.4). Figures 8.4 and 8.5 present the state response with different attack

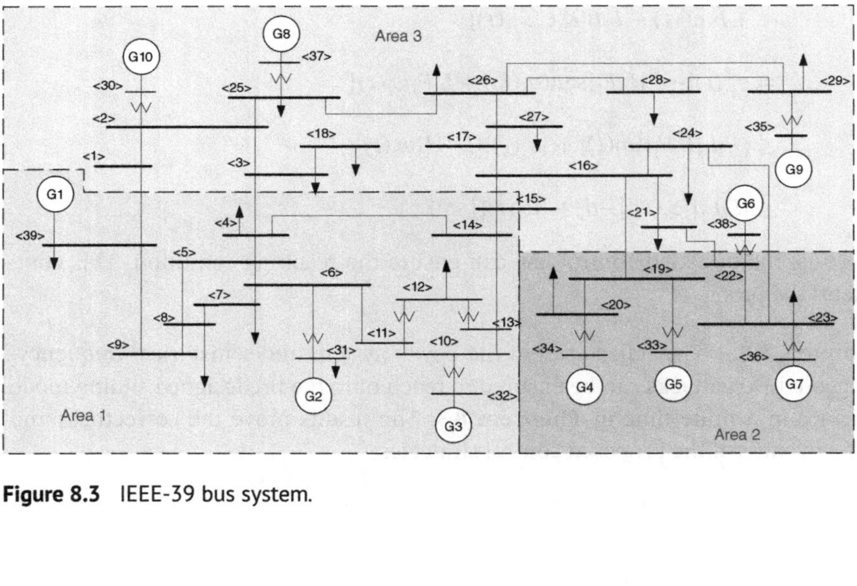

Figure 8.3 IEEE-39 bus system.

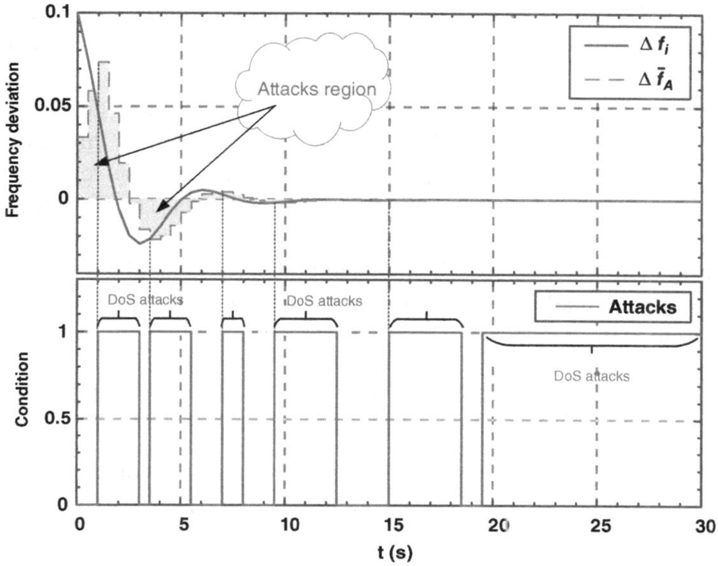

Figure 8.4 Historical-frequency-triggered DoS attacks with $m = 3$.

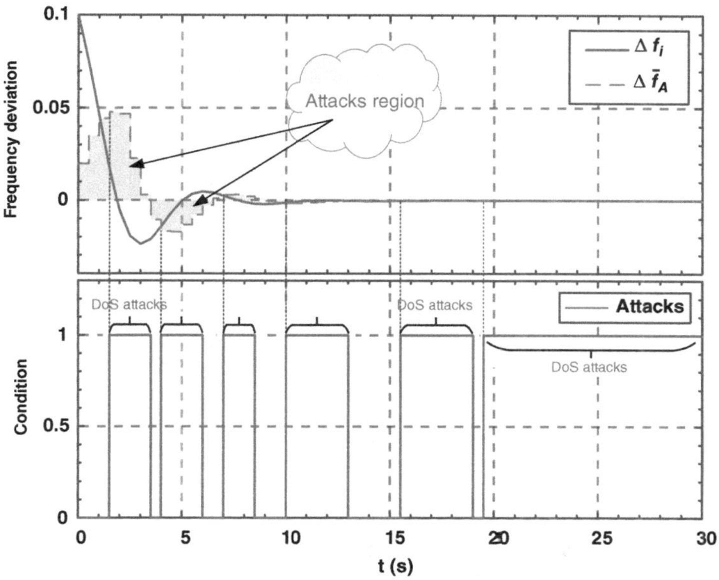

Figure 8.5 Historical-frequency-triggered DoS attacks with $m = 5$.

thresholds. Simulated results show that the characteristic of historical-frequency-triggered DoS attacks is that they only occur when the system is tending toward a stable state.

The sampling intervals of the attack frequency sensor are given as $h = 0.1s, 0.3s, 0.5s, 0.8s$. A smaller sampling interval brings the attack threshold closer to the real-time frequency of the power system. It can be observed from Figure 8.6 that a smaller sampling interval triggers more attacks. The triggering threshold $\Delta\bar{f}_A$ is closer to the system frequency deviation with $h = 0.1$ than the others. The system states approach the asymptotic value at about $t = 3s$. The results show that a smaller sampling interval causes attackers to pay more attention to the state of the system.

Example 8.2 *A three-area interconnected power system is provided to illustrate the usefulness of the proposed approach.*

Based on the IEEE 10-generator 39-bus system, the state responses of Δf_i, ΔP^i_{tie}, ΔP_{mi} are shown in Figure 8.7. As can be seen in Figure 8.7 that Δf_i, ΔP^i_{tie}, and ΔP_{mi} under the proposed method can be stable at about $t = 15s$, $t = 30s$, and $t = 20s$, respectively. Simulated results prove that with the intervention of the controller actions, the system under frequency-based deception attacks gradually becomes stable. The result verifies the effectiveness of the proposed approach.

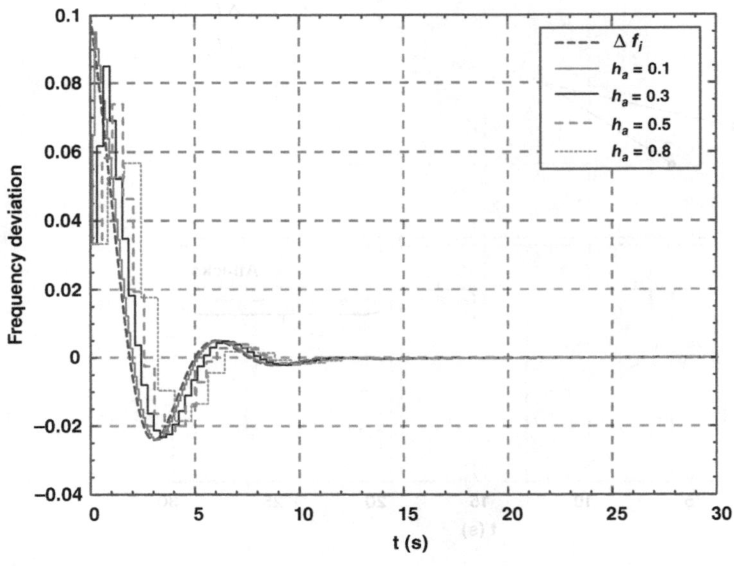

Figure 8.6 Changing of attacks threshold $\Delta\bar{f}_A$ with different sampling interval.

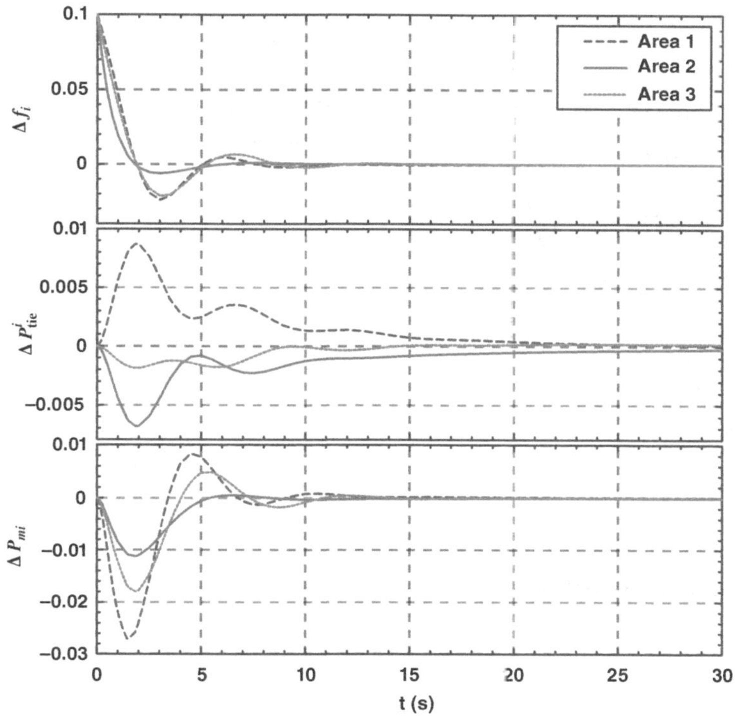

Figure 8.7 State response of three-area 10-generator 39-bus system.

Example 8.3 *The difference among different control strategies is shown in this example. The performance of the power system under different sizes of λ is described in this section.*

To validate the superiority of the proposed method, a comparison is made among the proposed method, output SMC in [31], SMC in [28], and PI-SMC in [39].

Under the controller's action, all four control methods are able to make the power system reach a stable state, as shown in Figure 8.8. Due to the application of system state and output in the controller design, the performance of the proposed strategy is better than that of other control methods. Simulation results prove the superiority of the proposed method.

Then, to illustrate the impact of λ in **Theorem 8.2**, the power system is tested with $\lambda_1 = 1 \times 10^{-7}$, $\lambda_2 = 1 \times 10^{-5}$ and $\lambda_3 = 1 \times 10^{-2}$, respectively.

The results in Figure 8.9 prove that with the condition of $\lambda \to 0$, the stable performance changes for the better.

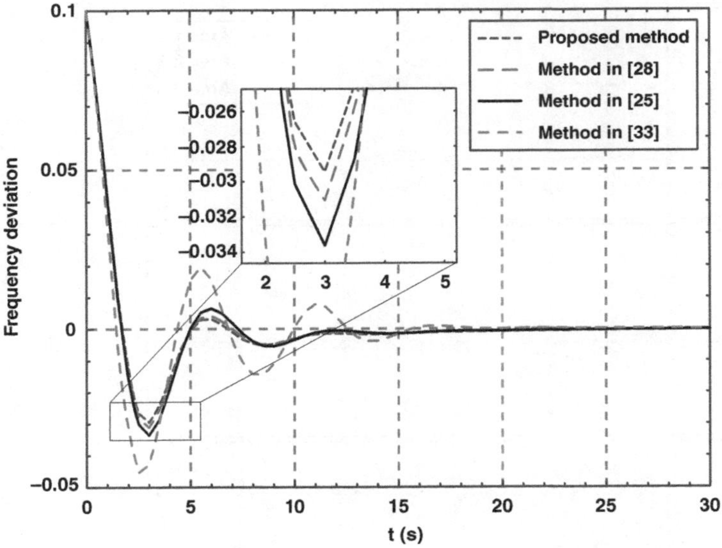

Figure 8.8 State response with different strategies.

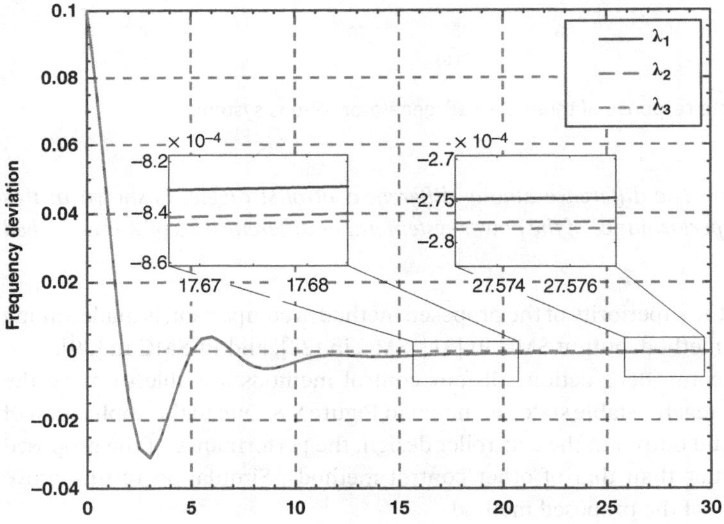

Figure 8.9 State response with different λ.

8.5 Conclusion

To describe the actual attack behavior, a historical-frequency-triggered DoS attack model has been developed by combining power system historical information, memory-event-triggered scheme, and the attributes of DoS attacks. A MOSMLFC strategy has been investigated to ensure the security performance of power systems under historical-frequency-triggered DoS attacks. The controller has been designed by applying the state and multiple output information of the power system. The security performance of power systems under historical-frequency-triggered DoS attacks has been ensured by Lyapunov's theorem. Numerical examples have been tested on the IEEE 10-generator 39-bus system, and the simulated results have shown the usefulness and effectiveness of the proposed strategy.

References

1 K. Gruber, T. Gauster, G. Laaha, et al. Profitability and investment risk of Texan power system winterization. *Nature Energy*, 7(5):409–416, 2022.

2 J. W. Li, T. Zhou, and H. Y. Cui. Brain-inspired deep meta-reinforcement learning for active coordinated fault-tolerant load frequency control of multi-area grids. *IEEE Transactions on Automation Science and Engineering*, 21(3):2518–2530, 2024.

3 X. H. Liu, Q. M. Jiao, S. W. Qiao, et al. A hybrid monotonic neural network approach for multi-area power system load frequency control against FGSM attack. *IEEE Transactions on Circuits and Systems II: Express Briefs*, 71(8):3780–3784, 2024.

4 M. Zhang, S. L. Dong, Z. G. Wu, et al. Reliable event-triggered load frequency control of uncertain multiarea power systems with actuator failures. *IEEE Transactions on Automation Science and Engineering*, 20(4):2516–2526, 2022.

5 J.-N. Li, H. Feng, Y. Wang, and G.-Y. Liu. A novel failure-distribution-dependent non-fragile H_∞ fault-tolerant load frequency control for faulty multi-area power systems. *IEEE Transactions on Power Systems*, 39(2):2936–2946, 2024.

6 B. Khokhar and K. P. S. Parmar. Utilizing diverse mix of energy storage for LFC performance enhancement of a microgrid: A novel MPC approach. *Applied Energy*, 333:120639, 2023.

7 X. Chen, M. Zhang, Z. Wu, et al. Model-free load frequency control of nonlinear power systems based on deep reinforcement learning. *IEEE Transactions on Industrial Informatics*, 20(4):6825–6833, 2024.

8 J. L. Liu, E. Y. Gong, L. J. Zha, et al. Observer-based security fuzzy control for nonlinear networked systems under weighted try-once-discard protocol. *IEEE Transactions on Fuzzy Systems*, 31(11):3853–3865, 2023.

9 H. Y. Chen, G. D. Zong, F. Z. Gao, and Y. Shi. Probabilistic event-triggered policy for extended dissipative finite-time control of MJSs under cyber-attacks and actuator failures. *IEEE Transactions on Automatic Control*, 68(12):7803–7810, 2023.

10 S. L. Hu, X. H. Ge, W. Zhang, and D. Yue. DoS-resilient load frequency control of multi-area power systems: An attack-parameter-dependent approach. *IEEE Transactions on Information Forensics and Security*, 19:3423–3434, 2024.

11 K. D. Lu and Z. G. Wu. Resilient event-triggered load frequency control for cyber-physical power systems under DoS attacks. *IEEE Transactions on Power Systems*, 38(6):5302–5313, 2023.

12 L. J. Zha, R. F. Liao, J. L. Liu, et al. Outlier-resistant distributed filtering over sensor networks under dynamic event-triggered schemes and DoS attacks. *IEEE Transactions on Automation Science and Engineering*, 22:1152–1162, 2025. https://doi.org/10.1109/TASE.2024.3360718.

13 D. D. Li, Q. Q. Cai, D. Marelli, et al. Stabilization of networked switched systems under DoS attacks. *IEEE Transactions on Cybernetics*, 54(8):4859–4866, 2024. https://doi.org/10.1109/TCYB.2023.3348192.

14 L. Yao, X. Huang, Z. Wang, and H. Shen. Memory-based event-triggered control of Markov jump systems under hybrid cyber attacks: A switching-like adaptive law. *IEEE Transactions on Automation Science and Engineering*, 21(4):6347–6357, 2024. https://doi.org/10.1109/TASE.2023.3324649.

15 Y. C. Wang, Z. Y. Wang, J. Gan, et al. Switched observer-based adaptive event-triggered load frequency control for networked power systems under aperiodic DoS attacks. *IEEE Transactions on Smart Grid*, 14(6):4816–4826, 2023.

16 X. L. Chen, S. L. Hu, Y. Li, et al. Co-estimation of state and FDI attacks and attack compensation control for multi-area load frequency control systems under FDI and DoS attacks. *IEEE Transactions on Smart Grid*, 13(3):2357–2368, 2022.

17 P. C. Chen, D. Zhang, L. Yu, and H. C. Yan. Dynamic event-triggered output feedback control for load frequency control in power systems with multiple cyber attacks. *IEEE Transactions on Systems, Man, and Cybernetics: Systems*, 52(10):6246–6258, 2022.

18 X. Wang, E. G. Tian, W. X. Zheng, and X. P. Xie. Important-data-based DoS attack mechanism and resilient H_∞ filter design for networked T-S fuzzy systems. *IEEE Transactions on Cybernetics*, 54(5):3352–3362, 2024. https://doi.org/10.1109/TCYB.2023.3285526.

19 A. Amulya, K. S. Swarup, and R. Ramanathan. Multivariate spectral analysis and hypothesis testing-based robust attack detection for multiarea frequency control. *IEEE Transactions on Systems, Man, and Cybernetics: Systems*, 53(8):4636–4648, 2023. https://doi.org/10.1109/TSMC.2023.3254212.

20 G. D. Zong, D. Yang, J. Lam, and X. Q. Song. Fault-tolerant control of switched LPV systems: A bumpless transfer approach. *IEEE/ASME Transactions on Mechatronics*, 27(3):1436–1446, 2021.

21 H. Y. Chen, G. D. Zong, X. Liu, et al. A sub-domain-awareness adaptive probabilistic event-triggered policy for attack-compensated output control of Markov jump CPSs with dynamically matching modes. *IEEE Transactions on Automation Science and Engineering*, 21(3):4419–4431, 2024. https://doi.org/10.1109/TASE.2023.3296259.

22 J. Yang, Q. S. Zhong, K. B. Shi, and S. M. Zhong. Distributed coordination LFC approach for interconnected power systems under detection and compensation mechanism targeting DoS attacks. *IEEE Transactions on Industrial Informatics*, 19(11):11008–11018, 2023. https://doi.org/10.1109/TII.2023.3242587.

23 M. Zhang, S. L. Dong, P. Shi, et al. Distributed observer-based event-triggered load frequency control of multiarea power systems under cyber attacks. *IEEE Transactions on Automation Science and Engineering*, 20(4):2435–2444, 2022.

24 X. H. Li and D. Ye. Security-based event-triggered fuzzy control for multiarea power systems under cross-layer DoS attacks. *IEEE Transactions on Circuits and Systems I: Regular Papers*, 70(7):2995–3004, 2023. https://doi.org/10.1109/TCSI.2023.3268309.

25 A. Wang, M. Fei, Y. Song, et al. Secure adaptive event-triggered control for cyber-physical power systems under denial-of-service attacks. *IEEE Transactions on Cybernetics*, 54(3):1722–1733, 2024.

26 Y. R. Fan, Y. X. Wu, J. B. Yuan, et al. Event-triggered sliding-mode control for a discrete-time muscle-driven musculoskeletal system. *IEEE Transactions on Automation Science and Engineering*, 21(4):5866–5880, 2024. https://doi.org/10.1109/TASE.2023.3319503.

27 F. Yang, Y. Shen, D. Li, et al. Fractional-order sliding mode load frequency control and stability analysis for interconnected power systems with time-varying delay. *IEEE Transactions on Power Systems*, 39(1):1006–1018, 2024.

28 S. W. Qiao, X. H. Liu, Y. R. Liang, et al. Event-triggered sliding mode load frequency control of multiarea power systems under periodic denial-of-service attacks. *IEEE Systems Journal*, 17(2):2803–2814, 2023.

29 V. Van Huynh, P. T. Tran, C. S. T. Dong, et al. Sliding surface design for sliding mode load frequency control of multiarea multisource power system. *IEEE Transactions on Industrial Informatics*, 20(5):7797–7809, 2024. https://doi.org/10.1109/TII.2024.3359445.

30 J. H. Zhang and Y. Q. Xia. Design of static output feedback sliding mode control for uncertain linear systems. *IEEE Transactions on Industrial Electronics*, 57(6):2161–2170, 2009.

31 J. L. Chang. Dynamic output integral sliding-mode control with disturbance attenuation. *IEEE Transactions on Automatic Control*, 54(11):2653–2658, 2009.

32 J. Xu, Y. G. Niu, and Y. Y. Zou. Finite-time consensus for singularity-perturbed multiagent system via memory output sliding-mode control. *IEEE Transactions on Cybernetics*, 52(9):8692–8702, 2021.

33 A. Farnam and R. M. Esfanjani. Improved linear matrix inequality approach to stability analysis of linear systems with interval time-varying delays. *Journal of Computational and Applied Mathematics*, 294:49–56, 2016.

34 S. Qiao, X. Liu, Y. Wang, et al. H_∞ load frequency control of power system integrated with EVs under DoS attacks: Non-fragile output sliding mode control approach. *IEEE Transactions on Intelligent Transportation Systems*, 25(5): 4565–4577, 2024.

35 H. Y. Chen, G. D. Zong, X. D. Zhao, et al. Secure filter design of fuzzy switched CPSs with mismatched modes and application: A multidomain event-triggered strategy. *IEEE Transactions on Industrial Informatics*, 19(10):10034–10044, 2023.

36 A. Saxena, R. Shankar, C. Kumar, and S. K. Parida. A resilient frequency regulation for enhancing power system security against hybrid cyber-attacks. *IEEE Transactions on Industry Applications*, 60(3):4583–4597, 2024. https://doi.org/10.1109/TIA.2024.3354228.

37 R. J. Ma, P. Shi, and L. G. Wu. Dissipativity-based sliding-mode control of cyber-physical systems under denial-of-service attacks. *IEEE Transactions on Cybernetics*, 51(5):2306–2318, 2020.

38 W. Qi, G. Zong, and H. R. Karimi. Finite-time observer-based sliding mode control for quantized semi-Markov switching systems with application. *IEEE Transactions on Industrial Informatics*, 16(2):1259–1271, 2019.

39 Y. Mi, Y. Fu, C. S. Wang, and P. Wang. Decentralized sliding mode load frequency control for multi-area power systems. *IEEE Transactions on Power Systems*, 28(4):4301–4309, 2013.

40 H. Y. Li, J. H. Wang, and P. Shi. Output-feedback based sliding mode control for fuzzy systems with actuator saturation. *IEEE Transactions on Fuzzy Systems*, 24(6):1282–1293, 2015.

9

Security Concern and Fuzzy Output Sliding Mode Load Frequency Control of Power Systems

After reading this chapter, you should be able to:

- Figure out the characteristics of motion-trends-based deception attacks.
- Explain the promotion of LMI-based fuzzy logic in fuzzy controller.
- Identify the uniqueness of the fuzzy output sliding mode control approach.

9.1 Background

Given the pervasive integration of the power system into all facets of our society, ensuring an uninterrupted supply of high-quality electricity has become a primary focus of investigation among researchers [1]. The load frequency control (LFC) has been critical to ensure the frequency stability and power balance of multi-area power systems [2, 3]. In [4], the distributed fuzzy LFC problem was investigated for an event-based power system with switching topology. In [5], a tube-based distributed model predictive control method was proposed in handling the strong uncertainties coming from high wind power penetration. In [6], a memory event-triggered device was investigated for the normal operation of various renewable power systems to alleviate the data transfer pressure. Notably, the efficient operation of modern power systems is inextricably linked to communications networks.

The issue of system security has always been a major concern for researchers. To improve the security performance of the system, some researchers have focused their attention on the safe operation of critical equipment. In [7], a multi-hop graph pooling adversarial network based on distributed federated learning was investigated for the remaining useful life prediction of equipment. To accurately estimate the state of health without recognizing the systematic mathematical model and to extend the lifetime, the authors in [8] proposed a lifetime extension approach based on the Levenberg–Marquardt backpropagation neural network and power routing of interleaved DC-DC boost conversion systems. Some other researchers have paid their attention on the security performance of the entire system. To enhance the integrity and confidentiality of transmitted information, a decentralized and tamper-proof nature of blockchain approach was added among the nodes in [9]. The resilience evaluation approach was proposed for the power system that considers the damage of cyberattacks, so that the resilience of the system under cyberattacks can be accurately quantified in [10]. In recent years, frequent cyberattacks have gradually become the largest threats to the security performance of power systems [11]. To improve the security performance of the system [12], a detection mechanism using Byrnes–Isidori normal form was designed to secure the system under zero-dynamics attack.

On the other hand, various effective countermeasures have been investigated to reduce or eliminate the contamination caused by cyberattacks. Cyberattacks come in various forms, with denial-of-service (DoS) attacks and deception attacks being the most considered in the LFC power system [13, 14]. In [15], the observer-based predictive event-triggered LFC problem was investigated for a multi-area power system in a smart grid under DoS attacks. In [16], a decentralized H_∞ secure defense approach was investigated for a networked power system to against deception attack. In [17], a dynamic event-triggered device was investigated for a LFC system under malicious DoS attacks. In addition to investigate the dangers of different cyberattacks, it is also worth exploring the damage of different attack models of the same attack. Different types of cyberattack models have been proposed to search for the actual attack behavior. In [18], the cyberattacks model was designed based on the Bernoulli distribution. In [19], the happening of cyberattacks was determined by the predefined periodic. In [20], the duration and frequency of the cyberattack were given. In [21], an important data-based mechanism was used in attack model to enhance the damage of the cyberattacks. The variety of cyberattack models will make it more difficult for us to develop a defensive approach. It is therefore necessary to anticipate new models of cyberattack that may emerge so that appropriate strategies can be developed. Inspired by this question, we consider a new cyberattack model to explore the actual behavior of cyberattacks, and investigate a corresponding effective countermeasure to defend against the malicious deception attacks.

In this chapter, a motion-trends-based deception attack model is proposed to describe the malicious threats from the communication network. A clever attacker may notice the system's changing tendencies from one moment to the next. Usually we think that a system tending toward a stable state means that the state of the system is developing in a positive direction. From an attacker's perspective, it is necessary to inject cyberattacks to disrupt the trend of system improvement. Based on this consideration, we propose a more destructive attack model that determines the occurrence of attacks by monitoring the operational state of the system. Unlike most considered deception attack plants, the investigated attack model utilizes the system motion-trends-based mechanism to evaluate its future trends and determine whether an attack will occur. The proposed attack mechanism consists of the system state and its derivative. The main object of the attack mechanism is to determine whether the current state is worth attacking by determining its future operational trends. It is noteworthy that there is an implicit assumption in a large amount of the existing literature that the cyberattack decides attacks indiscriminately [21]. In the proposed attack mechanism, if the state of the system is converging toward a stable state, then this moment is considered to be worth attacking. Conversely, the system state is considered to be deteriorating and no attacks occur. In this attack mechanism, the target of the deception attack is focused on the time when the system state tends to stabilize. This means that a cyberattack will only destabilize the system or prevent it from converging to a stable state. At the same time, this is more in line with the purpose of cyberattacks, which is to disrupt the stable operation of the system.

System information is often applied in controller design. Cybersecurity relates to whether the information transmitted can support normal system operations [22, 23]. This means that cyberattacks can result in poor control effectiveness or even controller failure. Sliding mode control (SMC) has been investigated to address unexpected disturbances [24]. In [25], an event-triggered adaptive SMC approach was designed for a discrete uncertain nonlinear system to mitigate the contamination of false data injection attack. In [26], the authors proposed quantized feedback SMC method for a cyber-physical system under malicious attack. In [27], an event-triggered SMC strategy was investigated for multi-area LFC power systems under periodic DoS attacks. Researchers have found that the combination of fuzzy logic and SMC improves the control performance of SMC. Fuzzy control has been widely attempted in various fields since its introduction. In [28], two fuzzy integral switching surfaces were built in the sliding mode controller. In [29], a SMC scheme was constructed for a T-S fuzzy system under cyberattack. In this chapter, a fuzzy output sliding mode load frequency control (FOSMLFC) approach is developed to solve the security issue for the power systems under motion-trends-based deception attack.

When using a fuzzy controller, the controller gains or member functions are usually determined by trial-error or experience, which may result in a poor control performance [28, 29]. However, few studies discuss how to set the member functions. To improve the fuzzy control performance, a linear matrix inequality

(LMI)-based approach is investigated. Initial fuzzy controller gain and member functions are designed according to the result of LMI. Based on the Lyapunov theorem, the controller gain obtained using LMI ensures system stability. On the basis of the obtained controller gain, it is easy to determine the coordinate range of the member function. The trial-error method or experience filled with uncertainty is replaced by LMIs, which helps to determine member functions more quickly and accurately. This saves us the cost of trial-error and extra expense. The use of LMI in fuzzy controller design ensures control performance while making the system quickly stable.

In this study, we focus on the security issue for the power system under motion-trends-based deception attacks. Malicious threats from communication networks are described as motion-trends-based deception attacks. A FOSMLFC approach is investigated to resolve the unexpected influence of motion-trends-based deception attacks on the power system. By using the LMI, a controller gain is obtained that ensures a stable power system. Fuzzy logic is then used to modify the resulting controller gain parameters so that the controlled system can be brought to a stable state more quickly. A LMI-based design method is presented for fuzzy logic member functions.

Main contributions are summarized as follows.

1. A novel motion-trends-based deception attack model is proposed to describe the actual behavior of cyberattacks, which helps to formulate defense strategies in advance. Compared to the deception attack models in [14, 16], this innovative attack model can give a more complex attack behavior with the larger threat to power system.
2. A fuzzy output SMC approach is proposed for the LFC power system under motion-trends-based deception attacks. Due to the use of fuzzy logic and unique sliding mode surface design, the proposed method achieves a better control performance than that of [26].
3. A LMI-based fuzzy logic member function design approach is presented to further improve the power system performance, which can provide a more rapid convergence than the trial-error method or experience [28–30].

9.2 Problem Formulation

Figure 9.1 is given as the simplified multi-area power system model, which consists of the power generator, turbine, governor, and so on.

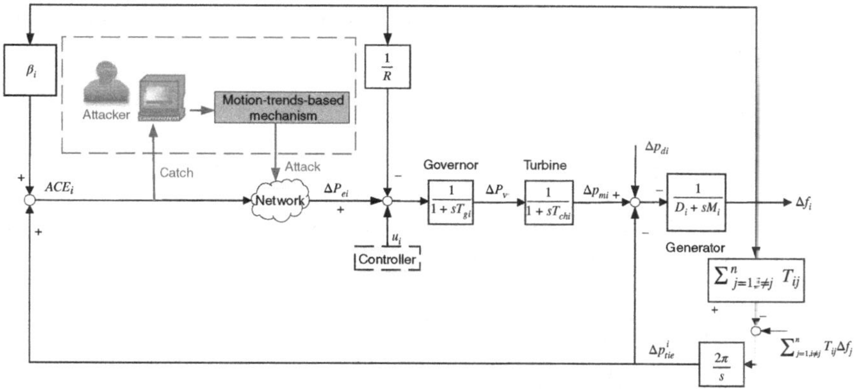

Figure 9.1 i-th area power system simplified model.

9.2.1 Multi-area Power System Model

The dynamic equations of the system based on the system model and information transfer are shown as

$$\begin{cases} \Delta \dot{f}_i(t) = \dfrac{\Delta P_{mi}(t)}{M_i} - \dfrac{D_i \Delta f_i(t)}{M_i} - \dfrac{\Delta P^i_{\text{tie}}(t)}{M_i} - \dfrac{\Delta P_{di}(t)}{M_i} \\ \Delta \dot{P}_{mi}(t) = -\dfrac{\Delta P_{mi}(t)}{T_{\text{chi}}} + \dfrac{\Delta P_{vi}(t)}{T_{\text{chi}}} \\ \Delta \dot{P}_{vi}(t) = -\dfrac{\Delta f_i(t)}{R_i T_{gi}} - \dfrac{\Delta P_{vi}(t)}{T_{gi}} - \dfrac{\Delta P_{ei}(t)}{T_{gi}} \\ \Delta \dot{P}^i_{\text{tie}}(t) = 2\pi \sum\limits_{j=1,i\neq j}^{n} T_{ij}(\Delta f_i(t) - \Delta f_j(t)) \\ ACE_i(t) = \Delta \dot{P}_{\text{tie}i}(t) + \beta_i \Delta f_i(t) \end{cases} \tag{9.1}$$

where Δf_i is the frequency deviation, M_i is inertia constant of the generator, ΔP_{mi} is the mechanical output deviation, D_i is the damping coefficient, ΔP^i_{tie} is the deviation of tie-line power exchange between areas, ΔP_{di} is the load perturbations, T_{chi} is the time constant of the turbine, ΔP_{vi} is the deviation value of governor position, R_i is the droop coefficient, T_{gi} is the time constant of the governor, ΔP_{ei} is the integral of ACE_i, T_{ij} is the synchronization coefficient of the tie-line power, Δf_j represents the frequency deviation of the neighbor area, ACE_i is the area control error, and β_i is the frequency deviation coefficient.

Define $x_i(t) = [\Delta P_{mi}\ \Delta P_{vi}\ \Delta P^i_{\text{tie}}\ \Delta f_i]^\mathsf{T}, y_i(t) = ACE_i(t)$. Hence, the state vector space function can be represented as

$$\begin{cases} \dot{x}_i(t) = A_i x_i(t) + B_i u_i(t) + F_i w_i(t) \\ y_i(t) = C_i x_i(t) \end{cases} \tag{9.2}$$

where in $w_i(t)$ is the bounded system disturbance, which is generally defined as $w_i(t) = \Delta P_{di} < d_i$, d_i is a positive constant, and

$$A_i = \begin{bmatrix} -\dfrac{1}{T_{\text{chi}}} & \dfrac{1}{T_{\text{chi}}} & 0 & 0 \\ 0 & -\dfrac{1}{T_{gi}} & 0 & -\dfrac{1}{R_i T_{gi}} \\ 0 & 0 & 0 & 2\pi \displaystyle\sum_{j=1,j\neq i}^{n} T_{ij} \\ \dfrac{1}{M_i} & 0 & -\dfrac{1}{M_i} & -\dfrac{D_i}{M_i} \end{bmatrix}, \quad \begin{aligned} B_i &= \begin{bmatrix} 0 & \dfrac{1}{T_{gi}} & 0 & 0 \end{bmatrix}^\mathsf{T}, \\ F_i &= \begin{bmatrix} 0 & 0 & 0 & -\dfrac{1}{M_i} \end{bmatrix}, \\ C_i &= \begin{bmatrix} 0 & 0 & 1 & \beta_i \end{bmatrix}. \end{aligned} \tag{9.3}$$

9.2.2 Motion-trends-based Deception Attacks Model

Cyberattacks mainly disrupt the stability performance of the power system by altering the normal transmission of information [23]. For an ideal LFC power system, stable performance can be ensured when $x_i(t) = 0$. An intelligent attacker may be concerned about the operational state of the power system. Therefore, a motion-trends-based deception attack model is proposed in this chapter. Figure 9.2 provides an example to illustrate the characteristics of the motion-trends-based mechanism.

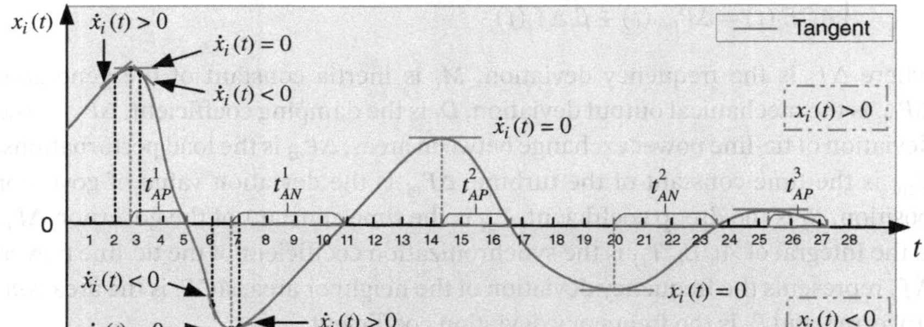

Figure 9.2 Example of motion-trends-based mechanism.

The transmitted data is assumed to be intercepted by the attackers and replaced with attack information [31]. The trend of system state changes can be determined through $\dot{x}_i(t)$. According to Figure 9.2, we have the following findings

a. For $x_i(t) = 0$, the power system state is stable at this moment.
b. For $x_i(t) > 0$, $\dot{x}_i(t) \leq 0$ indicates that the system state is approaching stability.
c. For $x_i(t) > 0$, $\dot{x}_i(t) > 0$ indicates that the power system state is far away from stability.
d. For $x_i(t) < 0$, $\dot{x}_i(t) \geq 0$ indicates that the power system state is approaching stability.
e. For $x_i(t) < 0$, $\dot{x}_i(t) < 0$ indicates that the power system state is far away from stability.

Remark 9.1: The trend of system state changes can be determined by the value of $\dot{x}_i(t)$. From the attacker's perspective, cyberattacks should prevent the system from moving to a better operating state. Therefore, an attack occurs when the power system tends to stability. The proposed motion-trends-based deception attacks are designed to prevent the power system from successfully approaching a stable state.

Based on our current understanding of cyberattacks, the goal is to destroy the stable state of the target system. Therefore, conditions (a), (b), and (d) should be the target of cyberattacks. Then, the motion-trends-based mechanism for the deception attacks is given as

$$\begin{cases} \begin{cases} \dot{x}_i(t) \leq 0, \quad x_i(t) > 0 \\ x_i(t) = 0 \\ \dot{x}_i(t) \geq 0, \quad x_i(t) < 0 \end{cases}, \quad \text{Attacks} \\ \text{Otherwise, Without Attacks} \end{cases} \tag{9.4}$$

For further, (9.4) can be rewritten as

$$\begin{cases} \dot{x}_i(t) \times x_i(t) \leq 0, \quad \text{Attacks} \\ \dot{x}_i(t) \times x_i(t) > 0, \quad \text{Without Attacks} \end{cases} \tag{9.5}$$

The attack time is defined as t_A, and $t_A \in T_A$. $T_A = t_{AP} \cup t_{AN}$, where $t_{AP} = \{t_{AP}^1, t_{AP}^2, \cdots\}$, and $t_{AN} = \{t_{AN}^1, t_{AN}^2, \cdots\}$.

The attack signal is given as

$$x_a(t) = f_a(x_i(t_A)) + x_i(t) \tag{9.6}$$

The attack signals $f_a(x_i(t_A))$ are assumed to meet

$$\|f_a(x_i(t_A))\|_2 \leq \|G_i x_i(t)\|_2 \tag{9.7}$$

Then, we have

$$\hat{x}_i(t) = \begin{cases} x_a(t), & t \in T_A \\ x_i(t), & \text{otherwise} \end{cases} \tag{9.8}$$

A flowchart is proposed in Figure 9.3 to show the process of injecting the motion-trends-based deception attacks.

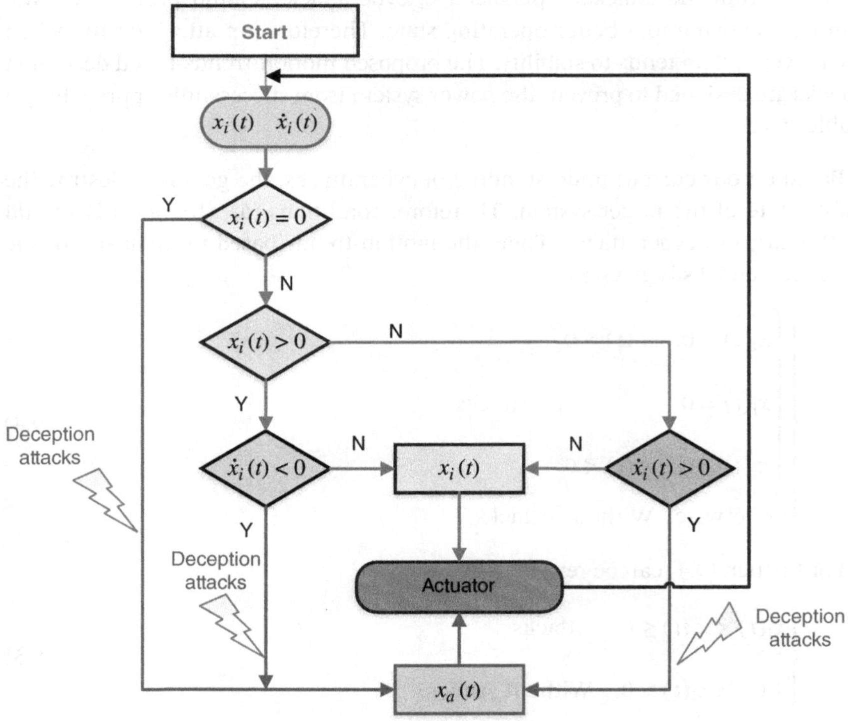

Figure 9.3 Flowchart of motion-trends-based deception attacks.

9.2.3 Fuzzy Output SMC Approach

A FOSMLFC approach is investigated to address the security issue, where the fuzzy logic is applied to modify the controller gain. Based on fuzzy logic, the controller gain calculated by the LMIs can be continuously adjusted.

A fuzzy output SMC is developed by using the output information for the power systems. Since $\Delta P_{ei}(t) = \int_0^t ACE_i(s)ds$, and $y_i(t) = ACE_i(t)$, we have

$$\Delta \dot{P}_{ei}(t) = ACE_i(t) = y_i(t). \tag{9.9}$$

Considering the impact of deception attacks (9.8), the fuzzy output sliding mode surface is developed as

$$\sigma = L_i x_i(t) - \int_0^t L_i A_i x_i(s)ds - \int_0^t L_i B_i K_i \hat{x}_i(s)ds - K_f^j \Delta P_{ei}(t), \tag{9.10}$$

where K_f^j is the fuzzy logic controller gain, $j = 1, \cdots, m$ is represented as the number of fuzzy rules, $m \in \mathbb{N}^+$, and L_i is selected to ensure $L_i B_i$ nonsingular.

Remark 9.2: In the design of the sliding surface, the output information of the power system is taken into account. The use of system status and output helps to improve control performance [32]. It should be noted that there are discernible distinctions between the output sliding mode and output feedback sliding mode surfaces. In [33], the output feedback sliding mode surface is designed as $\sigma = y_i(t)$ and its derivative is $\dot{\sigma} = \dot{y}_i(t) = C_i \dot{x}(t)$. This increases the complexity of the controller in comparison to the fuzzy output sliding mode surface, which is defined as (9.10). This means that the complexity is reduced while using the output information.

Fuzzy gain K_f^j is modified by the fuzzy rules by observing the output deviation and its derivative. System output error represents the difference between the last output $y_i(t-1)$ and current output $y_i(t)$, which is given as

$$e_i(t) = y_i(t) - y_i(t-1) \tag{9.11}$$

and its derivative is $\dot{e}_i(t)$.

Table 9.1 is given as the fuzzy logic rules. The IF-THEN plants rules are described as

- **IF** $e_i(t)$ is PB and $\dot{e}_i(t)$ is PB, **Then** K_f^j is PB.

Then, the output sliding mode surface (9.10) is reformulated in the following fuzzy form.

Table 9.1 Fuzzy rules of the proposed controller gain $\hat{K}_f^j$.

$e_i(t)$	$\dot{e}_i(t)$						
	NB	**NM**	**NS**	**Z**	**PS**	**PM**	**PB**
NB	NB	NB	NB	NB	NM	NS	Z
NM	NB	NB	NB	NM	NS	Z	PS
NS	NB	NB	NM	NS	Z	PS	PM
Z	NB	NM	NS	Z	PS	PM	PB
PS	NM	NS	Z	PS	PM	PB	PB
PM	NS	Z	PS	PM	PB	PB	PB
PB	Z	PS	PM	PB	PB	PB	PB

NS, negative small; NM, negative medium; NB, negative big; Z, zero; PB, positive big; PM, positive medium; PS, positive small.

- Rule 1: **IF** K_f^j is K_f^1, **Then**

$$\sigma_1 = L_i x_i(t) - \int_0^t L_i A_i x_i(s) ds - \int_0^t L_i B_i K_f^1 \hat{x}_i(s) ds - K_f^1 \Delta P_{ei}(t), \tag{9.12}$$

- Rule m: **IF** K_f^j is K_f^m, **Then**

$$\sigma_m = L_i x_i(t) - \int_0^t L_i A_i x_i(s) ds - \int_0^t L_i B_i K_f^m \hat{x}_i(s) ds - K_f^m \Delta P_{ei}(t), \tag{9.13}$$

A proper fuzzy sliding mode surface should satisfy $\sigma = \dot{\sigma} = 0$. Take the derivative on (9.10), we have

$$L_i \dot{x}_i(t) - L_i (A_i x_i(t) + B_i K_f^j \hat{x}_i(t)) - K_f^j y_i(t) = 0. \tag{9.14}$$

Combining with (9.2), the equivalent fuzzy output sliding mode controller is derived as

$$u_{eq}(t) = \hat{K}_{\text{FUZ}}^j \hat{x}_i(t) - (L_i B_i)^{-1} L_i F_i w_i(t), \tag{9.15}$$

where $\hat{K}_{\text{FUZ}}^j = K_f^j + (L_i B_i)^{-1} K_f^j C_i$.

Based on the above analysis, a generic fuzzy power system state space function can be written as

- Rule 1: **IF** Controller gain is $\hat{K}_{\text{FUZ}}^1$, **Then**

$$\begin{cases} \dot{x}_i(t) = A_i x_i(t) + B_i \hat{K}_{\text{FUZ}}^1 \hat{x}_i(t) + \hat{F}_i w_i(t) \\ y_i(t) = C_i x_i(t) \end{cases} \tag{9.16}$$

- Rule j: **IF** Controller gain is $\hat{K}^{j}_{\mathrm{FUZ}}$, **Then**

$$\begin{cases} \dot{x}_i(t) = A_i x_i(t) + B_i \hat{K}^{j}_{\mathrm{FUZ}} \hat{x}_i(t) + \hat{F}_i w_i(t) \\ y_i(t) = C_i x_i(t) \end{cases} \tag{9.17}$$

where $\hat{F}_i = F_i - B_i(L_iB_i)^{-1}L_iF_i$.

Remark 9.3: We define $\hat{K}^{1}_{\mathrm{FUZ}}$ as the initial value of controller gains. It is worth noting that $\hat{K}^{1}_{\mathrm{FUZ}}$ serves two purposes: one is to help us determine what value to regulate from, and the other is to provide a range of horizontal coordinates for the member function.

Remark 9.4: Normally, the initial value is determined through trial-error, experience, or intelligence algorithms [28–30]. However, trial-error and experience rely on long-term relevant research and cannot guarantee accuracy and effectiveness. The effectiveness of method intelligence algorithms depends on the choice of algorithm. These methods all require a long time to determine $\hat{K}^{1}_{\mathrm{FUZ}}$. Differently, this chapter provides a solution for quickly determining $\hat{K}^{1}_{\mathrm{FUZ}}$ by utilizing LMI.

9.3 Main Results

This part derives a sufficient condition to ensure the security performance of FOSMLFC power systems under cyberattacks. By using LMIs, a controller gain that ensures the security performance of power system is obtained. Then, we propose a design method for fuzzy logic member functions based on the obtained controller gain.

Lemma 9.1 [34] *For a given constant h, a symmetric matrix $P > 0$, and a symmetric matrix U with appropriate dimension, there exist*

$$-h \int_{t-h}^{t} \dot{x}_i^{\mathrm{T}}(s) P \dot{x}_i(s) ds \leq - \begin{bmatrix} \xi_1 \\ \xi_2 \end{bmatrix}^{\mathrm{T}} \begin{bmatrix} -P & U \\ U^{\mathrm{T}} & -P \end{bmatrix} \begin{bmatrix} \xi_1 \\ \xi_2 \end{bmatrix},$$

where $\xi_1 = x_i(t)$, $\xi_2 = x_i(t-h)$.

Lemma 9.2 [34] *For a given constant h, and a positive definite symmetric matrix P, there exist*

$$-\frac{h^2}{2} \int_{t-h}^{t} \int_{t+\alpha}^{t} \dot{x}_i^{\mathrm{T}}(s) P \dot{x}_i(s) ds d\alpha \leq \begin{bmatrix} h x_i(t) \\ \int_{t-h}^{t} x_i(s) ds \end{bmatrix}^{\mathrm{T}} \begin{bmatrix} P & -P \\ -P & P \end{bmatrix} \begin{bmatrix} h x_i(t) \\ \int_{t-h}^{t} x_i(s) ds \end{bmatrix},$$

9.3.1 Stability Analysis

Theorem 9.1 *For given scalars $h > 0$ and $\gamma > 0$, power system (9.17) is stable if there exist $P_i > 0$, $Q_1 > 0$, $Q_2 > 0$, $R_1 > 0$, $R_2 > 0$, $S_1 > 0$, and matrix U with appropriate dimensions, such that*

$$\begin{bmatrix} \Theta_{11} & \Theta_{21}^{\mathrm{T}} \\ \Theta_{21} & \Theta_{22} \end{bmatrix} < 0 \tag{9.18}$$

$$\begin{bmatrix} -R_2 & U^{\mathrm{T}} \\ U & -R_2 \end{bmatrix} > 0 \tag{9.19}$$

where $\Theta_{11} = [\eta_{5\times5}]$,

$$\begin{aligned}
\eta_{1,1} &= A_i^{\mathrm{T}} P_i + P_i A_i + \hat{K}_{\mathrm{FUZ}}^{j\mathrm{T}} B_i^{\mathrm{T}} P_i + P_i B_i \hat{K}_{\mathrm{FUZ}}^{j\mathrm{T}} + Q_1 + h^2 R_1 - R_2 - h^2 S_1 \\
&\quad + C_i^{\mathrm{T}} C_i + G_i^{\mathrm{T}} P_i G_i, \\
\eta_{2,1} &= -2U, \eta_{2,2} = -Q_1 - R_2, \eta_{3,1} = \hat{K}_{\mathrm{FUZ}}^{j\mathrm{T}} B_i^{\mathrm{T}} P_i, \eta_{3,3} = -P_i, \eta_{4,1} \\
&= P_i + Q_2 + hS_1, \\
\eta_{4,2} &= -P_i - Q_2, \eta_{4,4} = -R_1 - R_2 - S_1, \eta_{5,1} = \hat{F}^{\mathrm{T}} P_i, \eta_{5,5} = -\gamma, \\
\Theta_{2,1} &= \mathrm{col}[h\phi, \frac{h^2}{2}\phi], \Theta_{2,2} = -\mathrm{diag}\{R_2, S_1\}, \\
\phi &= [A_i^{\mathrm{T}} P_i + \hat{K}_{\mathrm{FUZ}}^{j\mathrm{T}} B_i^{\mathrm{T}} P_i, 0, \hat{K}_{\mathrm{FUZ}}^{j\mathrm{T}} B_i^{\mathrm{T}} P_i, 0, \hat{F}_i^{\mathrm{T}} P_i].
\end{aligned}$$

Proof. The Lyapunov candidate function is chosen as

$$V(t) = V_1(t) + V_2(t) + V_3(t) + V_4(t), \tag{9.20}$$

wherein

$$V_1(t) = x_i^{\mathrm{T}}(t) P_i x_i(t) + \int_{t-h}^{t} x_i^{\mathrm{T}}(s) ds P_i \int_{t-h}^{t} x_i(s) ds$$

$$V_2(t) = \int_{t-h}^{t} x_i^{\mathrm{T}}(s) Q_1\, x_i(s) ds + \int_{t-h}^{t} x_i^{\mathrm{T}}(s) ds Q_2 \int_{t-h}^{t} x_i(s) ds,$$

$$V_3(t) = h\int_{-h}^{0}\int_{t+\alpha}^{t} x_i^{\mathrm{T}}(\alpha)R_1x_i(\alpha)dsd\alpha + h\int_{-h}^{t}\int_{t+\alpha}^{t} \dot{x}_i^{\mathrm{T}}(\alpha)R_2\dot{x}_i(u)dsd\alpha,$$

$$V_4(t) = \frac{h^2}{2}\int_{-h}^{t}\int_{\alpha}^{0}\int_{t+v}^{t} \dot{x}_i^{\mathrm{T}}(v)S_1\dot{x}_i(v)dsd\alpha dv.$$

Taking deviation on $V_i(t)$, we have

$$\dot{V}_1(t) = \mathrm{sym}\left\{\dot{x}_i^{\mathrm{T}}P_ix_i(t) + \left(x_i^{\mathrm{T}}(t) - x_i^{\mathrm{T}}(t-h)\right)P_i\int_{t-h}^{t} x_i(s)ds\right\},$$

$$\begin{aligned}\dot{V}_2(t) &= x_i^{\mathrm{T}}(t)Q_1x_i(t) - x_i^{\mathrm{T}}(t-h)Q_1x_i(t-h)\\ &\quad + \left(x_i^{\mathrm{T}}(t) - x_i^{\mathrm{T}}(t-h)\right)Q_2\int_{t-h}^{t} x_i(s)ds,\end{aligned}$$

$$\begin{aligned}\dot{V}_3(t) &= h^2x_i^{\mathrm{T}}(t)R_1x_i(t) - h\int_{t-h}^{t} x_i^{\mathrm{T}}(s)R_1x_i(s)ds\\ &\quad + h^2\dot{x}_i^{\mathrm{T}}(t)R_2\dot{x}_i(t) - h\int_{t-h}^{t} \dot{x}_i^{\mathrm{T}}(s)R_2\dot{x}_i(s)ds\end{aligned}$$

$$\dot{V}_4(t) = \frac{h^4}{4}\dot{x}_i^{\mathrm{T}}(t)S_1\dot{x}_i(t) - \frac{h^2}{2}\int_{-h}^{0}\int_{t+\alpha}^{t} \dot{x}_i^{\mathrm{T}}(s)S_1\dot{x}_i(s)dsd\alpha$$

Combining with system under motion-trends-based deception attacks **(9.17)**, $\dot{V}_1(t)$ can be written as

$$\begin{aligned}\dot{V}_1(t) = \mathrm{sym}&\left\{\left[(A_i + B_i\hat{K}_{\mathrm{FUZ}}^{j})x_i(t) + B_i\hat{K}_{\mathrm{FUZ}}^{j}f_a(x_i(t_A)) + F_iw_i(t)\right]^{\mathrm{T}}P_ix_i(t)\right.\\ &\left. + \left(x_i^{\mathrm{T}}(t) - x_i^{\mathrm{T}}(t-h)\right)P_i\int_{t-h}^{t} x_i(s)ds\right\},\end{aligned}$$

Focusing on $\dot{V}_3(t)$, the following can be derived

$$-h\int_{t-h}^{t} x_i^{\mathrm{T}}(s)R_1x_i(s)ds \leq -\int_{t-h}^{t} x_i^{\mathrm{T}}(s)dsR_1\int_{t-h}^{t} x_i(s)ds \tag{9.21}$$

By applying **Lemma 9.1**, we have

$$\begin{aligned}-h\int_{t-h}^{t} \dot{x}_i^{\mathrm{T}}(s)R_2\dot{x}_i(s)ds &\leq -x_i^{\mathrm{T}}(t)R_2(t) + x_i^{\mathrm{T}}(t)U^{\mathrm{T}}x_i(t-h)\\ &\quad + x_i^{\mathrm{T}}(t-h)Ux_i(t) - x_i^{\mathrm{T}}(t-h)R_2x_i(t-h)\end{aligned} \tag{9.22}$$

Utilizing **Lemma 9.2**, we have

$$
\begin{aligned}
-\frac{h^2}{2}\int_{-h}^{0}\int_{t+\alpha}^{t} \dot{x}_i^{\mathsf{T}}(s)S_1\dot{x}_i(s)dsd\alpha &\leq -h^2x_i^{\mathsf{T}}(t)S_1x_i(t) + hx_i^{\mathsf{T}}(t)S_1\int_{t-h}^{t} x_i(s)ds \\
&+ h\int_{t-h}^{t} x_i^{\mathsf{T}}(s)dsS_1x_i(t) - \int_{t-h}^{t} x_i^{\mathsf{T}}(s)dsS_1\int_{t-h}^{t} x_i(s)ds
\end{aligned}
\tag{9.23}
$$

Noticing the impact of motion-trends-based deception attacks, Equation (9.7) can be written as

$$
x_i^{\mathsf{T}}(t)G^{\mathsf{T}}P_iG_ix_i(t) - f_a^{\mathsf{T}}(x_i(t_A))P_if_a(x_i(t_A)) > 0 \tag{9.24}
$$

For $w_i(t) \neq 0$, the following definition is introduced to show the attenuation level $\gamma > 0$, then

$$
\dot{V}_i(t) - y_i^{\mathsf{T}}(t)y_i(t) + \gamma^2w_i^{\mathsf{T}}(t)w_i(t) \leq -y_i^{\mathsf{T}}(t)y_i(t) + \gamma^2w_i^{\mathsf{T}}(t)w_i(t). \tag{9.25}
$$

Defining the following

$$
\Psi^{\mathsf{T}}(t) = \left[x_i(t), x_i(t-h), f_a(x_i(t_A)), \int_{t-h}^{t} x_i(s)ds, w_i(t)\right]^{\mathsf{T}}.
$$

Combining with $\dot{V}_i(t)$, (9.21)–(9.25) one has

$$
\Psi^{\mathsf{T}}(t)(\Theta_{11} - \Theta_{21}^{\mathsf{T}}\Theta_{22}\Theta_{12})\Psi(t) \leq 0. \tag{9.26}
$$

Under zero initial conditions, we have

$$
\int_0^{\infty} y_i^{\mathsf{T}}(s)y_i(s)ds \leq \gamma^2w_i^{\mathsf{T}}(s)w_i(s)ds. \tag{9.27}
$$

Thus, $\|y_i(t)\| \leq \gamma\|w_i(t)\|$ exists for nonzero $w_i(t) \in L_2[0, +\infty)$.

Therefore, the FOSMLFC power system under motion-trends-based deception attacks has proved to be stable. This completes the proof.

In **Theorem 9.1**, the stable performance of power system under motion-trends-based deception attacks is proved. Then, a controller gain that ensures the stable performance of power system shall be obtained in **Theorem 9.2**.

Remark 9.5: In **Theorem 9.1**, the FOSMLFC power system under motion-trends-based deception attacks is proved to be stable with condition (9.18). The result shows that there exists a suitable fuzzy output sliding mode controller gain to help the power system reach stability. The following theorem explains the process of obtaining the controller gain.

9.3.2 Fuzzy Gains

The following theorem is provided to obtain the initial fuzzy gain $\hat{K}^1_{\mathrm{FUZ}}$. A gain that ensures the system stable can be obtained by unitizing LMI. Then we take it as the initial fuzzy gain $\hat{K}^1_{\mathrm{FUZ}}$.

To obtain the $\hat{K}^1_{\mathrm{FUZ}}$, we introduce a fuzzy power model for analysis.

- Rule 1: **IF** Controller gain is $\hat{K}^1_{\mathrm{FUZ}}$, **Then**

$$\begin{cases} \dot{x}_i(t) = A_i x_i(t) + B_i \hat{K}^1_{\mathrm{FUZ}} \hat{x}_i(t) + \hat{F}_i w_i(t) \\ y_i(t) = C_i x_i(t) \end{cases} \tag{9.28}$$

Theorem 9.2 *For given scalars $h > 0$ and $\gamma > 0$, power system (9.28) is stable if there exist $X_i > 0$, $\tilde{Q}_1 > 0$, $\tilde{Q}_2 > 0$, $\tilde{R}_1 > 0$, $\tilde{R}_2 > 0$, $\tilde{S}_1 > 0$, and matrices $\tilde{U}$ Y_i with appropriate dimensions, such that the following LMIs hold*

$$\left[\tilde{\eta}\right]_{9\times 9} \prec 0 \tag{9.29}$$

$$\begin{bmatrix} -\tilde{R}_2 & \tilde{U}^{\mathrm{T}} \\ \tilde{U} & -\tilde{R}_2 \end{bmatrix} \succ 0 \tag{9.30}$$

where

$$\begin{aligned}
\tilde{\eta}_{1,1} &= X_i A_i^{\mathrm{T}} + A_i X_i + Y_i^{\mathrm{T}} B_i^{\mathrm{T}} + \tilde{Q}_1 + h^2 \tilde{R}_{_} - \tilde{R}_2 - h^2 \tilde{S}_1, \tilde{\eta}_{2,1} = -\tilde{U}, \\
&\quad \tilde{\eta}_{2,2} = -\tilde{Q}_1 - 4\tilde{R}_2, \\
\tilde{\eta}_{3,1} &= Y_i^{\mathrm{T}} B_i^{\mathrm{T}}, \tilde{\eta}_{3,3} = -X_i, \tilde{\eta}_{4,1} = X_i + \tilde{Q}_2 + h\tilde{S}_1, \tilde{\eta}_{4,2} = -X_i - \tilde{Q}_2, \\
&\quad \tilde{\eta}_{4,4} = -\tilde{R}_1 - \tilde{S}_1, \\
\tilde{\eta}_{5,1} &= \hat{F}^{\mathrm{T}}, \tilde{\eta}_{5,5} = -\gamma^2, \tilde{\eta}_{6,1} = X_i C_i^{\mathrm{T}}, \tilde{\eta}_{6,6} = -I, \tilde{\eta}_{7,1} = X_i G_i^{\mathrm{T}}, \tilde{\eta}_{7,7} = -X_i, \\
\tilde{\eta}_{8,1} &= hX_i A_i^{\mathrm{T}} + hY_i^{\mathrm{T}} B_i^{\mathrm{T}}, \tilde{\eta}_{8,3} = hY_i^{\mathrm{T}} B_i^{\mathrm{T}}, \tilde{r}_{8,5} = h\hat{F}_i^{\mathrm{T}}, \tilde{\eta}_{8,8} = -\tilde{R}_2, \\
\tilde{\eta}_{9,1} &= \frac{h^2}{2} X_i A_i^{\mathrm{T}} + \frac{h^2}{2} Y_i^{\mathrm{T}} B_i^{\mathrm{T}}, \tilde{\eta}_{9,3} = \frac{h^2}{2} Y_i^{\mathrm{T}} B_i^{\mathrm{T}}, \tilde{\eta}_{9,5} = \frac{h^2}{2} \hat{F}_i^{\mathrm{T}} \tilde{\eta}_{9,5} = \frac{h^2}{2} \hat{F}_i^{\mathrm{T}}, \\
&\quad \tilde{\eta}_{9,9} = -\tilde{S}_1.
\end{aligned}$$

Proof. To solve the complex terms in **Theorem 9.1**, pre- and post-multiplying both sides with diag$\{X_i, X_i, X_i, X_i, X_i, X_i, X_i, X_i, X_i, X_i\}$. The constant controller gain can be calculated by $\hat{K}^1_{\text{FUZ}} = Y_i X_i^{-1}$. In the following, $\hat{K}^1_{\text{FUZ}}$ is applied to the fuzzy controller.

Remark 9.6: The initial gain of the fuzzy output sliding controller $\hat{K}^1_{\text{FUZ}}$ can be calculated by LMIs. Based on the obtained $\hat{K}^1_{\text{FUZ}}$, a design method for membership functions is provided. Controller gain $\hat{K}^1_{\text{FUZ}}$ is then applied to modify $\hat{K}^j_{\text{FUZ}}$ to improve control performance.

Remark 9.7: The choice of initial controller gain and membership function is mostly determined by experience or trial-error. In fact, an inappropriate choice of controller parameters will prolong the convergent time and lead to higher consumption. Control signal level and control consumption are guaranteed by LMI. The control speed is modified by LMI-based membership functions. Overall, the proposed LMI-based fuzzy logic controller obtains a better control performance than the fuzzy control method [28–30].

The power system state is defined as $x_i(t) = [\Delta P_{mi} \Delta P_{vi} \Delta P^i_{\text{tie}} \Delta f_i]^{\text{T}}$. Fuzzy controller gain, $\hat{K}^j_{\text{FUZ}} = [k_1, k_2, k_3, k_4]$, which makes the power system reach stability, is proved in **Theorem 9.2**. However, if LMI finds a gain $\hat{K}^1_{\text{FUZ}}$ that makes system to be stable, the calculation will be stopped. Therefore, the abscissa of membership functions can be determined by $\hat{K}^1_{\text{FUZ}}$.

The design of the membership functions is displayed in Figure 9.4. The obtained controller gain can help us to determine the coordinate range and provide an appropriate membership function. For positive a_1, a_2, a_3, there exist $a_1 < a_2 < |k_b| < a_3$, $b = 1, 2, 3, 4$. The specific situation is shown in simulation part.

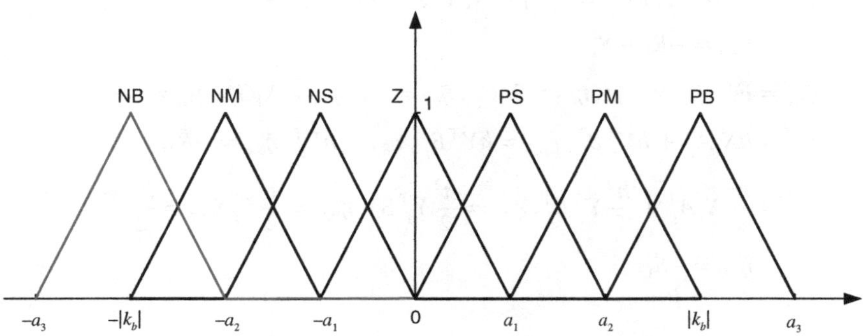

Figure 9.4 Member function design for fuzzy controller gains.

Remark 9.8: Mostly, the member functions coordinate range is designed based on trial-error method or experience [28–30]. This process is complex and cannot guarantee the control performance. Differently, the coordinate range can be determined directly by using LMI. This saves us the cost of trial-error and extra expense.

Remark 9.9: The obtained controller gain ensures the stable performance of the power system. However, the LMI toolbox calculation will stop when it finds a gain that is feasible. Due to the limitations of LMI, we cannot guarantee that the gain obtained is optimal. Therefore, fuzzy logic is applied to modify the controller gain to improve control performance.

9.3.3 Reachability Analysis

Theorem 9.3 *Power system (9.17) trajectories are ensured to be on the predefined fuzzy sliding surface in a finite time with the following control law*

$$u_i(t) = \hat{K}^j_{\text{FUZ}}\hat{x}_i(t) \quad +(L_iB_i)^{-1}\hat{K}^j_{\text{FUZ}}C_i\hat{x}_i(t) - d_i(L_iB_i)^{-1}\|L_iF_i\|\text{sgn}(\sigma_i(t)) \tag{9.31}$$

where $d_i > 0$, sgn(·) *denotes the switch function.*

Proof. The Lyapunov function is chosen as

$$V_s = \frac{1}{2}\sigma_i^{\text{T}}(t)\sigma_i(t) \tag{9.32}$$

Take the derivative on V_s, we have

$$\begin{aligned}
\dot{V}_s &= \sigma_i^{\text{T}}(t)\dot{\sigma}_i(t) \\
&= \sigma_i^{\text{T}}(t)[L_i\dot{x}_i(t) - L_i(A_ix_i(t) + B_i\hat{K}^j_{\text{FUZ}}\hat{x}_i(t)) - K^j_{\text{FUZ}}y_i(t)] \\
&= \sigma_i^{\text{T}}(t)\{L_iA_ix_i(t) + L_iB_iu_i(t) + L_i\hat{F}_iw_i(t) - L_i(A_ix_i(t) + B_i\hat{K}^j_{\text{FUZ}}\hat{x}_i(t)) \\
&\quad - \hat{K}^j_{\text{FUZ}}C_ix_i(t)\} \\
&= \sigma_i^{\text{T}}(t)\{-L_iB_i\hat{K}^j_{\text{FUZ}}C_i\hat{x}_i(t) - d_i\|L_i\hat{F}_i\|\text{sgn}(\sigma_i(t)) + \hat{F}_iw_i(t) - L_iB_i\hat{K}^j_{\text{FUZ}}C_i\hat{x}_i(t)\} \\
&= \sigma_i^{\text{T}}(t)[-d_i\|L_iF_i\|\text{sgn}(\sigma_i(t)) + L_iF_iw_i(t)] \\
&\le \{-d_i\|L_iF_i\|\|\sigma_i(t)\| + \|\sigma_i(t)\|\|L_iF_i\|\|w_i(t)\| \\
&\le \|L_iF_i\|\|\sigma_i(t)\|[-d_i + \|w_i(t)\|] < 0.
\end{aligned}$$

Thus, the designed control law can ensure the reaching condition. This completes the proof.

9.4 Numerical Examples

Numerical cases are tested on an IEEE 10-generator 39-bus system, as presented in Figure 9.5. The system parameters as shown in Table 9.2 are largely the same as those in [35].

The fuzzy controller member functions can be built by using gain matrices. According to **Theorem 9.2**, we have $\tilde{K}^1_{\text{FUZ}} = [-0.0034, 0.2004, 0.0031, -0.0361]$ for area-1. Based on $\tilde{K}^1_{\text{FUZ1}}$, the member functions for $\tilde{K}^j_{\text{FUZ}}$ are given in Figure 9.6.

Example 9.1 *The characteristics of the motion-trends-based deception attacks are discussed in this example.*

The deception attack signals are given as $x_{a1} = 1.5 * x_i(t)$, $x_{a2} = 3 * x_i(t)$, and $x_{a3} = 5 * x_i(t)$, respectively. As the power system performance can be reflected by real-time frequency. Simulation results are shown as the changes of Δf_i.

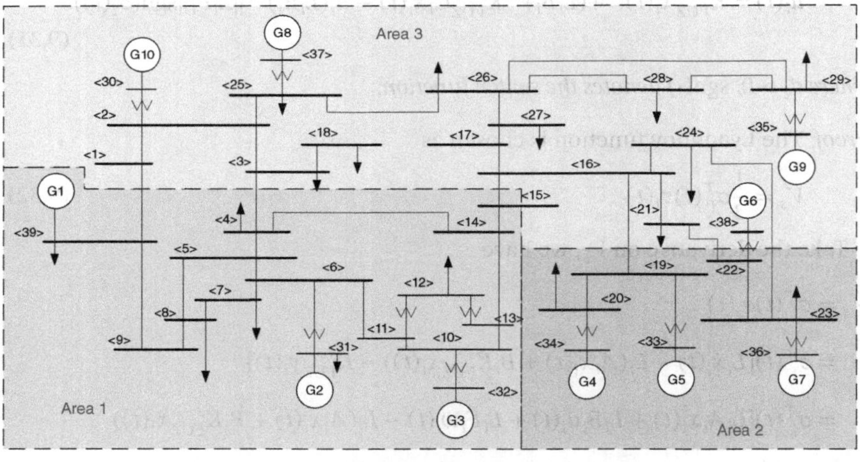

Figure 9.5 IEEE 10-generator 39-bus system.

Table 9.2 Parameters of power system.

Area-1	Parameters	T_{ch1}	T_{g1}	R_1	T_{12}	M_1	D_1
	Value	0.7	0.7	0.5	0.06*π	2	1
Area-2	Parameters	T_{ch2}	T_{g2}	R_2	T_{13}	M_2	D_2
	Value	1.2	0.9	1.5	0.06*π	2	1.5
Area-3	Parameters	T_{ch3}	T_{g3}	R_3	T_{23}	M_3	D_3
	Value	0.9	1.5	1.5	0.06*π	2	3

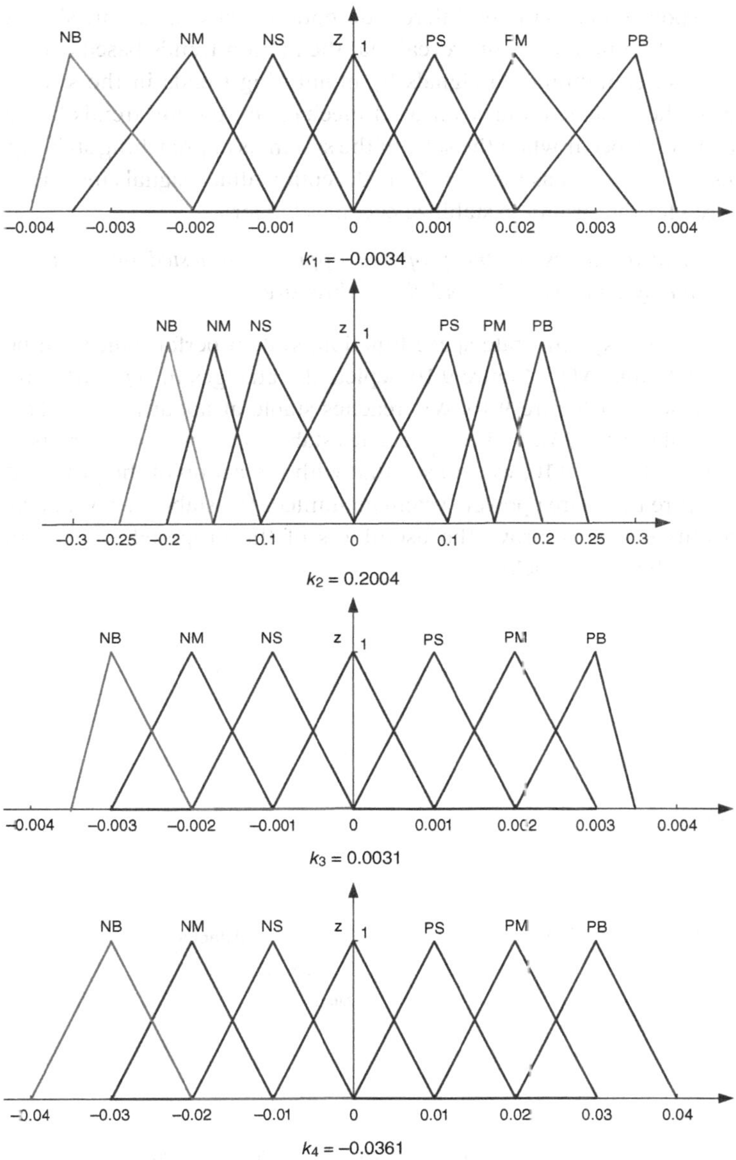

Figure 9.6 Member functions for $\hat{K}^{j}_{\mathrm{FUZ}}$.

The state responses of Δf_i under different deception attack signals are shown in Figures 9.7–9.9. Simulation results reveal that the motion-trends-based deception attacks release contamination signals by monitoring trends in the system state. Caused by the designed attack released mechanism, motion-trends-based deception attacks only occur when the state of the system is approaching stability. Simulation results show that as the size of the deception attack signals increases, the time required for the system to stabilize becomes longer.

Example 9.2 *The usefulness of the proposed approach is tested on an IEEE 10-generator 39-bus system, which is divided into three areas.*

As defined in power system state space function, system performance can be shown by Δf_i, ΔP_{vi} and ΔP^i_{tie}. Figure 9.10 depicts the changes in Δf_i, ΔP_{vi} and ΔP^i_{tie}. As can be seen in Figure 9.10, Δf_i reaches stable at about $t = 15s$, ΔP_{vi} reaches stable at about $t = 15s$, and ΔP^i_{tie} reaches stable at about $t = 30s$, respectively. According to Figure 9.10, we can see that with the action of the proposed approach, the entire three-area power system guarantees the stable performance. Simulation results evidently prove the usefulness of the proposed strategy in defending against the cyberattacks.

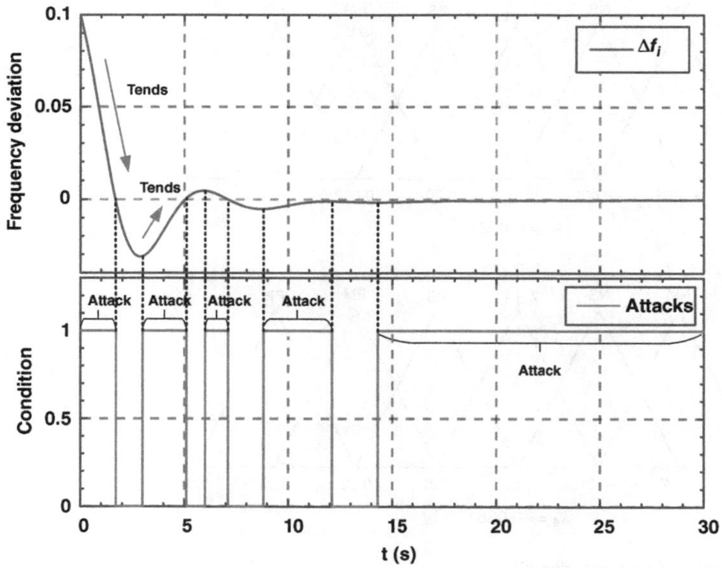

Figure 9.7 Frequency deviation under deception attacks x_{a1}.

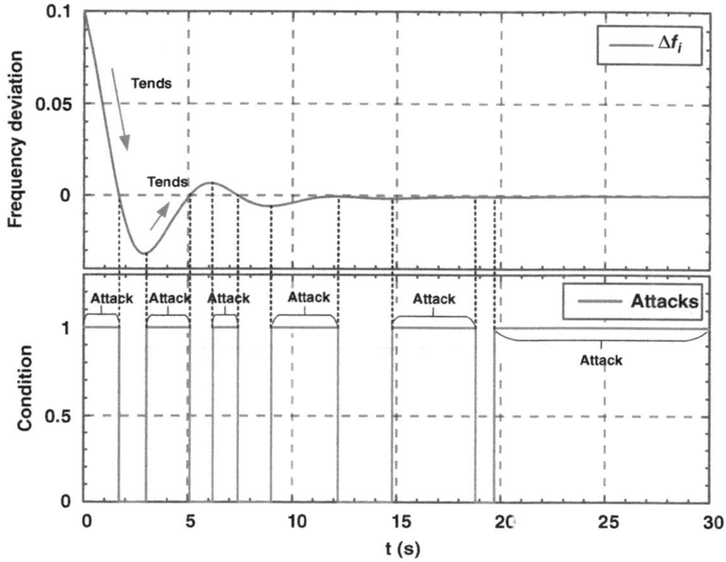

Figure 9.8 Frequency deviation under deception attacks x_{a2}.

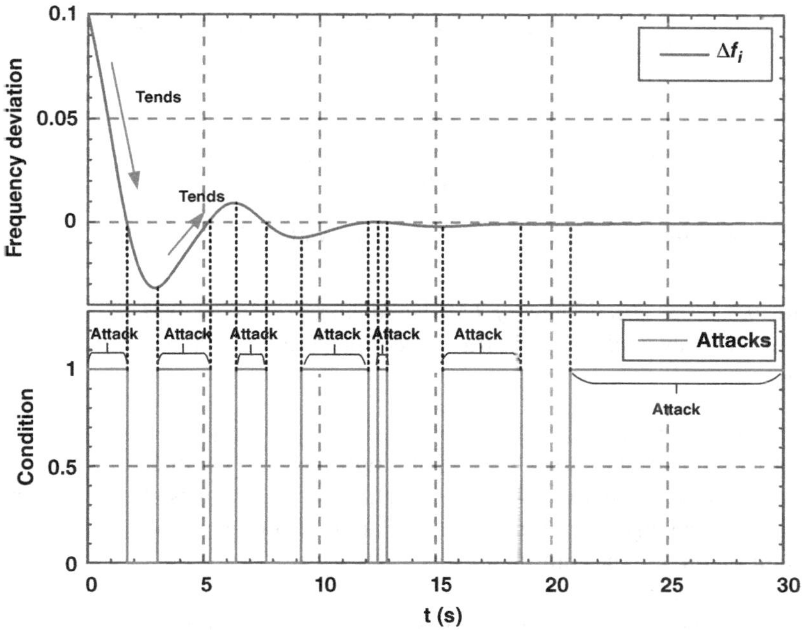

Figure 9.9 Frequency deviation under deception attacks x_{a3}.

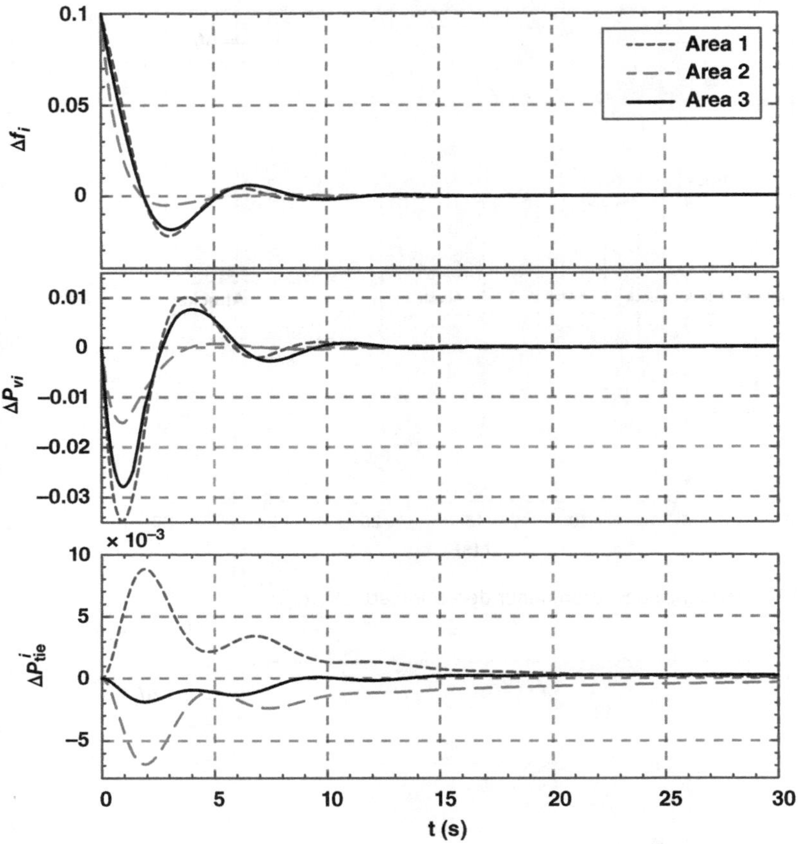

Figure 9.10 State response of three-area power system with FOSMLFC.

Example 9.3 *The effectiveness of LMI-based fuzzy controller is demonstrated in this part. This example also illustrates the difference between different control strategies.*

The fuzzy controller gain is defined as $\hat{K}^{j}_{\mathrm{FUZ}} = [k_1, k_2, k_3, k_4]$.

The moderation effect of fuzzy logic on the controller gain parameters is shown in Figure 9.11. Due to the use of the system output and its changes, the controller gain parameters change synchronously. Simulation results show that fuzzy logic is useful for modifying the controller gains.

The following example illustrates the need for fuzzy logic in controller design. Comparisons are made between the proposed method and output SMC [33].

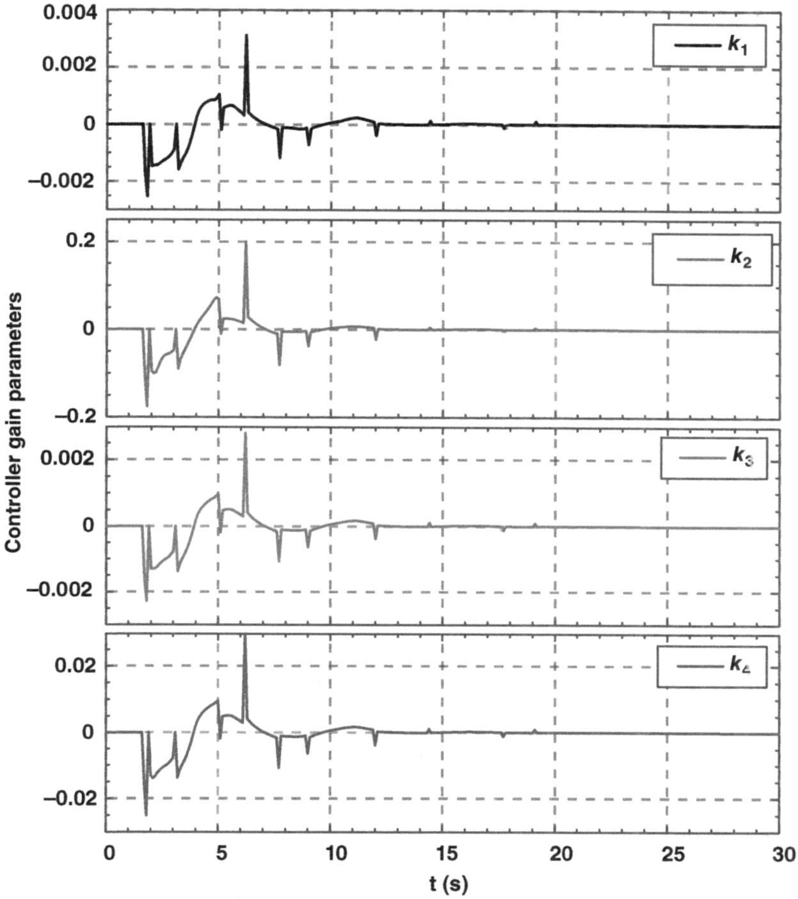

Figure 9.11 Fuzzy logic for controller gain $\hat{K}^j_{\text{FUZ}}$.

The comparison of the proposed approach and the output SMLFC is displayed in Figure 9.12. In can be observed that Δf_i, ΔP_{vi}, and ΔP_{mi} become stable at about $t = 15s$, $t = 20s$ and $t = 30s$, respectively. However, the system state remains unstable under output SMLFC. Comparison results show that the use of fuzzy logic can effectively improve system performance. Simulation results confirm the superiority of the proposed approach. The results clearly demonstrate the necessity and effectiveness of fuzzy logic in the controller design.

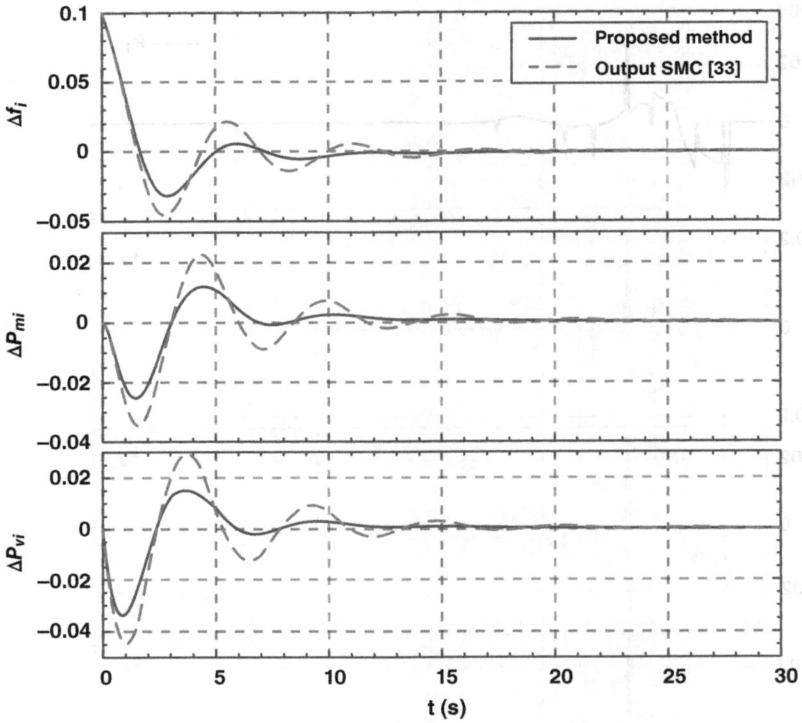

Figure 9.12 State response of different control methods.

9.5 Conclusion

In this chapter, a new motion-trends-based deception attack model has been proposed to describe potentially malicious threats from communication networks. Recognizing that intelligent attackers may monitor the operational state of the target system to release attacks, the occurrence of deception attacks has been determined by the motion-trends-based condition. A fuzzy output SMC method has been discussed, where the system state and output have been added in the sliding mode surface design. Later, we have analyzed the stability of the FOSMLFC power system in the presence of motion-trends-based deception attacks. Using the LMI, a controller gain has been obtained that ensures the stability of the power system. However, the search process stops when LMI finds a feasible solution to stabilizes the system. To further improve the power system performance, a LMI-based approach has been proposed for the fuzzy output SMC. A design method for the fuzzy logic member functions has been investigated using the obtained controller gain. Simulation results have confirmed the usefulness of the proposed approach in defending cyberattacks. Besides, we have verified the feasibility of LMI-based fuzzy logic in controller modifying.

References

1 M. Zeyringer, J. Price, B. Fais, et al. Designing low-carbon power systems for Great Britain in 2050 that are robust to the spatiotemporal and inter-annual variability of weather. *Nature Energy*, 3:395–403, 2018.

2 S. Kuppusamy, Y. H. Joo, and H. S. Kim. Fault-tolerant load frequency control for DFIG-based interconnected wind power systems. *Information Sciences*, 582:73–88, 2022.

3 L. Shanmugam and Y. H. Joo. Investigation on stability of delayed TS fuzzy interconnected systems via decentralized memory-based sampled-data control and validation through interconnected power systems with DFIG-based wind turbines. *Information Sciences*, 580:934–952, 2021.

4 X. H. Li and D. Ye. Event-based distributed fuzzy load frequency control for multiarea nonlinear power systems with switching topology. *IEEE Transactions on Fuzzy Systems*, 30:4262–4272, 2022.

5 X. Liu, C. Wang, X. Kong, et al. Tube-based distributed MPC for load frequency control of power system with high wind power penetration. *IEEE Transactions on Power Systems*, 39(2):3118–3129, 2024.

6 X. X. Lv, F. Qiao, Y. H. Sun, et al. Improved memory event-triggered load frequency control in multi-area power system with renewable energy. *IEEE Transactions on Circuits and Systems II: Express Briefs*, 71:311–315, 2024.

7 J. S. Zhang, J. L. Tian, P. F. Yan, et al. Multi-hop graph pooling adversarial network for cross-domain remaining useful life prediction: A distributed federated learning perspective. *Reliability Engineering & System Safety*, 244:109950, 2024.

8 J. S. Zhang, J. L. Tian, A. M. Alcaide, et al. Lifetime extension approach based on Levenberg-Marquardt neural network and power routing of DC-DC converters. *IEEE Transactions on Power Electronics*, 38:10280–10291, 2023.

9 S. Nasiri, H. Seifi, and H. Delkhosh. A secure power system distributed state estimation via a consensus-based mechanism and a cooperative trust management strategy. *IEEE Transactions on Industrial Informatics*, 20:3002–3014, 2024.

10 X. X. Kong, Z. G. Lu, X. Q. Guo, et al. Resilience evaluation of cyber-physical power system considering cyber attacks. *IEEE Transactions on Reliability*, 73:245–256, 2024.

11 Z. Zhang, X. N. Song, X. L. Sun, and V. Stojanović. Hybrid-driven-based fuzzy secure filtering for nonlinear parabolic partial differential equation systems with cyber attacks. *International Journal of Adaptive Control and Signal Processing*, 37:380–398, 2023.

12 Z. Wang, H. Zhang, X. H. Cao, et al. Modeling and detection scheme for zero-dynamics attack on wind power system. *IEEE Transactions on Smart Grid*, 15:934–943, 2024.

13 M. J. Yang, H. Zhang, C. Peng, and Y. L. Wang. A penalty-based adaptive secure estimation for power systems under false data injection attacks. *Information Sciences*, 508:380–392, 2020.

14 J. W. Xing, C. Peng, Z. R. Cao, and W. B. Xie. Security-based control for networked interval type-2 fuzzy systems with multiple cyber-attacks: An improved dynamic event-triggered scheme. *IEEE Transactions on Fuzzy Systems*, 31:2747–2760, 2024.

15 M. M. Hossain and C. Peng. Observer-based event triggering H_∞ LFC for multi-area power systems under DoS attacks. *Information Sciences*, 543:437–453, 2021.

16 S. Yan, Z. Gu, J. H. Park, et al. Probability-density-dependent load frequency control of power systems with random delays and cyber-attacks via circuital implementation. *IEEE Transactions on Smart Grid*, 13:4837–4847, 2022.

17 G. P. Liu, J. H. Park, C. C. Hua, and Y. F. Li. Hybrid dynamic event-triggered load frequency control for power systems with unreliable transmission networks. *IEEE Transactions on Cybernetics*, 53:806–817, 2022.

18 J. L. Liu, Y. H. Dong, L. J. Zha, et al. Event-based security tracking control for networked control systems against stochastic cyber-attacks. *Information Sciences*, 612:306–321, 2022.

19 K. D. Lu and Z. G. Wu. Resilient event-triggered load frequency control for cyber-physical power systems under DoS attacks. *IEEE Transactions on Power Systems*, 38(6):5302–5313, 2023.

20 X. Y. Liu and W. W. Che. Event-based distributed secondary voltage tracking control of microgrids under DoS attacks. *Information Sciences*, 608:1572–1590, 2022.

21 X. Wang, E. G. Tian, W. X. Zheng, and X. P. Xie. Important-data-based DoS attack mechanism and resilient H_∞ filter design for networked T-S fuzzy systems. *IEEE Transactions on Cybernetics*, 54(5):3352–3362, 2024. https://doi.org/10.1109/TCYB.2023.3285526.

22 X. N. Song, N. N. Wu, S. Song, and V. Stojanović. Switching-like event-triggered state estimation for reaction–diffusion neural networks against DoS attacks. *Neural Processing Letters*, 55:8997–9018, 2023.

23 G. Falco, M. Eling, D. Jablanski, et al. Cyber risk research impeded by disciplinary barriers. *Science*, 366:1066–1069, 2019.

24 Q. Y. Zhang, X. N. Song, S. Song, and V. Stojanović. Finite-time sliding mode control for singularly perturbed PDE systems. *Journal of the Franklin Institute*, 360:841–861, 2023.

25 H. Zhang, J. Hu, G. Liu, and X. Yu. Event-triggered secure control of discrete systems under cyber-attacks using an observer-based sliding mode strategy. *Information Sciences*, 587:587–606, 2022.

26 B. C. Zheng, Y. Y. Zhao, Y. A. Zhu, and T. Li. Sliding-mode control under denial-of-service attacks: An event-triggered quantized state feedback approach. *IEEE Transactions on Circuits and Systems II: Express Briefs*, 70:2949–2953, 2023.

27 S. W. Qiao, X. H. Liu, Y. R. Liang, et al. Event-triggered sliding mode load frequency control of multiarea power systems under periodic denial-of-service attacks. *IEEE Systems Journal*, 17(2):2803–2814, 2023.

28 Z. Echreshavi, M. Farbood, and M. Shasadeghi. Fuzzy event-triggered integral sliding mode control of nonlinear continuous-time systems. *IEEE Transactions on Fuzzy Systems*, 30:2347–2359, 2021.

29 Y. Yang, B. Y. Wen, X. J. Su, et al. Sliding mode fuzzy control of stochastic nonlinear systems under cyber-attacks. *IEEE Transactions on Cybernetics*, 54(5):2347–2359, 2023.

30 W. H. Liu, Q. Ma, Y. Lu, and S. Y. Xu. Adaptive fixed-time event-triggered fuzzy control for time-delay nonlinear systems with disturbances and quantization. *IEEE Transactions on Fuzzy Systems*, 31:2848–2860, 2023.

31 H. Y. Zhang, N. Zhao, S. Y. Wang, and R. K. Agarwal. Improved event-triggered dynamic output feedback control for networked T–S fuzzy systems with actuator failure and deception attacks. *IEEE Transactions on Cybernetics*, 53:7989–7999, 2023.

32 V. Stojanović. Fault-tolerant control of a hydraulic servo actuator via adaptive dynamic programming. *Mathematical Modelling and Control*, 3:181–191, 2023.

33 J. H. Zhang and Y. Q. Xia. Design of static output feedback sliding mode control for uncertain linear systems. *IEEE Transactions on Industrial Electronics*, 57(6):2161–2170, 2009.

34 A. Farnam and R. M. Esfanjani. Improved linear matrix inequality approach to stability analysis of linear systems with interval time-varying delays. *Journal of Computational and Applied Mathematics*, 294:49–56, 2016.

35 J. Li, H. Feng, Y. Wang, and G. Liu. A novel failure-distribution-dependent non-fragile H_∞ fault-tolerant load frequency control for faulty multi-area power systems. *IEEE Transactions on Power Systems*, 39(2):2936–2946, 2024.

10

Future Work

At present, the stability problem and communication security problem with respect to renewable energy interconnected power systems are well worth noting. Integration and optimization of multiple renewable energy sources represents a major challenge facing future research. Furthermore, the development of innovative technologies such as cloud computing, big data, and artificial intelligence shows great potential for solving load frequency control (LFC)-related problems. The current challenges and future research directions are discussed below.

1. Renewable energy: With the deepening of research on renewable energy, certain sources of renewable energy that could not be used before have been gradually developed for power systems. The diverse power output of renewable energy sources connected to the power system increases the complexity and variability in terms of power generation and adds uncertainty. This presents challenges for the stabilization of LFC power systems integrated with renewable energy sources. In addition, the integration of multiple renewable energy sources means that more electronic equipment needs to be used. The resulting need to ensure matching between devices, efficient information transmission, and other issues are unavoidable challenges facing LFC power systems.
2. Cybersecurity: Cybersecurity problems are directly related to the normal operation of power systems. Compared with a single type of cyberattack, hybrid cyberattacks can have greater impact. Furthermore, new types of cyberattacks on LFC power systems deserve attention. According to our investigation, most of the cyberattacks considered in the context of LFC power systems can be modeled using the queueing model or stochastic model. However, the types of cyberattack are increasing. One possible development direction of cyberattacks involves smart design. For example, attackers may use devices to obtain information about the power system and determine vulnerable nodes in the network according to the obtained information. Cyberattacks are expected to remain an important problem in LFC power systems in the future.

3. Control capability: In recent years, a large number of intelligent algorithm-based control methods have been researched for LFC power systems. Multiple issues around renewable energy, cyberattacks, and other issues are focusing greater development space and attention on intelligent control methods. How to improve control capability is becoming an important issue. To achieve this goal, researchers need to choose their research direction; possibilities include combining traditional control methods with modern control methods, using intelligent control methods, and developing new intelligent algorithms.
4. Flexibility: A highly flexible and secure interconnected power system is a necessity due to the high penetration and networking requirements of renewable energy sources. The transition of the current energy production mix can be aided by hosting a larger proportion of renewable energy sources. In this way, multi-area power systems can become more flexible and reliable, resulting in the development of novel intelligent algorithms along with control, storage, and market approaches.
5. Future: Cybersecurity research will focus on AI-enhanced intrusion detection, leveraging machine learning to identify anomalous behavior and enable dynamic responses. It will integrate with blockchain to establish decentralized identity authentication and ensure signal integrity, thereby defending against cyberattacks. Additionally, future research will adopt quantum cryptographic algorithms and quantum key distribution to counter quantum computing threats. The synergy of these three technologies forms a powerful defense system, though challenges remain in practical application.

Index

b

c

t

Printed in the USA/Agawam, MA
August 4, 2026

908586.012